The Atmosphere and Ocean

The Atmosphere and Ocean

A PHYSICAL INTRODUCTION
THIRD EDITION

Neil C. Wells

University of Southampton, UK

A John Wiley & Sons, Ltd., Publication

This edition first published 2012
© 2012 by John Wiley & Sons, Ltd

Wiley-Blackwell is an imprint of John Wiley & Sons, formed by the merger of Wiley's global Scientific, Technical and Medical business with Blackwell Publishing.

Registered office: John Wiley & Sons, Ltd, The Atrium, Southern Gate, Chichester, West Sussex, PO19 8SQ, UK

Editorial offices: 9600 Garsington Road, Oxford, OX4 2DQ, UK
The Atrium, Southern Gate, Chichester, West Sussex, PO19 8SQ, UK
111 River Street, Hoboken, NJ 07030-5774, USA

For details of our global editorial offices, for customer services and for information about how to apply for permission to reuse the copyright material in this book please see our website at www.wiley.com/wiley-blackwell.

Library of Congress Cataloging-in-Publication Data

Wells, Neil.
   The atmosphere and ocean : a physical introduction / Neil Wells. – 3rd ed.
       p. cm.
   Includes bibliographical references and index.
   ISBN 978-0-470-69469-5 (cloth) – ISBN 978-0-470-69468-8 (pbk.)
   1. Atmospheric physics. 2. Oceanography. I. Title.
   QC861.3.W45 2011
   551.5–dc23

                                                    2011016838

A catalogue record for this book is available from the British Library.

This book is published in the following electronic formats: ePDF 9781119994596; Wiley Online Library 9781119994589; ePub 9781119979845; Mobi 9781119979852

Typeset in 11/13pt Palatino-Roman by Laserwords Private Limited, Chennai, India

First Impression 2012

Contents

Series Foreword

Advances in Weather and Climate

Meteorology is a rapidly moving science. New developments in weather forecasting, climate science and observing techniques are happening all the time, as shown by the wealth of papers published in the various meteorological journals. Often these developments take many years to make it into academic textbooks, by which time the science itself has moved on. At the same time, the underpinning principles of atmospheric science are well understood but could be brought up to date in the light of the ever increasing volume of new and exciting observations and the underlying patterns of climate change that may affect so many aspects of weather and the climate system.

In this series, the Royal Meteorological Society, in conjunction with Wiley-Blackwell, is aiming to bring together both the underpinning principles and new developments in the science into a unified set of books suitable for undergraduate and postgraduate study as well as being a useful resource for the professional meteorologist or Earth system scientist. New developments in weather and climate sciences will be described together with a comprehensive survey of the underpinning principles, thoroughly updated for the 21st century. The series will build into a comprehensive teaching resource for the growing number of courses in weather and climate science at undergraduate and postgraduate level.

Series Editors
Peter Inness, University of Reading, UK
William Beasley, University of Oklahoma, USA

Preface to the Third Edition

The third edition of *The Atmosphere and Ocean* is a major revision of the material in previous editions to reflect the very significant changes in the subject. In particular chapters on mathematical modelling and climate change have been added to the new edition. Furthermore, problems and their solutions have been provided for each chapter, which some readers may find useful.

The book is not an exhaustive account of the subject but reflects my interests over the last 40 years in teaching and research of the ocean and atmosphere. I hope it may continue to instil enthusiasm and fascination for a subject which continues to challenge humankind in the 21st century.

I wish to thank the following people: Kate Davis for the drafting and improvement of the figures, Helen Wells for suggesting improvements to the earlier chapters, and Jenny Wells for careful checking of the final manuscript.

Finally I would like to thank the team at John Wiley, both in Chichester, U.K. and Singapore, who have provided a high level of professional support during the development and the production of this edition.

1

The Earth within the Solar System

1.1 The Sun and its constancy

Any account of the Earth's atmosphere and ocean cannot be regarded as complete without a discussion of the Sun, the solar system and the place of the Earth within this system. The Sun supplies the energy absorbed by the Earth's atmosphere-ocean system. Some of Sun's energy is converted directly into thermal energy, which drives the atmospheric circulation. A small portion of this energy appears as the kinetic energy of the winds which, in turn, drives the ocean circulation. Some of the intercepted solar energy is transformed by photosynthesis into biomass, a large proportion of which is ultimately converted into heat energy by chemical oxidation within the bodies of animals and by the decomposition and burning of vegetable matter. A very small proportion of the photosynthetic process produces organic sediments which may eventually be transformed into fossil fuels. It is estimated that the solar radiation intercepted by the Earth in seven days is equivalent to the heat that would be released by the combustion of all known reserves of fossil fuels on the Earth. The Sun, therefore, is of fundamental importance in the understanding of the uniqueness of the Earth.

The Sun is a main sequence star in the middle stages of its life and was formed 4.6×10^9 years ago. It is composed mainly of hydrogen (75% by mass) and helium (24% by mass); the remaining 1% of the Sun's mass comprises the elements oxygen; nitrogen; carbon; silicon; iron; magnesium and calcium. The emitted energy of the Sun is 3.8×10^{26} W and this energy emission arises from the thermonuclear fusion of hydrogen into helium at temperatures around 1.5×10^7 K in the core of the Sun.

In the core, the dominant constituent is helium (65% by mass) and the hydrogen content is reduced to 35% by mass as a direct result of its consumption in the fusion reactions. It is estimated that the remaining hydrogen in the Sun's core is sufficient to maintain the Sun at its present luminosity and size for a further 4×10^9 years. At this stage it is expected that the Sun will

The Atmosphere and Ocean: A Physical Introduction, Third Edition. Neil C. Wells.
© 2012 John Wiley & Sons, Ltd. Published 2012 by John Wiley & Sons, Ltd.

expand into a red giant and engulf all of the inner planets of the solar system (i.e. Mercury, Venus, Earth and Mars).

There exists a high-pressure gradient between the core of the Sun and its perimeter, and this is balanced by the gravitational attraction of the mass of the Sun. In the core, the energy released by the thermonuclear reaction is transported by energetic photons but, because of the strong absorption by peripheral gases, most of these photons do not penetrate to the surface. This absorption causes heating in the region outside the core. In contrast, the outer layers of the Sun are continually losing energy by radiative emission into space in all regions of the electromagnetic spectrum. This causes a large temperature gradient to develop between the surface and the inner region of the Sun. This large temperature gradient produces an unstable region and large scale convection currents are set up that transfer heat to the surface of the Sun. The convection currents are visible as the fine grain structure, or granules, in high resolution photos of the Sun's surface. It is thought that the convection currents have a three-tier structure within the Sun. The largest cells, 200×10^3 km in diameter are close to the core. In the middle tier the convection cells are about 30×10^3 km in diameter and at the surface they are 1×10^3 km. The latter cells have a depth of 2000 km. In each cell, hot gas is transported towards the cooler surface, whilst the return flow transports cooler gases towards the interior.

Almost all of the solar radiation emitted into space, approximately 99.9%, originates from the visible disc of the Sun, known as the photosphere. The photosphere is the region of the Sun where the density of the solar gas is sufficient to produce and emit a large number of photons, but where the density of the overlying layers of gas is insufficient to absorb the emitted photons. This region of the Sun has a thickness of approximately 500 km but no sharp boundaries can be defined. The radiative spectrum of the Sun, when fitted to a theoretical black body curve, gives a black body temperature of 6000 K, although the effective temperature, deduced from the total energy emitted by the Sun, gives a lower temperature of 5800 K.

The photosphere is not uniform in temperature. The lower regions of the photosphere have temperatures of 8000 K, whilst the outer regions have temperatures of 4000 K. Furthermore, the convection cells produce horizontal temperature variations of 100 K between the ascending and descending currents of solar gas. Larger convection cells also appear in the photosphere and they have diameters of 30 000 km. They appear to originate from the second tier of convection within the Sun. The appearance of sunspots gives rise to horizontal variations of 2000 K within the photosphere. The inner regions of the sunspots have black body temperatures of 4000 K. Sunspots have diameters of 10 000–150 000 km and they may last for many weeks. It is thought that the 'sunspot' causes a localised suppression of the convection and therefore leads to a reduction in the transfer of heat into the photosphere from the interior. However, although the sunspot features are dramatic, they occupy less than 1% of the Sun's disc and therefore the effect on the luminosity of the Sun is small.

Beyond the photosphere lies the chromosphere where the temperature decreases to a minimum of 4000 K at 2000 km above the photosphere and then increases sharply to a temperature of 10^6 K at a height of 5000 km in the region of the corona. However, because of the low density of the gas in this region, the radiation emitted from the chromosphere and the corona only amount to 0.1% of the total radiation from the Sun.

These different temperature zones of the Sun can be observed in the solar spectrum (Figure 1.1). The visible and infra-red radiation emitted from the photosphere follow reasonably closely the black body curve for a temperature of 5800 K. However, substantial deviations from the theoretical curve occur in the X-ray and radio wavebands, and lesser deviations occur in the ultraviolet spectrum. The high temperature of the corona is responsible for the intense X-ray band, whilst the high radio frequency energy is associated with the solar wind and solar activity. However, the energy in these wavebands is a negligible fraction of the total emitted energy and therefore these very variable regions of the spectrum have little direct influence on the total solar energy received on the Earth. The depletion of energy in the ultraviolet spectrum is the result of emission at a lower temperature than the photosphere and therefore probably originates in the temperature minimum of the lower

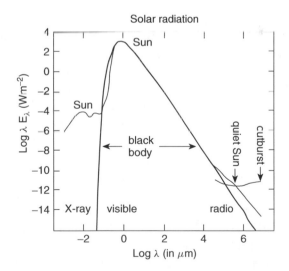

Figure 1.1 Solar spectrum and blackbody curve: energy distribution of the Sun and a black body at 5800 K. Reproduced, with permission, from Allen, C.W., 1958, Quarterly Journal of the Royal Meteorological Society, 84: page 311, figure 3

chromosphere. Recent observations have shown that the ultraviolet energy is not constant and shows considerable variability in the short-wave part of the spectrum amounting to 25% of its average value at 0.15 μm, and 1% at 0.23 μm. Again, the amount of energy in this band is relatively small and contributes less than 0.1% of the total energy. Furthermore, satellite observations of the solar spectrum since 1978 have demonstrated a 0.1% variation of the solar output over the 11 year sunspot cycle. These variations are rather small when compared to those produced by orbital variations of the Earth around the Sun.

1.2   Orbital variations in solar radiation

Let S be the total solar output of radiation in all frequencies. At a distance, r, from the centre of the Sun, imagine a sphere of radius r on which the flux of radiation will be the same (assuming the radiation from the Sun is equal in all directions). If the flux of radiation per unit area at a distance r is given by $Q(r)$, then the total radiation on the imagined sphere is $4\pi r^2 Q(r)$.

In the absence of additional energy sources to the Sun

$$S = 4\pi r^2 Q(r) \tag{1.1}$$

Rearranging:

$$Q(r) = \frac{S}{4\pi r^2} \tag{1.2}$$

In practice, the solar radiation cannot be measured at the Sun but it can be measured by satellites above the Earth's atmosphere. Recent determinations of the flux of radiation per unit area, Q, give a value of $1360\,\mathrm{W\,m^{-2}}$. Given that the Earth is approximately 1.5×10^{11} m away from the Sun, S can be calculated to be 3.8×10^{26} W.

Though S is the true solar constant, in meteorology Q is defined as the solar constant of the Earth. Table 1.1 shows the value of the solar constant obtained for other planets in the solar system. It is noted that the dramatic changes in the solar constant between the Earth and our nearest planetary neighbours, Mars and Venus, merely serve to highlight the uniqueness of the position of the Earth in the solar system. At the radius of Pluto (some 39 Earth-Sun distances), the flux of the radiation from the Sun is less than $1\,\mathrm{W\,m^{-2}}$.

The radiation incident on a spherical planet is not equal to the solar constant of the planet. The planet intercepts a disc of radiation of area πa^2, where a is the planetary radius, whereas the surface area of the planet is $4\pi a^2$. Hence the solar radiation per unit area on a spherical planet is

$$\frac{Q\pi a^2}{4\pi a^2} = \frac{Q}{4} \tag{1.3}$$

Table 1.1   Radiative properties of terrestrial planets. Q is the solar irradiance at distance r from the Sun, α is the planetary albedo, T_e is the radiative equilibrium temperature and T_s is the surface temperature

	$r(10^9 m)$	$Q(W\,m^{-2})$	α	$T_e(K)$	$T_s(K)$
Venus	108	2623	0.75	232	760
Earth	150	1360	0.30	255	288
Mars	228	589	0.15	217	227

The average radiation on the Earth's surface is $340\,W\,m^{-2}$. The above discussion assumes the total absence of an atmosphere and that the Earth is a perfect sphere in a spherical orbit.

The three geometrical factors which determine the seasonal variation of solar radiation incident on the Earth are shown in Figure 1.2. The Earth revolves around the Sun in an elliptical orbit, being closest to the Sun, i.e. at perihelion, about 4 January and farthest from the Sun, i.e. at aphelion, on 4 July. The eccentricity, represented by the symbol e, of the present day orbit is 0.017. At aphelion, it can be shown that

$$r_{aphelion} = (1 + e)\bar{r}$$

where $\bar{r}$ is the mean distance between the Earth and the Sun.

As the incident radiation is inversely proportional to the square of the Earth's distance from the Sun, it can also be shown that, at aphelion, the radiation received by the Earth is 3.5% less than the annual average, whilst

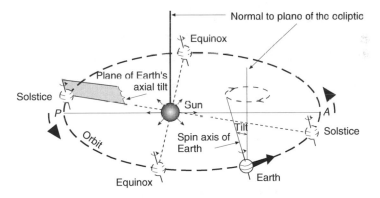

Figure 1.2   Geometry of the Earth–Sun system. The Earth's orbit, the large ellipse with major axis AP and the Sun at one focus, defines the plane of the ecliptic. The plane of the Earth's axial tilt (obliquity) is shown passing through the Sun corresponding to the time of the southern summer solstice. The Earth moves around its orbit in the direction of the solid arrow (period one year) whilst spinning about its axis in the direction shown by the thin curved arrows (period one day). The broken arrows shown opposite the points of aphelion (A) and perihelion (P) indicate the direction of the very slow rotation of the orbit. Reproduced, with permission, from Pittock, A.B. et al. eds, 1978, Climate Change and Variability: A Southern Perspective, Cambridge University Press: page 10, figure 2.1

at perihelion it is 3.5% greater than the annual average. Therefore the total radiation incident on the Earth over the course of one year is independent of the eccentricity of the Earth's orbit. The long-term variation in eccentricity indicates changes of 0.005–0.060 occurring with a period of 100 000 years. This would produce changes of up to 10% in the radiation incident on the surface of the Earth at perihelion and aphelion.

However, the angle of tilt (ε) of the Earth with respect to the plane of rotation of the Earth's orbit, i.e. the obliquity of the ecliptic, is the dominant influence on the seasonal cycle of solar radiation (Figure 1.3). It not only determines the march of the seasons, but it is also important in determining the latitudinal distribution of the climatic zones. At the present time, the angle is 23.5°, which is similar to the obliquity of Mars. It can be shown that the amplitude of the seasonal variation in the solar radiation is directly proportional to the obliquity. If the obliquity were reduced, there would be a reduction in the amplitude of the seasonal solar radiation cycle, and if

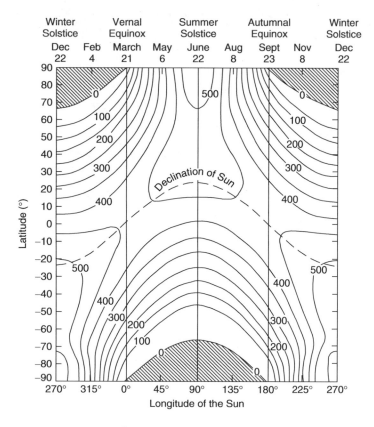

Figure 1.3   Solar radiation in W m^{-2} arriving at the Earth's surface, in the absence of an atmosphere, as a function of latitude and time of year at 2000 A.D. Reproduced, with permission, from Hess, S.L., 1959, Introduction to Theoretical Meteorology, Holt, Rinehart & Winston: page 132, figure 9.1

the obliquity were zero, the seasonal variation would depend solely on the eccentricity of the orbit.

The obliquity has varied between 22° and 24.5° over the past million years and this variation has a total period of 41 000 years. Such variations cannot influence the total radiation intercepted by the Earth, but they will exert an influence on the seasonal cycle.

The third parameter which affects the seasonal cycle is the longitude of the perihelion, measured relative to the vernal equinox. It precesses by 360° in a period of 21 000 years. Therefore, in about 10 500 years' time, the Earth will be closest to the Sun in July and farthest away in January.

The sensitivity of the global temperature to the seasonal cycle, which arises from ice, cloud and ocean feedback processes, indicates that these changes may be sufficient to initiate long term changes in climate and, in the most extreme cases, ice ages (Figure 1.4).

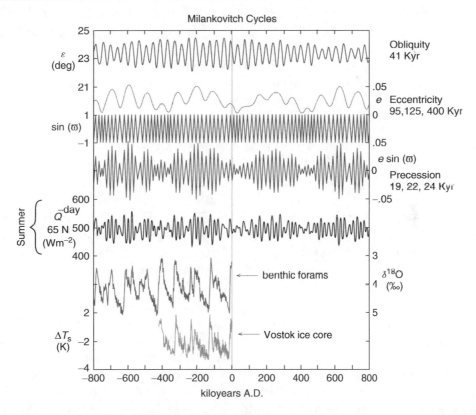

Figure 1.4   Past and future Milankovitch cycles ε is obliquity (axial tilt), e is eccentricity, ϖ is longitude of perihelion. $e\sin(\varpi)$ is the precession index, which together with obliquity, controls the seasonal cycle of insolation. $\overline{Q}^{day}$ is the calculated daily-averaged insolation at the top of the atmosphere, on the day of the summer solstice at 65 N latitude. Benthic forams and Vostok ice core show two distinct proxies for past global sea level and temperature, from ocean sediment and Antarctic ice respectively. Vertical gray line is current condition, at 2 ky A.D. See plate section for a colour version of this image

Milankovitch, a Serbian scientist, recognized the relationship between the variations in the Earth's orbit around the Sun and climate change over the last two million years. In particular, he determined the orbital conditions which would lead to the initiation of an ice age over the Northern Hemisphere continents. He reasoned that a reduction in solar radiation during the polar summer would lead to a reduction in summer melting of the snow cover. This, in turn, over many years, would lead to the build up of ice over the continents. The optimum orbital conditions for this reduction in the summer solar radiation are:

(i)   That aphelion occurs during the Northern Hemisphere summer.
(ii) That the eccentricity be large, to maximise the Earth–Sun separation at aphelion.
(iii) That the obliquity be small.

These three factors would act to decrease the amplitude of the solar radiation cycle and therefore reduce the solar radiation in summer at high latitudes in the Northern Hemisphere. However, all of these factors would tend to lead to an increase in solar radiation in the Northern Hemisphere winter. Comparison of the orbital parameters with long term temperature records, deduced from the dating of deep ocean sediment cores, indicate that there is a good statistical agreement with the theory.

1.3   Radiative equilibrium temperature

The inner planets of the solar system have one common attribute – the lack of a major source of internal energy. It is, of course, true that there is a steady geothermal flow of energy from the interior of the Earth as a result of radioactive decay in the Earth's core, but this energy flow is less than $0.1\,W\,m^{-2}$ and can be compared with an average solar heat flux of $340\,W\,m^{-2}$. What, then, is the fate of this steady input of solar energy? On average, about 30% is scattered back into space by clouds; snow; ice; atmospheric gases and aerosols, and the land surface, leaving 70% available for the heating of the atmosphere, the ocean and the Earth's surface. However, the annual average of the Earth's temperature has not changed by more than $1\,K$ over the past 100 years. Why, then, have the temperatures of the atmosphere and ocean not increased? The observation that both the temperatures of the atmosphere and the ocean have remained relatively constant implies that energy is being lost from the ocean, the atmosphere and the Earth's surface at approximately the same rate as it is being supplied by the Sun. The only mechanism by which this heat can be lost is by the emission of electromagnetic radiation from the Earth's atmosphere into space. If it is assumed that the Earth can be represented by a black body, then it is possible to apply Stefan's law to the electromagnetic emission of the Earth. Stefan's law states that the total emission of radiation from a black body over all wavelengths is proportional to the fourth power of the temperature, i.e. $E = \sigma T^4$, where σ is the Stefan – Boltzmann constant. For the emission by the Earth of an amount of radiation equal to that received from the Sun, there can be defined a radiative equilibrium temperature, T_e.

If $Q/4$ is the energy flux from the Sun and α is the fraction of solar radiation reflected back into space, known as the planetary albedo, then for the Earth's system in thermal equilibrium:

$$\left(\frac{Q}{4}\right)(1 - \alpha) = \sigma T_e^4 \tag{1.4}$$

where $\sigma = 5.67 \times 10^{-8}\,\mathrm{W\,m^{-2}\,K^{-4}}$. By inserting $\alpha = 0.3$ and $Q/4 = 340\,\mathrm{W\,m^{-2}}$, a radiative temperature of 255 K or $-18°C$ can be calculated.

This temperature is lower than might be expected. The mean surface temperature of the Earth is 15°C, or 288 K, and so it is clear that the radiative temperature bears little direct relationship to the observed surface temperature. This is because most of the planetary radiation is emitted by the atmosphere, whilst only a small fraction originates from the surface of the Earth. The temperature of the atmosphere decreases by $6.5\,\mathrm{K\,km^{-1}}$ from the surface to the tropopause, 10 km above the Earth's surface, and therefore the radiative equilibrium temperature corresponds to the temperature at a height of 5 km.

The Wien displacement equation enables the calculation of the wavelength of maximum radiation, λ_{max}, and it states that

$$\lambda_{max} T = 2897\,\mathrm{\mu m\,K} \tag{1.5}$$

If the radiative temperature, $T_e = 255\,\mathrm{K}$, is substituted in this equation, then λ_{max} is calculated to be approximately 11 μm, which lies in the middle part of the infra red spectrum. In this region of the electromagnetic spectrum, water and carbon dioxide have large absorption bands so that all substances containing water are particularly good absorbers of radiation in the middle infra red. As far as emission is concerned, water has an emissivity of 0.97, which means that it emits radiation at the rate of 97% of the theoretical black body value. Hence clouds, which are composed mainly of water droplets, are good infra red emitters. The prevalence of liquid water and water vapour in the Earth's atmosphere therefore leads to the strong absorption and re-emission of radiation. This ability of water to absorb and re-emit radiation back to the Earth's surface results in the higher observed mean surface temperature. If the atmosphere was transparent to the emitted planetary radiation, then the surface temperature would be close to the radiative equilibrium temperature.

The temperature of the surface of a planet without an atmosphere, such as Mercury, would be observed to have a surface equilibrium temperature equal to that of the radiative equilibrium temperature. Table 1.1 shows the radiative equilibrium temperatures for Mars and Venus, as well as their surface temperatures. The surface temperatures are higher than the radiative temperatures. On Mars the atmospheric mass is smaller than that of Earth by two orders of magnitude and therefore, though carbon dioxide absorbs and re-emits planetary radiation back to the surface, a surface warming of only 19 K is produced. However, on Venus the surface temperature of 760 K is sufficient to melt lead, in spite of the fact that the radiative equilibrium temperature is less than that of the Earth. This low radiative equilibrium temperature is caused by the reflection of 77% of the incident solar radiation back into space by the omnipresent clouds, which are not composed of water droplets as they are on Earth. Therefore, although the planet receives less net solar radiation than the Earth, it has a surface temperature 472 K higher. This is the result of the massive carbon dioxide atmosphere, which absorbs virtually all of the radiation emitted by the surface and re-emits it back to the surface. Furthermore, it is known that clouds of sulphuric acid also enhance this warming effect.

The ability of an atmosphere to maintain a surface temperature above the radiative equilibrium temperature is commonly known as the 'greenhouse effect'.

It is clear that, although the distance from the Sun determines the energy incident on an atmosphere, the mass and the constituents of that atmosphere are also important factors in the determination of the surface temperature of the planet.

1.4   Thermal inertia of the atmosphere

The thermal inertia of the atmosphere gives an indication of how quickly the atmosphere would respond to variations in solar radiation and it is therefore of importance in the understanding of climatic change. Consider an atmosphere having a mass M, per unit area, and a specific heat at constant pressure, C_p. The thermal inertia of the atmosphere is MC_p.

Initially, the atmosphere is in thermal equilibrium and therefore the solar radiation absorbed by the atmosphere is in balance with the emission of long wave planetary radiation into space (equation 1.4).

Now consider a situation where heat is suddenly added to the atmosphere, for example, as the result of burning fossil fuels or by a large thermonuclear explosion. How long would the atmosphere take to regain its former equilibrium? If all the heat were liberated at the same time, then the atmosphere's temperature would increase suddenly by ΔT. The planetary radiation emitted into space would now be $\sigma(T_e + \Delta T)^4$ and therefore more radiation would be emitted than received from the Sun. This deficit implies that the atmosphere would cool. The rate of temperature change, dT/dt, is proportional to the net difference between the emitted planetary radiation and the solar radiation and thus:

$$MC_p \frac{dT}{dt} = -\sigma(T_e + \Delta T)^4 + \frac{Q}{4}(1 - \alpha)$$

Rearranging and substituting from equation 1.4:

$$MC_p \frac{dT}{dt} = -\sigma T_e^4 \left(1 + \frac{\Delta T}{T_e}\right)^4 + \sigma T_e^4$$

Expanding by the binomial expansion

$$MC_p \frac{dT}{dt} = -\sigma T_e^4 \left(1 + \frac{4\Delta T}{T_e} + 6\left[\frac{\Delta T}{T_e}\right]^2 + 4\left[\frac{\Delta T}{T_e}\right]^3 + \left[\frac{\Delta T}{T_e}\right]^4\right) + \sigma T_e^4$$

Providing that $\Delta T/T_e \ll 1$, then the higher order terms in the above expression can be neglected. Therefore, retaining only the first order terms:

$$MC_p \frac{dT}{dt} = -\sigma T_e^4 \frac{4\Delta T}{T_e}$$

Since $T = T_e + \Delta T$, where T_e is independent of time, then:

$$\frac{dT}{dt} = \frac{d\Delta T}{dt}$$

Hence:

$$\frac{d\Delta T}{dt} = -\left(\frac{4\sigma T_e^3}{MC_p}\right)\Delta T \tag{1.6}$$

The solution of the above equation is:

$$\Delta T(t) = \Delta T_0 \exp(-t/\tau) \tag{1.7}$$

where ΔT_0 is the initial temperature perturbation at $t = 0$ and τ_R is known as the radiation relaxation time constant.

$$\tau_R = \frac{MC_p}{4\sigma T_e^3} \tag{1.8}$$

For Earth's atmosphere, where $M = 10.316 \times 10^3 \text{kg m}^{-2}$, $\sigma = 5.67 \times 10^{-8} \text{W m}^{-2} \text{K}^{-4}$, $C_p = 1004 \text{J kg}^{-1} \text{K}^{-1}$ and $T_e = 255 \text{K}$, τ_R can be calculated to be 2.7×10^6s or 32 days.

From equation 1.7:

$$\Delta T = \Delta T_0 e^{-1} \text{ at } t = \tau_R$$

$$\Delta T = \Delta T_0 e^{-2} \text{ at } t = 2\tau_R$$

$$\Delta T = \Delta T_0 e^{-3} \text{ at } t = 3\tau_R$$

For an initial temperature perturbation of 1 K, equivalent to the instantaneous burning of 160×10^9 t of coal, then after approximately 32 days, the temperature perturbation will be reduced to 0.37 K; after 65 days it will be 0.13 K and after 96 days it will be 0.05 K. For the Earth's atmosphere, therefore, the colossal excess of heat will be lost by planetary radiation to space within 100 days and the Earth's radiative equilibrium will be restored.

Although this particular problem is hypothetical, the concept is a useful one for understanding the response of the atmosphere to changes in the radiative balance caused by daily and seasonal variations in solar radiation. For instance, because the radiative time constant is much greater than 1 day, the lower part of the Earth's atmosphere (wherein is located the major part of the mass of the atmosphere) does not respond significantly to the daily variation in solar radiation. However, for seasonal variations, where the solar radiation changes are of a longer period than the radiative time constant, there will be a significant response in the radiation temperature of the atmosphere.

Table 1.2 shows a comparison of the radiative time constants of Mars, Venus and the Earth. On Mars, because of the small thermal inertia, the diurnal variation of temperature at the surface is very large and, as a result, large thermal tides are set up in the atmosphere.

This phenomenon is also found in the upper reaches of the Earth's atmosphere, where the atmospheric mass is small and the thermal capacity is reduced. However, on Venus the diurnal variation in temperature is only 1–2 K because of the very large radiative time constant, even though the Venusian day is 243 Earth days long.

It must, of course, be remembered that the seasonal change in temperature on Earth is modified by the thermal capacity of the ocean. The thermal capacity of the atmosphere is equivalent to 2.6 m of sea water, and therefore it can be seen that the thermal capacity of the ocean is equivalent to approximately

Table 1.2 Radiative time constant, τ_R for the terrestrial planets. C_p is the specific heat at constant pressure, p is the surface pressure and g is the acceleration due to gravity

	C_p (J kg^{-1} K^{-1})	p (hPa)	g (m s^{-2})	τ_R (days)
Venus	830	93 000	8.9	3550
Earth	1004	1013	9.8	32
Mars	830	6	3.7	1

1600 atmospheres. However, during the seasonal cycle, only the upper 100 m of the ocean is affected significantly and therefore the thermal capacity is equivalent to only 38 atmospheric masses, giving a thermal relaxation time of approximately 3.5 years. This simple sum suggests that the ocean is responsible for a major modification of the seasonal cycle of temperature in the atmosphere. An example of the oceanic influence is shown in Figure 1.5. The flux of the long wave radiation from the top of the atmosphere gives an indication of the seasonal change in temperature. In the Northern Hemisphere, where the largest ratio of continents to land exists (Figure 1.10 shows about 70% is land at 60°N), there is a large change of approximately 40 K in the radiation temperature during the seasonal cycle, whilst in the Southern Hemisphere between 30° and 60°S, where the continental area is less than 10%, there is only a small change in radiation temperature during the year. The most continental region benefits from amelioration in its seasonal climate

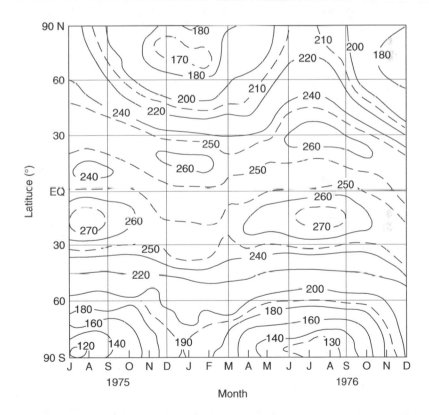

Figure 1.5 Seasonal variation of the zonal mean long-wave radiation flux, $W m^{-2}$, measured by satellite from July 1975 to December 1976. A flux of $260 \, W m^{-2}$ corresponds to a radiation temperature of 260 K, whilst a flux of $160 \, W m^{-2}$ is equivalent to a temperature of 230 K. Reproduced, with permission, from Jacobowitz, H. *et al.*, 1979, Journal of Atmospheric Science, 36: page 506, figure 6

brought about by the ocean. Even in Siberia, the winters are not as severe as they would be if the ocean did not exist.

1.5   Albedo

The planetary albedo is the ratio of the reflected (scattered) solar radiation to the incident solar radiation, measured above the atmosphere. As this reflected radiation is lost immediately from the Earth-atmosphere system, the accurate determination of planetary albedo is important in the calculation of the amount of solar radiation absorbed by the system, and therefore the climate of the Earth. On average, the global albedo is 30% ($\pm$2%). This value has been determined from a number of satellite measurements over a period of many years. The global albedo is not constant but varies on both seasonal and interannual time scales by approximately 2%. Observations have shown that the albedo is a maximum in January and a minimum in July. If the cloud and surface distributions were similar in both hemispheres, then no annual variation would be expected. However, because of the large seasonal cycle in snow extent over the Northern Hemisphere continents, particularly over the Eurasian land mass, and the seasonal cycle of cloud in the Northern Hemisphere, the global albedo has a seasonal component.

The latitudinal variation in albedo is determined by:

(i)   The elevation of the Sun.
(ii)   The distribution and type of cloud.
(iii) The surface albedo.

At latitudes less than 30°, the planetary albedo is relatively constant at 25% but it then increases with latitude to a maximum value of 70% over the Poles. However, because the polar regions occupy less than 8% of the Earth's surface, the value of the global albedo is weighted to the albedo of the middle and low latitudes. It is noted that the latitude distributions in the albedo in the Northern and Southern Hemispheres are similar, despite the difference in land and sea distribution between the two hemispheres. This indicates that the cloud distributions in both hemispheres have a dominant influence over the surface variations in albedo in determining the planetary albedo (see Figure 1.6).

The long term latitudinal variation in absorbed solar energy and the emitted planetary variation are depicted in Figure 1.7a. The absorbed solar energy has a maximum value of 300 W m^{-2} in low latitudes. The planetary radiation, which is dominated by emission from the lower troposphere, shows a decrease with latitude that occurs at a slower rate than the decrease in absorbed solar radiation. If the atmosphere were in radiative balance at each latitude, the

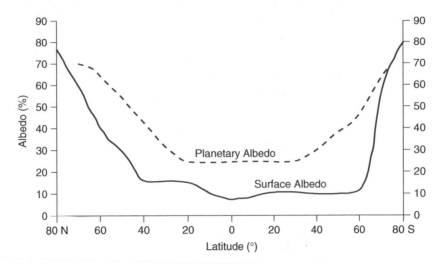

Figure 1.6 Planetary and surface albedo (%) expressed as a function of latitude. The planetary albedo is the fraction of incident solar energy reflected back into space, by the atmosphere and the Earth's surface

two curves should be identical. However, the solar absorption exceeds the planetary emission between 40°N and 40°S, and therefore there is a net excess in low latitudes and a net deficit at high latitudes (i.e. poleward of 40°). This latitudinal imbalance in radiation implies that thermal energy (heat) must be transferred from low to high latitudes by the circulation of the atmosphere and ocean. In effect, the atmosphere and oceanic circulations are maintaining higher atmospheric temperatures polewards of 40° than can be maintained by solar absorption.

To obtain the transport of heat by the atmosphere and ocean, the net radiation at each latitude, i.e. the difference between the absorbed solar radiation and the emitted planetary radiation, is integrated from pole to pole (Figure 1.7b). Thus, it is possible to determine the poleward transport of heat from satellite radiation measurements, but it is not possible to distinguish what proportion of heat is transferred by the atmosphere and what proportion by the ocean. It is noted that the maximum transfer of heat occurs between 30° and 40° of latitude and is approximately 5×10^{15} W.

Seasonal variations in the surface albedo are significant, particularly in the middle and high latitudes of the Northern Hemisphere, and arise from changes in snow and cloud cover. At 60°N the albedo varies from 65% in February to less than 40% in July, as a result of the seasonal variation in snow cover. In the Southern Hemisphere the seasonal cycle is not so pronounced but a minimum in albedo can be observed in March at 60°S, which is probably associated with the minimum sea ice extent around Antarctica. The emitted planetary radiation shows a strong asymmetry between the Southern and

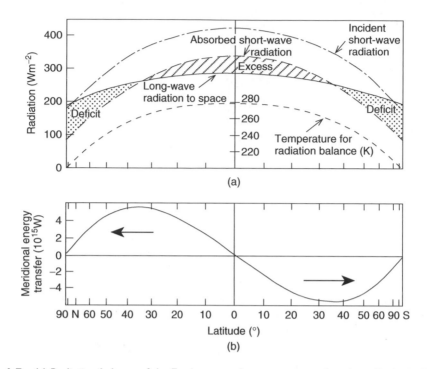

Figure 1.7 (a) Radiation balance of the Earth–atmosphere system as a function of latitude. The thin broken curve indicates an equivalent radiative temperature (centre scale) for each latitude band to be individually in balance with the absorbed solar radiation (heavy broken curve). The thin solid curve shows the actual observed long-wave radiation emitted to space as a function of latitude; (b) Poleward energy transfer within the Earth – atmosphere system needed to balance the resulting meridional distribution of net radiative gains (hatched) and losses (stippled) as indicated in (a). Note that the horizontal scale is compressed towards the poles so as to be proportional to the surface area of the Earth between latitude circles. Reproduced, with permission, from Pittock, A.B. *et al.* eds, 1978, *Climate Change and Variability: A Southern Ocean Perspective*, Cambridge University Press: page 12, figure 2.1.2

the Northern Hemispheres, which is the result of the different land-sea distributions in each hemisphere.

The net radiation budget, illustrated in Figure 1.8, indicates substantial seasonal variation as the result of the seasonal solar distribution. The zero contour varies from 15° to 70° latitude, thus showing that the zone for maximum heat transport will also vary during the seasonal cycle. The maximum net radiation gain varies between 30°S and 30°N, and is larger in the Southern Hemisphere than in the Northern Hemisphere.

The surface albedo, as has been shown, makes an important contribution to the planetary albedo although the cloud distribution is, without doubt, the dominant influence. However, the amount of solar radiation absorbed by the surface, and therefore the surface temperature, are strongly influenced by

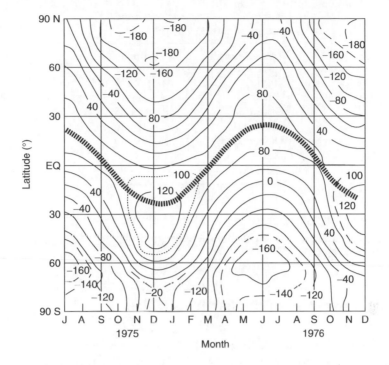

Figure 1.8   Seasonal variation of zonal mean net (solar–longwave) radiation budget, W m^{-2}, measured by satellite from July 1975 to December 1976. Reproduced, with permission, from Jacobowitz, H. *et al.*, 1979, Journal of Atmospheric Science, 36: page 506, figure 7

the albedo of that surface. The albedo of any particular surface depends on the following factors:

(i)   The type of surface.
(ii) The solar elevation and the geometry of the surface relative to the Sun.
(iii) The spectral distribution of the solar radiation and the spectral reflection.

Table 1.3 shows typical values of the surface albedo. For a calm tropical sea surface, where the Sun's elevation is high, the albedo can be as low as 2%, and therefore the oceans are generally very good absorbers of solar radiation in comparison with most types of land surface. Forests and wet surfaces generally have low albedos whilst deserts and snow-covered surfaces have high albedos. The surface albedo is also a function of the spectral reflectivity of the surface, and variations between visible and near infra red can be substantial. For instance, though freshly fallen snow has an albedo greater than 90% in the visible, at 1.5 μm, in the near infra red, its albedo is only 25%. Vegetation usually has a low albedo in the visible region but a high albedo in the near infra red. Although it is useful to be able to derive a single albedo for

Table 1.3 Albedos for different terrestrial surfaces

Type of surface	Albedo%
Ocean	2–10
Forest	6–18
Grass	7–25
Soil	10–20
Desert (sand)	35–45
Ice	20–70
Snow (fresh)	70–80

the solar spectrum, it must be remembered that the solar spectrum incident on the surface changes with solar elevation and cloud cover, and that the albedo will also change.

1.6   The topography of the Earth's surface

The Earth is a spheroid which is slightly distorted by rotation. This distortion amounts to approximately 0.3% of the Earth's mean radius, with the equatorial radius being greater than the polar radius, Because of the irregularities in the Earth's surface, geophysicists have defined a surface called the geoid. The geoid is a surface of equipotential gravity. The height of this surface varies with position relative to the centre of mass of the Earth but *along* the surface there is no gravitational force, i.e. the direction of the gravity force is everywhere perpendicular to the equipotential surface.

The ocean surface would be an equipotential surface if winds; tides; currents and density variations did not produce deformations in its surface. Oceanographers, therefore, have defined their reference level in terms of mean sea level, after the removal of tides and meteorological influences. The mean sea level has been obtained by long-term observations of the sea level at coastal stations around the world. With reference to the mean sea level, the continental surface has a mean altitude of 840 m, whilst the mean depth of the ocean is 3800 m. This implies that the average height of the Earth's crust is 2440 m below sea level.

The relative topography of the Earth's surface is depicted in Figure 1.9, where it is noted that the frequency distribution has two maxima; one corresponding to the mean height of the continents and the second corresponding to an ocean depth of approximately 4500 m. This diagram shows the most frequent height of the Earth's crust corresponds to the ocean bottom. The ocean occupies 70.8% of the Earth's surface, and 76% of this ocean has depths between 3000 m and 6000 m.

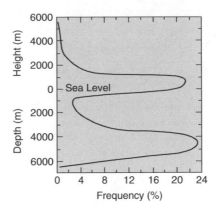

Figure 1.9 Frequency distribution of elevation intervals of the Earth's surface. Reproduced, with permission, from McLellan, H.J., 1965, Elements of Physical Oceanography, Pergamon Press: page 5, figure 1.2

These depths are known as the abyssal depths of the ocean. Of the remaining 24% of the ocean area, 7.5% is occupied by the continental shelf, which is roughly confined to depths less than 200 m, and 15.3% is occupied by the continental slope, which is the transition region between the continental shelf and the abyssal ocean. Only 1.1% of the ocean has depths greater than 6000 m and only 0.1% has depths greater than 7000 m. This small percentage is associated with ocean trenches, whose depths plunge down to 11 500 m in the Mindanao Trench.

The relative distributions of land and ocean as a function of latitude are shown in Figure 1.10. It is again noted that the land/sea distribution is asymmetrical, with over 80% of the total global land area being located in the Northern Hemisphere and 63% of the global ocean area in the Southern Hemisphere. The continental land masses of North America and Eurasia contribute to the dominance of land over ocean between 45° and 70°N. At corresponding latitudes in the Southern Hemisphere are regions where land occupies less than 3% of the total area, with essentially no land between 55° and 60°S. Poleward of 70° latitude, the situation is reversed, with the Northern Hemisphere being occupied by the Arctic Ocean and the Southern Hemisphere being occupied by the continent of Antarctica. It is believed that the presence of a continent in a polar region is a significant factor in causing glacial cycles over the past 60 million years. An interesting property of the land/ocean distribution is that 95% of all land points have antipodes in the ocean.

The distribution of the major topographic features of the globe is shown in Figure 1.11. The most extensive feature is the continental slope or escarpment which extends for more than 300 000 km, with an average slope of 4°. The continental slope marks the geological boundary between the older, relatively thick granitic continental crust and the young, basaltic ocean crust.

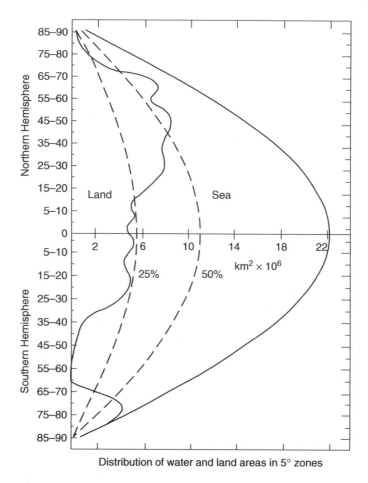

Distribution of water and land areas in 5° zones

Figure 1.10 Distribution of water and land areas in 5° zones for the Northern and Southern Hemispheres. Reproduced, with permission, from McLellan, H.J., 1965, Elements of Physical Oceanography, Pergamon Press: page 4, figure 1.1

The second most important feature is the mid-ocean ridge system which has a length of approximately 75 000 km. The typical width of the ridge is about 500 km and the height ranges from 1 to 3 km above the abyssal depths. The ridges are regions of newly formed ocean crust, with the crust moving outwards from the central part of the ridge system. A rift valley approximately 20–50 km wide, along the central part of the system, is the region of major crust formation and it is frequently associated with volcanic and seismic activity. The ocean ridges are of importance in the movement of deep water masses, because they split the oceans into smaller ocean basins.

In the Atlantic Ocean, the eastern abyssal basins are separated from the western abyssal basin by the mid-Atlantic ocean ridge. Connections between the basins, however, occur in the fracture zones which are orientated

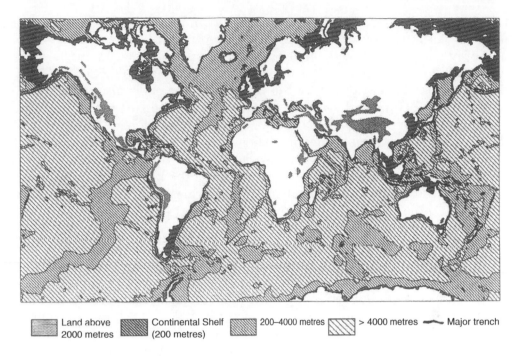

Land above 2000 metres · Continental Shelf (200 metres) · 200–4000 metres · > 4000 metres · Major trench

Figure 1.11   The major topographic features of the ocean and the continents

perpendicular to the axis of the ridge. Two important fracture zones in the Atlantic Ocean are the Romanche gap, close to the equator, which allows Atlantic Bottom Water to flow into the eastern basins, and the Gibbs fracture zone at about 53°N, which allows the North Atlantic Deep Water to flow westward into the Labrador basin. Furthermore, the ridge system has an influence on oceanic flows above the level of the ridge topography and in some regions, particularly in the Southern Ocean, its effect extends to the ocean surface currents.

The relatively low relief of the land masses has already been noted and only 11% of the land area is above 2000 m. However, there are two significant features associated with Tertiary mountain building that are worthy of further comment. These are the Western Cordillera of the American continent, which has an altitude of over 6 km above sea level, and the Himalaya mountain chain, with altitudes in excess of 8 km. These mountain ranges are very important in their influence on the general circulation of the atmosphere. Although the atmosphere extends well beyond the highest peaks, 50% of the atmospheric mass occurs below 5 km, and therefore these large mountain chains have a major disruptive effect on the flow of the lower atmosphere. The Western Cordillera has a predominantly north-south orientation and this disrupts the westerly flow for over 5000 km downstream. The Himalayas have an east-west orientation which results in an intensification of the zonal flow and an inhibition of the northward penetration of the Asian monsoon. Not only is the flow directly affected by the mountain barriers but, because most of the water vapour in the atmosphere is confined to the lowest 5 km, the geographical distribution of the hydrological cycle is also significantly affected. Furthermore, the release of latent heat by condensation into the atmosphere provides a major source of energy for the general circulation of the atmosphere and the mountain chains will influence the location of heat sources in the atmosphere and the associated low-level atmospheric flow.

A feature worthy of note is the comparison of the horizontal scales of the ocean and the atmosphere with their vertical scales. The ocean basins have horizontal dimensions of 5000 km in the Atlantic Ocean to over 15 000 km in the Pacific Ocean, whilst the average depths are approximately 3.8 km. Thus the vertical scale is small compared with the horizontal scale and the ratio between them, known as the aspect ratio, is at most 1 part in 1000 and, more typically, 1 part in 10 000. The atmosphere has 99% of its mass between the sea surface and a height of 30 km, and compared with the circumference of the globe, the aspect ratio is again less than 1 in 1000. Thus both atmosphere and ocean are very thin when compared with their horizontal dimensions.

It has been shown that the topography of the land masses has a significant effect on the circulation of the lower atmosphere. In a different way, the shape and size of the ocean basins have profound effects on the circulation of water masses. Except for the region lying between 50° and 60°S, the ocean is bounded by continental land masses and therefore zonal (i.e. west-east) flows are restricted to the Southern Ocean. The presence of continental margins produces intense meridional (i.e. north-south) flows in the western regions of the basins, as evidenced by the Gulf Stream in the North Atlantic and the Kuroshio current in the North Pacific. All the ocean basins are connected via the circumpolar flow of the Southern Ocean. In contrast, the flow into the Arctic Ocean from the Pacific Ocean is inhibited by the shallowness of the Bering Sea, whilst from the Atlantic there are 'sills' which control the flow between Greenland and Iceland, and between Iceland and Scotland. A sill is a relatively shallow region which connects either two deep ocean basins or a deep marginal sea with an ocean basin.

There is a deep-water connection between the Indian Ocean and the Pacific Ocean, through the Indonesian Archipelago, which has important influences on the flow around the Australian continent, and on the exchange of surface and intermediate water masses between these two ocean basins.

The width of the ocean basins has a marked effect on the ocean-atmosphere interaction. The relatively narrow North Atlantic Ocean is exposed to drier air from the surrounding continents than the Pacific Ocean, and therefore evaporation is approximately twice as large in the North Atlantic Ocean as in the Pacific Ocean. This high evaporation rate is a factor in the larger heat loss from the North Atlantic Ocean than from the Pacific Ocean, and hence dense, cold water masses are formed in the North Atlantic Ocean rather than in the Pacific Ocean. Thus, the deep-water vertical circulation in the North Atlantic Ocean is much more intense than the sluggish vertical circulation observed in the Pacific Ocean. It will be shown in Chapter 5 that this has consequences for the heat transport in the ocean. In the context of evaporation, the marginal seas, such as the Mediterranean Sea and the Red Sea, are important regions for the formation of dense water masses because of their high evaporation rates.

The continental shelf, though occupying a relatively small area of the ocean, is of considerable importance both to human kind and to life in the ocean. The intense flows, associated with winds and tides in the continental shelf region, result in the mixing of the nutrients from continental shelf sediment into the water column. These nutrients, in conjunction with readily available oxygen and sunlight in the shallow sea areas, produce an abundance of phytoplankton. These regions of high primary production are very important in the maintenance of marine food chains and contain the most intensive fisheries in the world. The sedimentation rates on the continental shelves are also high, being of the order of 1 m per 1000 years in comparison with 0.1–10 mm per 1000 years in the deep ocean. Much of the sediment deposited is derived from river-borne material and this contains large quantities of phosphorus which is an element necessary for the maintenance of productivity in the ocean.

In conclusion, it has been shown that the surface topography and the shape and interlinkage of the ocean basins have a significant effect on the circulation of both the atmosphere and ocean. This, in turn, affects the biological productivity of the land and ocean. In the geological past, different ocean geometry and surface topography would have been partly responsible for the many varied climatic patterns that have occurred since the formation of the Earth.

2

Composition and Physical Properties of the Ocean and Atmosphere

2.1 Evolution of the atmosphere and ocean

Direct evidence of the form and composition of the early Earth have long since been destroyed and clues can only be gleaned from an understanding of geochemical processes. Various radioactive elements can be used to evaluate the timing of the development of the atmosphere and understanding of the behaviour of certain elements allows estimation of the atmospheric and oceanic composition through time.

The Earth and other planets were formed some 4.6×10^9 years ago (4.6 Ga) via an accretionary process involving collisions between cosmic dust through to bombardment of planetismals by meteorites, eventually resulting in the solar system we know today. This accretion is inferred to have continued for some 150 million years and the extensive bombardment of the young Earth continued until at least 3.8 Ga before present.

The accepted paradigm for oceanic evolution involves precipitation from a transitory steam atmosphere created by the continued impact and heating of the proto-Earth surface during planetary accretion. The ocean precipitated out as the Earth cooled and has remained liquid throughout the intervening geological time. This has allowed development of life on Earth and puts lower limits on any climatic variations inferred to have taken place, since the oceans have never frozen completely. Upper constraints to surface temperatures and atmospheric structure are imposed by the presence of a large hydrosphere on Earth. The present day vertical atmospheric temperature structure means that water vapour precipitates in the upper troposphere and the concentration of water vapour in the stratosphere is very low. This retains water within the

The Atmosphere and Ocean: A Physical Introduction, Third Edition. Neil C. Wells.
© 2012 John Wiley & Sons, Ltd. Published 2012 by John Wiley & Sons, Ltd.

lower atmosphere and limits the passage of H_2O to the upper atmosphere, where extensive photodissociation would occur and the H_2 evolved would escape into space due to its low molecular mass.

Comparison of the rare, inert gas composition of the atmosphere with that of the Sun and average solar system abundancies demonstrates that the Earth lost its primordial atmosphere during the accretionary process, which involved heating and melting of the new planetary mass. The current atmosphere formed as a consequence of degassing of the cooling planet Earth once accretion ceased. Again, rare gas composition can be used to determine the timing of the atmospheric development; 80–85% of the present day rare gas content degassed from the Earth's interior in the first million years of Earth's history, whilst the remainder has leaked constantly out during the intervening eons.

Volcanism is the major route of degassing of volatile material from the inner Earth. Present day volcanic activity produces H_2O, CO_2, SO_2, N_2, H_2 and Cl_2 in substantial quantities. The composition of early volcanic gases is a matter of debate and there may have been a substantial component of NH_3 and CH_4 depending on the extent of oxidation of source rocks in the upper mantle.

The early solar system was illuminated by a weak, young Sun that only delivered ~75% of the present day energy to the Earth's surface. To provide a non-frozen ocean there must have been an increased 'greenhouse' effect provided by CO_2. One-dimensional climate models require CO_2 levels of 500 times present atmospheric levels (PAL) at 2.75 Ga. However, recent estimates of atmospheric CO_2 levels from study of ancient soils (palaeosoils) suggest upper limits on Archaen partial pressure of CO_2 of 100 PAL. The additional planetary warming must be derived from other greenhouse gases such as NH_3 and CH_4 or albedo variations. The same palaeosoil samples allow estimation of the early atmospheric O_2 content as being approximately 1.4×10^{-4} PAL. This O_2 was mainly derived from the photodissociation of H_2O in the upper atmosphere.

The behaviour of CO_2 is crucial to our understanding of early Earth history. Early levels were elevated and, by the Cambrian era (~600 Ma), CO_2 levels were close to PAL. The main reason for the extraction of CO_2 out of the atmosphere was the development of life forms in the ocean that sequestered carbon in organic and later inorganic (calcium carbonate) forms and buried it in sedimentary formations on the sea floor. Life must have originated as the planetary bombardment slowed after 3.8 Ga and the oldest known fossils are blue-green algae dated 3.5 Ga. The precursors to these organisms were almost certainly high temperature (>70°C) chemosynthetic archaea living in specialized niches such as submarine hydrothermal vent systems in areas of underwater volcanic activity.

Photosynthesis involves the oxidation of liquid water and the reduction of carbon dioxide to carbohydrate and the release of oxygen;

$$6CO_2 + 6H_2O \rightarrow (CH_2O)_6 + 6O_2$$

The reverse of this process is respiration which is carried out by all life forms. There is extensive evidence from palaeosoils and Fe-rich sedimentary formations that atmospheric O_2 levels did not achieve appreciable levels until 2 Ga and only approached present-day levels by 1.5 Ga. Until then, all O_2 produced was consumed in the oceans by respiration and reduction by dissolved chemical species such as Fe(II). Once O_2 started to accumulate in the atmosphere, the production of ozone (O_3) could occur. Ozone plays a vital role in the upper atmosphere by shielding the Earth's surface from damaging ultra violet (UV) radiation, allowing life to move from a purely oceanic realm and begin to colonize the land mass. The evolution of land plants and the terrestrial biosphere would extract more CO_2 from the atmosphere and provide an additional source of O_2.

The present day atmosphere is dominated by chemical species that demonstrate the existence of life on Earth. In the modern system, 99.8% of the present day carbon is locked up in sedimentary rocks of mainly biogenic origin, whereas only 1.0×10^{-3}% of carbon is in the atmosphere as CO_2 (Table 2.1). The atmospheric O_2 level of 21% represents the standing crop of photosynthetic products maintained by global primary production. This contrasts starkly with the early Earth atmosphere.

Comparison of Earth with its nearest neighbours in the solar system is illuminating. The starting products for Venus, Earth and Mars were likely to be similar as all three planets were formed by the same accretionary process. Venus has not developed the atmospheric temperature inversion between troposphere and stratosphere essential for maintaining liquid water. Venus has a dense CO_2 rich atmosphere with a small component of water vapour and

Table 2.1   The carbon budget for the Earth

Reservoir	Mass of carbon (10^{12} kg)
Atmosphere	748
Ocean	36 930
Land:	
Vegetation	625
Soil	2700
Sediments mainly carbonate rocks	65 500 000
Fossil fuels	5000

extremely high surface temperatures of 460°C. Studies of the heavy isotope of hydrogen, deuterium, suggest a large proportion of the early Venus water budget was lost to space via photodissociation in the upper atmosphere. In contrast, Mars has a very thin, CO_2 dominated atmosphere and cold surface temperatures, with solid CO_2 icecaps. There is some evidence from fluvial-type deposits that running water was in existence on the surface of Mars at 3.7 Ga, but it has been absent in liquid form since this time.

Water ice has recently been discovered in the Martian polar ice cap regions. Again, photodissociation in the upper atmosphere and escape of H_2 to space is invoked as a mechanism for water loss. The O_2 evolved by this process has oxidized the Fe(II) in the surface rocks to produce the red colour characteristic of Fe(III).

As already intimated, the atmospheric composition of greenhouse gases plays a major role in the Earth's climate. Evidence from fossils deposited on the sea floor allows estimation of sea-water temperatures through geological time. The mean surface temperature has decreased throughout the first few billion years of Earth's history. The first geological evidence for glaciation occurred between 2.5 and 2.3 Ga. However, earlier glaciations cannot be ruled out due to the paucity of geological evidence for this period. Climate variations have occurred throughout geological time and the geological record becomes more and more complete as the modern era is approached.

The Earth has undergone a major cooling episode over the last 65 Ma. This is associated with a decrease in CO_2 levels and the development of permanent ice sheets in the polar regions. Approximately 2 Ma ago the Earth went into a series of glacial-interglacial cycles, the most recent glaciation ending about 10 000 years ago. The pace-maker behind these climate oscillations has been attributed to variations in orbital parameters that determine surface insolation named Milankovitch cycles as previously discussed in Chapter 1.

Whatever drives the timing of glacial-interglacial cycles, it is apparent from evidence recorded in ice cores; deep-sea sediments; corals; varved lacustrine sediments and climate modelling that there have been changes in atmospheric CO_2 levels over the past 1 Ma. Atmospheric CO_2 levels were as low as 0.7 PAL during the last glacial maximum. Much effort has gone into understanding atmospheric and oceanic variability over the last few hundred thousand years in an attempt to understand the short term climate variability. This may allow prediction of the consequences of fossil fuel burning and deforestation that act to increase atmospheric CO_2 levels by 0.5% per annum.

2.1.1   The development of the ocean

Once accretion of the Earth stopped, the surface geological process of crustal generation and plate tectonics that we are familiar with today could begin.

Zircons are resistant mineral grains from continental crust that can withstand intense weathering and metamorphic processes. The oldest zircons, from Western Australia, have been dated at 4.1 to 4.3 Ga, giving a limit to the onset of continental crust generation. In contrast, the oldest oceanic crust is of the order of 200 Ma in age, a mere 0.4% of Earth's history. New oceanic crust is continually created at spreading centres located along the mid-ocean ridges and is destroyed at the convergent margins, where the ocean crust is subducted into the Earth's mantle. The rate of spreading is typically a few centimetres each year, with the fastest rates at the East Pacific Rise and the slowest at the mid-Atlantic ridge.

The amount of water in the hydrosphere is in steady-state balance between the input of juvenile, magma-derived water from the creation of new crust and the loss by photodissociation in the upper atmosphere. Sea level has varied through time by several hundred metres as varying amounts of water were locked up in the ice caps. In general, the rule of isostatic equilibrium means that the oceans fill the oceanic basins and flooding of continental crust is not extensive.

2.2 Present-day composition of sea water

Dissolved salts comprise about 3.5% by volume of sea water. The salts are present because of the relative ease with which ionic compounds dissolve in water. Originally it was believed that the salts dissolved in the ocean were derived from the weathering of continental rocks, the products of which were then transported by rivers to the ocean. Robert Boyle (1627–1691) made measurements of the composition of river water and he showed that salts could be detected. However, with more accurate methods of analysis, it was found that, although the river waters contained potassium; magnesium; calcium and sodium ions, they were deficient in dissolved carbon dioxide (i.e. carbonate and bicarbonate/hydrogen carbonate); chlorides; bromides and iodides. In fact, if the ocean salts had been derived from this one riverine source, their composition would have to be totally different from that observed (Table 2.2). The ions missing from land weathering products originated from volcanic and hot spring discharges, and atmospheric sources.

The observed distribution is not, however, simply determined by the inputs from weathered rock and volcanic emissions because ions are simultaneously being lost from the water column by sedimentation on the ocean floor and continental shelves, and by transfer to the atmosphere and biosphere. The present-day composition is, therefore, a geochemical balance between the sources and sinks of the ions. It is believed the ocean has been in equilibrium for 100 Ma years and that the total salt content of the ocean has not changed significantly in that time.

Table 2.2 Composition of standard ocean water (salinity $= 34.7\,psu$)

Constituent ion	Symbol	Mass of ion in grams for 1 kg of sea water	Percentage of global salt
Chloride	Cl^-	19.215	54.96
Sodium	Na^+	10.685	30.58
Sulphate	SO_4^{2-}	2.693	7.70
Magnesium	Mg^{2+}	1.287	3.69
Calcium	Ca^{2+}	0.410	1.17
Potassium	K^+	0.396	1.13
Bicarbonate	HCO_3^-	0.142	0.41
Bromide	Br^-	0.067	0.19
Boric acid	H_3BO_3	0.026	0.07

This deduction is based on the fact that the pH of the ocean must have remained reasonably constant to enable the present life forms to evolve. It is noted that a 1% decrease in the sodium ion concentration would result in a strongly acidic ocean with a pH of 2.6. However, the ratios of the other constituent salts in the ocean have shown considerable fluctuations. For example, the Na^+/K^+ ratio was approximately 1 in the pre-Cambrian period (1 Ga) and now it is 28. The changes in chemical composition are related to the evolution of organisms which extracted materials such as calcium and silicon for shell building and so altered the steady state balance.

There is recent evidence that the enhanced atmospheric CO_2, due to anthropogenic emissions, is starting to reduce the pH of the present day ocean and is leading to a slight acidification.

The first comprehensive analysis of the composition of sea water was made by William Dittmar who analysed 77 samples of sea water collected by HMS Challenger between 1872 and 1877. The samples were representative of all of the major ocean basins and were taken at depths ranging from the surface to abyssal layers. Dittmar analysed the samples for the eight major constituents (see Table 2.2) and showed that, within narrow limits, the ratios of these major constituents were in constant proportion. Together with boric acid, these eight constituents comprise 99.9% of the total dissolved salts in the ocean. The remaining 0.1% represents virtually all of the stable elements in the Periodic Table and these elements do not generally obey the rule of constant proportions. The most prominent of the minor constituents are strontium; silicon; iron; lithium; phosphorus and iodine as well as the dissolved gases oxygen and nitrogen, all of which are fundamental to the chemical and biological processes in the ocean.

Dittmar's experimental proof that the major constituents were present in constant proportions to each other throughout the world's oceans, including the Mediterranean and Baltic Seas, allowed oceanographers to define the salt content of the ocean as a single physical quantity, known as the salinity. From a determination of the salinity and the rule of constant proportions, the concentration of any one of the major constituents can be determined. Furthermore, in order to obtain a value of the salinity of sea water, it is only necessary to measure one of its major constituents. The most easily measured constituent is the chloride content of sea water. In practice the chlorinity (Cl) is determined, which is the amount of chlorine (grams) in 1 kg of sea water, where both bromine and iodine are replaced by chlorine. The relationship between salinity and chlorinity is given by:

$$\text{Salinity} = 1.80655\,\text{Cl}$$

By titration against silver nitrate, chlorinity can be measured and the salinity obtained to an accuracy of ± 0.01.

Since the 1980s, the titration method has been replaced by the direct measurement of the electrical conductivity of sea water. The definition of salinity has also been redefined to be the ratio of the electrical conductivity of sea water to a standard solution of potassium chloride at standard temperature and pressure. This ratio is expressed in Practical Salinity Units (psu) where $1\,\text{psu} = 1.00510 \times \text{salinity}$. The measurement accuracy of ± 0.001 was a considerable improvement over the titration method.

In 2009 a new definition of salinity has been adopted called the absolute salinity. This absolute salinity allows for spatial differences in sea water composition. The rule of constant proportions for seawater proposed by Dittmar is no longer strictly valid for the global ocean.

To avoid confusion salinity will be given as a simple ratio without units.

2.3   Introduction to gases and liquids

The relationship between pressure, p, temperature, T, and the density, ρ, for an ideal gas is:

$$p = \rho\,RT \tag{2.1}$$

where R is the gas constant for the constitutional gas having molecular weight m.

$$R = R^*/m \quad \text{where } R^* \text{ is the universal gas constant.}$$

The derivation of the ideal gas equation depends on the assumption that the molecules in an ideal gas are so far apart that intermolecular forces can be disregarded except during the very brief collisions. The strong intra-molecular forces within ideal gas molecules allow them to be treated as hard spheres and their collisions can be considered to be inelastic. Many gases, including the main atmospheric constituents, behave as ideal gases. The only exception is water vapour.

Such simple assumptions are invalid when dealing with the liquid and the solid state. In a liquid, the intermolecular forces cannot be neglected or generalized, since they vary from one liquid to another. It is not, therefore, possible to produce an ideal equation of state for an ideal liquid. The long-range attractive intermolecular forces in a liquid allow a much more efficient packing of the molecules than is found in a gas. Hence, liquids have densities approximately 100–1000 times greater than the density of a gas. For instance, air at normal atmospheric pressure and 0°C has a density of $1.28\,kg\,m^{-3}$, whilst sea water (salinity = 35) has a density of $1028\,kg\,m^{-3}$.

At very short ranges, of the order of 0.1 nm, the intermolecular repulsive forces become dominant and there is, therefore, a limit to the close packing of molecules. Solids may have densities approximately 15% greater than liquids, but this increase is achieved by more efficient long-range packing arrangements rather than smaller intermolecular separation. In this respect, the densities of water and ice are anomalous. Ice has a density which is 9% less than liquid water and thus ice floats on water.

The liquid state has properties similar to the solid state on a short time scale in that small clusters of molecules are ordered in a similar way to the solid state lattice. However, on a large time scale the liquid resembles a gas in that it is highly disordered. A liquid may therefore be considered, in some respects, as being an intermediate state between the extreme ordering of the solid state and the randomness of the gaseous state. On this longer time scale, a liquid and a gas will not support a shear stress and, as a result, they will yield and flow. In this case we may call the gas or the liquid a fluid and we can therefore mathematically describe properties of motion without referring to the substance as either a liquid or a gas.

Gases, furthermore, have the property that they will expand to fill the space available and therefore it is impossible for a gas to have a free surface. A liquid, on the other hand, always has a free surface, and therefore it is possible to define the interface between the atmosphere and ocean. Another important property of a gas is that it is readily compressible due to the large distances between molecules. However, because of the close packing in a liquid, it is difficult to compress a liquid without resorting to large pressures.

2.3.1   The equation of state for the atmosphere

The atmosphere can be described as a mixture of gases, predominantly oxygen; nitrogen; carbon dioxide; argon and water vapour (Table 2.3). Of these major constituents, only water vapour is a significantly variable constituent, and it is therefore considered separately from the other gases. The equation

Table 2.3   Composition and scale heights of terrestrial atmospheres

	Major constituents (% of molecules)	Mean molecular weight (amu[a]) (equation 2.2)	Gas constant (J kg^{-1} K^{-1})	Scale height (km) (equation 2.4)
Venus	95.5% CO_2 4% N_2	43.35	192	14.9[b]
Earth	78% N_2 21% O_2	28.97	287	7.4
Mars	96% CO_2 2.5% N_2 1.5% Ar	43.55	191	10.9

[a] a.m.u is atomic mass unit
[b] Lower atmosphere

of state for dry air can be obtained by summing the partial pressures of each constituent gas. If each gas occupies a similar volume, V, and has a mass of M_n, then the partial pressure p_n is given by:

$$p_n = \left(\frac{R^*}{m_n}\right)\frac{M_n}{V}T$$

where m_n is the molecular weight of the gas. The total pressure for N constituent gases is:

$$p = \sum_{n=1}^{N} p_n = \frac{R^* T}{V} \sum_{n=1}^{N} \frac{M_n}{m_n}$$

If the mean density of the mixture is $\overline{\rho} = \sum_{n=1}^{N}\left(\frac{M_n}{V}\right)$

Then the pressure for the mixture becomes: $p = \overline{\rho}\left(\frac{R^*}{\overline{m}}\right)T$

$$\text{where } \frac{1}{\overline{m}} = \frac{\sum_{n=1}^{N} M_n/m_n}{\sum_{n=1}^{N} M_n} \tag{2.2}$$

The term in the brackets, is R_D the gas constant for **dry air**. R_D has a value of 287 J K^{-1} kg^{-1}. The density of the mixture of gases is ρ.

The final equation of state for the mixture is gases is:

$$p = \rho R_D T \tag{2.3}$$

Comparing equation 2.3 with 2.1, it is noted that the equation of state for a mixture of gases is similar to a single constituent equation of state, provided that the molecular weight is replaced by a mean molecular weight for the mixture.

Table 2.4   Major constituents of the Earth's atmosphere by mass

Constituent	M_n	m_n
N_2	75.51	28.02
O_2	23.14	32.00
Ar	1.28	39.94
CO_2	0.02	44.01

Using an example for Earth with the values in Table 2.4.
For the major constituents:

$$\frac{1}{m} = \frac{75.51/28.02 + 23.14/32.00 + 1.28/39.94 + 0.02/44.01}{99.95}$$

Hence the mean molecular mass is 28.96.

2.3.2   The equation of state for sea water

The equation of state for liquid water is a function of three parameters: temperature, pressure and density. For sea water, the relationship involves the properties of all of the constituent salts and is therefore extremely complicated. However, as discussed in Section 2.2 the concept of salinity allows the individual constituents to be replaced by one variable to a good accuracy. As discussed earlier, there is no general statistical theory for all liquids and therefore the relationship has been determined by laboratory measurements and the results tabulated. The *in situ* density of sea water can be given as $\rho = \rho\,(T, P, S)$. Table 2.5 shows the value of *in situ* densities for extreme conditions in the ocean. It is noted that *in situ* density increases with both increasing pressure and salinity but decreases with temperature. However, the variations in density are less than 7% and it is therefore common practice to define

$$\sigma(T, P, S) = \rho(T, P, S) - 1000 \, \text{kg m}^{-3}$$

It is noted that sigma (σ) does not have any units.

Table 2.5   Density of ocean sea water for extreme ranges of temperature and pressure in the ocean

Depth (m)	Pressure (Pa)	Density (kgm^{-3}) ($T = 273\,\text{K}, \ S = 35\,\text{psu}$)	$T = 303\,\text{K}$ $S = 35\,\text{psu}$
0	10^5	1028	1022
10 000	10^8	1071	1060

The values of $\sigma(T, P, S)$ can be obtained from standard algorithms (e.g. UNESCO 1981; Thermodynamic Equation Of Seawater 2010). However, for many purposes, the effect of pressure is eliminated by using a reference pressure, i.e. mean sea level atmospheric pressure, to define density. This quantity is referred to as sigma-t or σ_t.

Figure 2.1 shows the relationship of σ_t as a function of temperature and salinity at atmospheric pressure. Although in all cases density increases with increasing salinity and decreasing temperature, the relationship is non-linear. At temperatures of 25°C–30°C, typical of the equatorial ocean, the isopycnals (lines of constant density) are reasonably linear and a simple relationship for σ_t can be used. However, at low temperatures a given salinity change produces a larger change of density than the corresponding change in salinity at high temperatures. This physical property of the equation of state results in a greater sensitivity of density changes to salinity at high latitudes and is an important influence on the formation of bottom water. Another interesting property of the equation of state is that two water masses with the same density but having different temperature and salinity (see Figure 5.18), when mixed, will produce a denser water than either of the two original water masses. Again, this is an important process in the formation of dense bottom waters in polar regions of the ocean. If the ocean was well mixed in temperature and salinity, because of the non-linearity of the equation of state, there would be an increase in density of the world's oceans. There would a corresponding decrease in ocean volume, because the ocean mass is close to constant. This would result in a decrease in sea level.

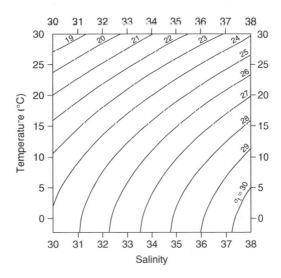

Figure 2.1 Density of sea water (σ_t) as a function of temperature and salinity. Reproduced, with permission, from Friedrich, H. and Levitus, S., 1972, Journal of Physical Oceanography, 2: page 516, figure 4

The anomalous behaviour of pure water compared with other liquids has already been briefly mentioned. In particular, the maximum density of liquid water at 4°C and the expansion of water on freezing are two of its remarkable properties. Other properties include the high dielectric constant, which results from the high electrical dipole with relatively small molecular volume. Because the interionic force of attraction varies inversely with the dielectric constant, the attraction between ions is diminished in water and hence ionic components are easily dissolved in water. The presence of dissolved salts in sea water increases the electrical conductivity by 10^6 over that of pure water. This property has been used to advantage in accurate electrical measurement of salinity in sea water, which is a reliable technique available to oceanographers. Additional remarkable features of water are its high specific heat, its high boiling point compared with other hydrogen based liquids (e.g. 213 K for H_2S and 240 K for NH_3) and its large latent heat of vaporisation of $2.3 \times 10^6 \, J \, kg^{-1}$, compared with $0.55 \times 10^6 \, J \, kg^{-1}$ for $H_2 S$ and $1.4 \times 10^6 \, J \, kg^{-1}$ for NH_3. This property allows water to be a major factor in the transfer of energy between the ocean and atmosphere.

The influence of salinity on the properties of water will be shown to be very important. Figure 2.2 shows the temperature of the maximum density of sea water and the freezing point as a function of salinity. For salinities less than 24.69, water reaches a maximum density before the freezing point is reached. Therefore between the temperature of maximum density and the freezing

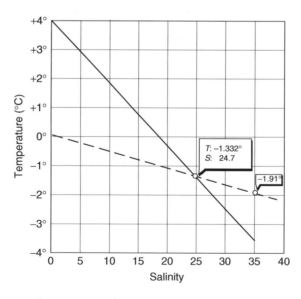

Figure 2.2 Changes in the maximum density temperature as a function of salinity (solid line). Changes in the freezing point as a function of salinity (dashed line). Reproduced, with permission, from Tchernia, P., 1980, Descriptive Regional Oceanography, Pergamon Press: page 162, figure 5.40

point there is an anomalous decrease in density with decreasing temperature. However, for a water mass having a salinity in excess of 24.69, which is true of all ocean water, the freezing point is reached before the maximum density of sea water. In this case, there is no anomalous behaviour of sea water and the densest water masses (for a given salinity) are associated with the lowest surface temperatures. For most ocean water masses, this temperature is −1.9°C. A second property of salinity is the effect on the specific heat of water. For pure water at 0°C the specific heat is $4218\,\mathrm{J\,kg^{-1}\,K^{-1}}$, whilst for sea water it is $3940\,\mathrm{J\,kg^{-1}\,K^{-1}}$ for a salinity of 35. Therefore sea water is 6% less efficient in storing heat than the equivalent mass of fresh water.

The propagation of sound in sea water is given by:

$$c = \sqrt{\frac{K}{\rho_0}}$$

where c is velocity of sound, ρ_0 is the mean density and K is the compressibility of sea water under adiabatic conditions. K is dependent on the temperature, pressure and salinity. The velocity of sound increases with temperature, pressure and salinity and typical values are shown in Table 2.6.

Figure 2.3 shows the variation in sound velocity for typical temperature and salinity profiles in the ocean. In the deep ocean (>1500 m), because of the uniformity of temperature and salinity (Figure 2.4), the sound velocity is dominated by pressure and therefore increases with depth. In the upper ocean, where large variations in temperature and, to a lesser extent, salinity

Table 2.6   Sound speed c in m/s and its dependency on pressure (decibars) and temperature °C for fresh water and ocean water for a salinity of 35. The pressure $p = 0$ corresponds to mean surface atmospheric pressure

Pressure (dB)	Temperature °C	Salinity	Sound Speed C (ms^{-1})
0	0	0	1402.4
1000	0	0	1418.0
4000	0	0	1468.0
10 000	0	0	1577.4
0	0	35	1449.1
1000	0	35	1465.5
4000	0	35	1516.5
10 000	0	35	1623.2
0	30	35	1545.6
1000	30	35	1562.4
4000	30	35	1612.5
10 000	30	35	1710.1

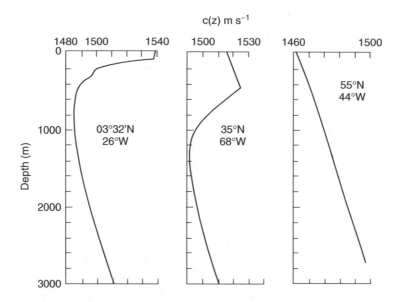

Figure 2.3 Velocity of sound as a function of depth, for tropical (left), temperate (centre) and sub-polar (right) oceans. Between the surface and 1000 m, temperature is the dominant effect, whilst pressure becomes dominant at depths greater than 1000 m. Reproduced, with permission, from Tolstoy, I. and Clay, C.S., 1966, Ocean Acoustics: Theory and Experiment in Underwater Sound, McGraw-Hill: page 7, figure 1.2

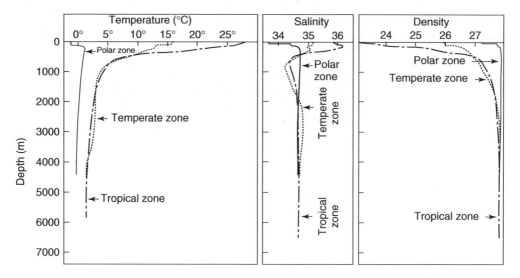

Figure 2.4 Temperature, salinity and density (σ_t) as a function of depth in the tropical, temperate and polar oceans. Reproduced, with permission, from Tchernia, P., 1980, Descriptive Regional Oceanography, Pergamon Press: page 35, figure 3.4

occur, the sound velocity displays more variability. The large decrease in temperature with depth, especially through the main thermocline, causes a decrease in velocity with depth which is larger than the opposing pressure effect. Hence, a minimum velocity is observed at a depth of approximately 1 km in profiles where the main thermocline is present. This minimum is very important for the propagation of sound and is known as the SOFAR channel. Sound waves emitted either upwards or downwards from this level will exhibit refraction into the sound channel because of the variation of the velocity of sound with depth. In this region, therefore, sound energy is trapped in a wave guide and a signal may propagate over hundreds and even thousands of kilometres before being significantly attenuated. This property has been used by oceanographers for the distant monitoring of SOFAR floats from research ships and shore bases.

2.4 Hydrostatic equilibrium

An observer descending form the outer reaches of the atmosphere to the Earth's surface with a barometer would notice an increase in pressure. The increase in pressure would not be linear but would show an exponential increase with depth. Though pressure itself is a scalar quantity, the vertical variation of pressure in the atmosphere produces a pressure force which acts in the direction of decreasing pressure (i.e. upwards) The observed pressure-gradient force would be able to accelerate a parcel of air upwards at ∼ 10 m s^{-2}. In the absence of other forces, this acceleration would produce a velocity of ∼5000 km hour^{-1} at a height of 100 km. Though the atmosphere is always in a state of motion, measured vertical velocities are rarely larger than 10 m s^{-1} and therefore this upward pressure gradient must be close to balance with the gravitational force acting towards the centre of the Earth. This balance between the vertical pressure gradient and the gravitational force is known as hydrostatic equilibrium. Ideally, the state requires no motion and therefore the atmosphere is never quite in hydrostatic balance. However, the approximation is very good for large scale motion and only breaks down in regions of large vertical acceleration associated with thunderstorm updrafts and mountain waves.

Consider a slab of atmosphere of thickness δz and cross-section area δA. The pressure gradient force (P.G.F.) across the slab is:

$$\left[p - \left(\frac{\partial p}{\partial z} \right) \delta z \right] \delta A - p \delta A$$

Hence P.G.F. $= - \left(\frac{\partial p}{\partial z} \right) \delta z \delta A$

The force due to gravity on the element of the atmosphere is $\rho \delta A \delta z g$, hence the balance is:

$$-\left(\frac{\partial p}{\partial z}\right) \delta z \delta A = \rho g \delta z \delta A$$

or

$$\frac{\partial p}{\partial z} = -\rho g \tag{2.4}$$

The hydrostatic equation 2.4 is a very important relationship in both meteorology and oceanography because it relates the pressure to the depth of fluid. In the atmosphere the density also varies with the height because of the compressibility of gases and hence equation 2.4 is not sufficient. Since the atmosphere is a very good approximation to an ideal gas, the ideal gas equation can be used to eliminate density in favour of temperature and pressure. The density from equation 2.3 can be substituted into equation 2.4 to obtain:

$$\frac{\partial p}{\partial z} = -\frac{pg}{R_D T}$$

If T is replaced by the vertical mean temperature of the atmosphere $\overline{T}$, the equation can be integrated from the surface ($z = 0$) to yield the pressure at a height z':

$$\int_{p_0}^{p'} \frac{\partial p}{p} = -\frac{g}{R_D \overline{T}} \int_0^{z'} \partial z$$

or

$$p' = p_0 \exp(-z'/H) \tag{2.5}$$

where p_0 is the surface pressure and $H = R_D \overline{T}/g$.

This equation is sometimes known as the barometer equation. The quantity H is known as the scale height and is a measure of the depth of the atmosphere. When $z' = H$, then $p' = p_0 e^{-1}$ or $p' = 0.37 p_0$. Therefore, H is the height at which the pressure is 37% of the surface atmospheric pressure. Furthermore the pressure decreases by a further 37% for each increase in height equal to the scale height. It is noted that since density is directly proportional to pressure in the ideal gas equation, the vertical variation of density will behave similarly to the vertical variation of pressure (equation 2.5).

Table 2.3 gives typical scale heights for the Earth's atmosphere and other planetary atmospheres, assuming the mean temperature is equal to either the radiative equilibrium temperature or the surface temperature (see Table 1.1). It is noted that the scale height is related to the mean molecular mass of the atmosphere and the acceleration due to gravity. On Mars, for instance, the gravity is approximately one-third of the Earth's gravity, but the atmosphere is composed of carbon dioxide with a molecular weight of 44, and hence there is a 50% increase in scale height compared with that on Earth. On Venus, the scale height is largest in the lowest layers of the atmosphere (Table 2.3), where the temperature approaches 760 K (Table 1.1).

As the reader will appreciate, the scale height is not a constant for a given atmospheric composition because of variations in the temperature of the atmosphere. To obtain accurate results from the equation, it is necessary to measure the temperature profile of the atmosphere as a function of height, and then to integrate the equation numerically. A simpler alternative to this procedure is to use a standard atmospheric temperature profile (Figure 3.7) to obtain the pressure height relationship.

In the ocean, the relationship between pressure and depth is very close to linear because of the small variation of density with depth. In the surface layers of the ocean the influence of pressure is small. However, because of the high pressures ($10^7 - 10^8$ Pa) existing in the deep water, compressibility does become an important factor when considering deep water masses (Table 2.5). An illustration of the importance of compressibility is that sea level would be tens of metres higher than it is if the ocean was assumed to be incompressible.

The pressure at a depth h in the ocean is obtained by integration of equation 2.4 from the surface to a depth h. This calculation is complicated by the fact that the ocean surface varies in height as a result of wind waves, tides and ocean currents. Locally, wind waves and swell can cause changes of many metres. On the scale of an ocean basin the tidal variations can be as large as 1m in deep ocean, and up to 10 m on the continental shelves where the tides are amplified. Secondly, the surface atmospheric pressure can vary by as much as 10% of the mean surface pressure, which leads to horizontal pressure variations in the oceans.

In the ocean, pressure is easier to measure than depth, and therefore a unit of depth is 1 decibar which is equivalent to 1 m of sea water. A bar (or 1 atmosphere) is 10^5 Pa. In Chapter 5 the measurement of temperature and salinity will be discussed.

2.5   Adiabatic changes and potential temperature

Consider a balloon that is thermally insulated, so that no heat is lost through the surface of the balloon. The balloon is filled with hydrogen and released into the atmosphere. At the moment of release, it is assumed that the hydrogen has the same temperature as the surrounding air. As the balloon ascends, it will expand because of the reduction in atmospheric pressure with height and, at the same time, the hydrogen in the balloon will become cooler. This reduction in temperature is due solely to the reduction in the internal energy of the gas, as a result of the energy used to expand the surface of the balloon. Since no heat is gained by or lost to the hydrogen gas, by thermal conduction across the surface or by thermal radiation from the atmosphere or Sun, the balloon is known as an adiabatic system and the reduction in temperature is known as adiabatic cooling. Of course, if there is no transfer of energy to its surroundings and the balloon is brought back down to the surface at the same initial temperature and pressure, then the temperature of the hydrogen gas will be the same as the initial temperature. Therefore a property of an adiabatic system is that it is reversible and no matter which path the balloon takes through the atmosphere, when it returns to its initial pressure level it will have the same initial temperature.

These properties of the adiabatic system can be understood by the application of the first law of thermodynamics:

$$dQ = C_v dT + p d\alpha \tag{2.6}$$

The first term represents the heat added across the boundaries of the system, which could be associated with heat conduction across the balloon's surface and radiational heating. The second term is the internal energy of the gas, which is associated with the average kinetic energy of the molecules and is proportional to the temperature of the gas. C_v is the specific heat of the gas at constant volume. The third term is the work done on the surroundings and, in our example, is associated with the expansion of the volume of the gas in the balloon. Hence, heat added to the system may either cause a change in volume of the gas, or it may result in a change in the temperature of the gas, or it may be a combination of both effects. As mentioned earlier, for an adiabatic system, there is no heat added across the boundaries of the system and therefore $dQ = 0$. Hence it can be seen from equation 2.6 that an expansion of the balloon ($d\alpha > 0$) will cause a reduction in the internal energy of the gas and a decrease in temperature.

Before this relationship can be usefully applied, it is necessary to substitute for the change in specific volume, $d\alpha$. By differentiating the ideal gas equation by parts, it can be shown that $\alpha dp + p d\alpha = R_D dT$. By substituting for $p d\alpha$ in equation 2.6 and noting $C_p = C_v + R_D$, where C_p is the specific heat at constant pressure, then $0 = C_p dT - \alpha dp$ and hence:

$$\frac{dT}{dp} = \frac{\alpha}{C_p} = \frac{1}{\rho C_p}$$

but, since ρ depends on pressure, this is not a particularly useful relationship. However, by rewriting the equation as:

$$\frac{dT}{dz}\frac{dz}{dp} = \frac{1}{\rho C_p}$$

and substituting for the hydrostatic equation (equation 2.4) we have:

$$\frac{dT}{dz} = -\frac{g}{C_p} \tag{2.7}$$

The variation of temperature with height in an adiabatic system is therefore determined solely from the acceleration due to gravity, g, and the specific heat of the gas at constant pressure. For the lower atmosphere ($<10\,\text{km}$), the change in g with height is small and therefore the adiabatic temperature gradient is, to a good approximation, a constant. Table 2.7 shows calculated adiabatic temperature gradients for the Earth and other planets. It is noted that both Earth and Venus have similar adiabatic temperature gradients, despite their different atmospheric compositions and masses.

It is necessary to emphasise that the adiabatic temperature gradient is only obtained when parcels of air do not gain or lose heat to their environment. In practice, mixing parcels of air, energy interchange by radiation and the condensation of water vapour produce a change in the total energy of the air parcel. The two latter effects are known as non-adiabatic processes or diabatic processes. Because of the diabatic processes, environmental temperature profiles are rarely adiabatic, though we shall see that the concept is still a useful one in understanding the vertical stability of the atmosphere. The average observed temperature gradient in the lowest 10 km of the atmosphere is $\sim 6.5\,\text{K}\,\text{km}^{-1}$, compared with an adiabatic temperature gradient of $-9.8\,\text{K}\,\text{km}^{-1}$.

The adiabatic temperature gradient for the ocean can also be derived from equation 2.6, but the final expression is not so simple because of the more complicated form of the equation of state. In this case, equation 2.6 becomes:

$$dQ = C_p\,dT - T\left(\frac{\partial \alpha}{\partial T}\right)_P dp$$

Table 2.7   Adiabatic temperature gradients for planetary atmospheres

	Earth	Mars	Venus	Jupiter
Adiabatic gradient temperature (K km^{-1})	−9.8	−4.5	−10.7	−20.2

Assuming $dQ = 0$ and substituting for the hydrostatic equation, it can be shown that $\frac{dT}{dz} = -\frac{g\beta T}{C_p}$ where $\beta = \left(\frac{1}{\alpha}\right)\left(\frac{\partial \alpha}{\partial p}\right)_p$, where β is the coefficient of heat expansion of sea water at constant pressure.

Both β and C_p for sea water are dependent on pressure, temperature and salinity. At atmospheric pressure, standard ocean salinity of 35 and a temperature of 0°C, the adiabatic temperature gradient is $-0.036\,\mathrm{K\,km}^{-1}$, whilst at a pressure of 5000 decibars (equivalent to a depth of 5000 m), the adiabatic temperature gradient in the ocean is $-0.094\,\mathrm{K\,km}^{-1}$. The latter is less than 1% that of the atmospheric adiabatic temperature gradient. However, because of the small variation in temperature in the deep-water masses, the adiabatic temperature gradient has to be taken into account when considering water masses at different depths.

We have seen that the adiabatic effect of pressure on temperature is significant in both atmosphere and ocean and therefore any vertical motions will produce an adiabatic temperature change in a parcel of fluid. We have also indicated that radiation, mixing and water vapour condensation (in the atmosphere) will also produce significant changes in the temperature of a parcel of air or sea water. It is useful, therefore, to separate the adiabatic process from diabatic processes in order to distinguish the two effects.

For example, a meteorologist may wish to decide if the air mass at the surface is similar to the air mass observed in the upper atmosphere. Similarly, an oceanographer may wish to determine the similarity of two water masses at two different depths in the ocean. For these processes, it is necessary to remove the adiabatic effect by bringing the air or water mass (adiabatically) to a reference pressure. This reference pressure is 1000 hPa for the atmosphere and surface atmospheric pressure of 1013 hPa for the upper ocean. For deep water masses reference levels of 1 km, 2 km, and 4 km are used. Because of the small adiabatic effect in the ocean, the effect of variations in atmospheric pressure is negligible. The temperature of the fluid parcel at the reference pressure is known as the potential temperature, θ.

Figure 2.5 shows the temperature change of an air parcel moved adiabatically a vertical distance of 1 km from the level of the reference pressure. The temperature at the initial position is 280 K, but the potential temperature is 298.8 K. Generally, since the reference pressure is close to sea-level pressure, the atmospheric potential temperatures are usually larger than the observed temperatures. In the ocean, the reference pressure is the minimum pressure of the ocean (i.e. surface) and therefore ocean potential temperatures are always lower than *in situ* temperatures.

Figure 2.6 shows the importance of reducing *in situ* temperatures to potential temperatures in the deep ocean. The measurements show temperatures increasing steadily towards the bottom of the Mindanao Trench, which suggests a strong vertical instability of the lower layer

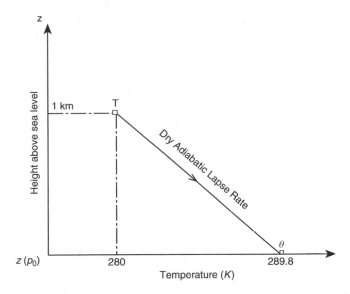

Figure 2.5 Relationship between temperature and potential temperature in the atmosphere as a function of height above mean sea level. ρ_0 is mean sea-level pressure

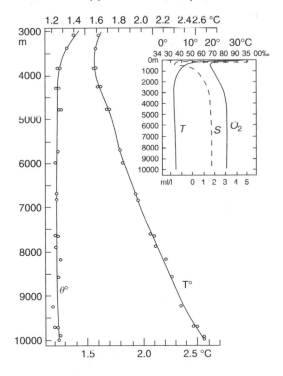

Figure 2.6 Observed temperature (T) and potential temperature (θ) in the Philippines Deep (Mindanao Trench). Observations from the Danish vessel *Galathea* in July–August 1951. Reproduced, with permission, from Tchernia, P., 1980, Descriptive Regional Oceanography, Pergamon Press: page 30, figure 3.2

(assuming salinity is not stabilising the temperature gradient). However, the potential temperature is in fact nearly constant with depth rather than increasing with depth. The apparent warm bottom water is, therefore, solely an artefact of the adiabatic effect.

An example of the reduction of an observed atmospheric temperature profile compared with a profile of potential temperature is given in Figure 2.7. It is again noted that the removal of the adiabatic influence substantially changes the profile, from one that decreases with height to a profile where the potential temperature increases with height. Usually it is only very close to the surface under strong heating conditions that the potential temperature decreases with height. The reason for this will be discussed in the following section.

2.6   Vertical stability of the ocean and atmosphere

We will first consider the stability of an incompressible fluid in the two situations shown in Figure 2.8b and d.

Initially, the parcel of fluid at point 0 is in hydrostatic equilibrium with its environment (i.e. the upward pressure gradient on the parcel is balanced by the downward force due to gravity). If the parcel of fluid in case (b) is moved *upwards* by an external force to a position A, and provided there is no mixing of the parcel with its environment, then the parcel will have a density greater than its environmental density at that level. The parcel is now heavier than its environment and will accelerate back towards its original

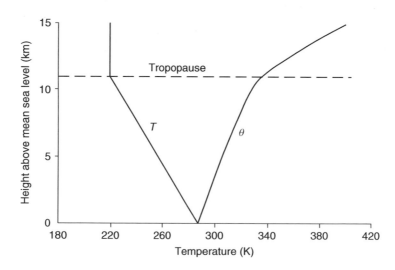

Figure 2.7   Mean vertical profile temperature (T) and potential temperature (θ) in the lowest 15 km of the atmosphere

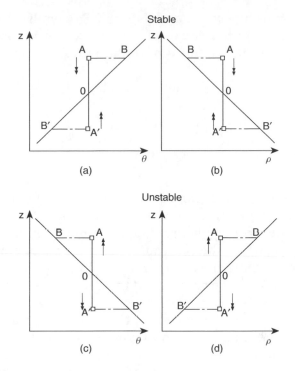

Figure 2.8   Stability of a parcel perturbed from equilibrium for atmosphere (a, c) and ocean (b, d)

level. In this case, we say the fluid system is *stable*. However, in case (d), the parcel at A will have a density less than its environment density at that level. In this case, the fluid is now lighter than its surroundings and the parcel will be subjected to an *upwards* force, which will accelerate the parcel away from its original position. Furthermore, as the parcel moves further away from its original position it will be subjected to a larger buoyancy force and hence subject to larger accelerations. In this case the fluid system is *unstable*. It should be appreciated that downward displacement of the parcels from their initial positions will produce similar behaviour. In the stable case, the fluid parcel will move back to its original position whilst in the unstable case it will accelerate away from its original position. In the situation where the density is uniform throughout the fluid, there will be no buoyancy forces and the system is said to be neutral. In this case, parcels will continue to move in their original direction until frictional forces slow them down.

The above criterion for stability can be applied to both the ocean and the atmosphere, provided adiabatic effects are taken into account. We have shown that a conservative property of an adiabatic fluid is the potential temperature and, in the atmosphere, the criterion for stability is simply based on the vertical gradient of the potential temperature. If the potential temperature *increases* with height, the atmosphere is *stable* (Figure 2.8a); whilst if it *decreases* with height, it is *unstable* (Figure 2.8c). It is left to the reader to consider the buoyancy forces on parcels displaced from equilibrium for these three cases, i.e.

$$\text{stable } \frac{\partial \vartheta}{\partial z} > 0$$

$$\text{neutral } \frac{\partial \vartheta}{\partial z} = 0$$

$$\text{unstable } \frac{\partial \vartheta}{\partial z} < 0 \tag{2.8}$$

Figure 2.9 shows the diurnal variation in potential temperature in the lowest 1 km over land when heat is added to the lower layer (diabatic process) by solar heating. In the morning there is a neutral layer close to the surface. This is the result of unstable conditions caused by solar heating of the surface. It is only very close to the ground (typically the lowest 1m) where a negative potential temperature can be present and this can not be seen in the figure. In this layer the solar heating can not be removed quickly enough by convective mixing. However, above this layer, vertical convection and the resulting mixing is a very efficient process and therefore a mixed layer with uniform potential temperature is found. The mixed layer is capped by a stable layer known as inversion. As mid-day approaches, the surface heating by solar

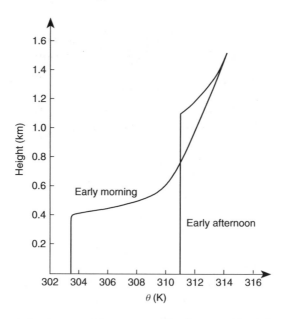

Figure 2.9   Diurnal variation in potential temperature in the atmospheric boundary layer over land in summer. Reproduced, with permission, from Stull, R.B., 1973, Journal of Atmospheric Science, 30: page 1097, figure 5

radiation reaches a maximum and the convective eddies become deeper and entrain stable air into the boundary layer. Therefore the region of uniform potential temperature, known as the mixed layer, will increase in depth.

During the night, the ground cools by loss of planetary radiation more effectively than the air above it, and therefore the potential temperature gradient becomes positive and the air becomes stable. In those circumstances the air is quiescent and smoke and other pollutants will tend to be trapped close to the ground. Only by artificially adding buoyancy to the pollutants can they be dispersed effectively.

Vertical convection is a process which occurs throughout the troposphere and will be discussed in Chapters 3 and 4.

In the ocean, the situation is complicated by the effect of salinity; hence the vertical gradient of the potential temperature does not always indicate the stability of the ocean. Generally we have to use a quantity known as the potential density, σ_θ, which is the density a parcel of sea water would have if it were moved adiabatically to the sea surface. For deep-water masses, a reference level of 1 km, 2 km or 4 km is used.

The stability of the ocean is then determined using the vertical gradient of the potential density, i.e.

$$\text{stable } \frac{\partial \sigma_\vartheta}{\partial z} > 0$$

$$\text{neutral } \frac{\partial \sigma_\vartheta}{\partial z} = 0$$

$$\text{unstable } \frac{\partial \sigma_\vartheta}{\partial z} < 0 \tag{2.9}$$

Unstable density gradients are rarely observed in the ocean. The reason for this is that the convective instability process is very efficient in mixing the unstable density gradient to produce a neutral profile. The presence, therefore, of a well-mixed ocean layer is a good indication of a convective instability process. The principal cause of the convective instability in the ocean is strong cooling of the surface during winter, which causes the formation of a relatively dense surface layer which will be unstable with respect to the deeper water and cause overturning. The surface cooling has a dual effect in destabilizing the water. It will not only remove heat from the water column but increased evaporation will also tend to increase the surface salinity and thus increase the density. The depth to which convection occurs will depend on the vertical density structure and the intensity of the surface cooling. In the North Pacific, the depth of winter convection is limited to the upper 150 m because of the presence of a very stable halocline below this depth. However, in the Labrador Sea, deep convection has been observed to a depth of 1500 m and, in the western Mediterranean in the Gulf of Lyons, deep convection occurs to the bottom.

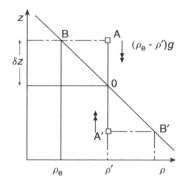

Figure 2.10   Stable layer and its relationship to buoyancy frequency (see text)

2.6.1   *Buoyancy frequency*

Let us reconsider the way in which a parcel of fluid returns to its equilibrium position (Figure 2.10) in a stable system. If there is no mixing of the parcel with its environment and no frictional drag on the parcel, then the parcel will accelerate towards its equilibrium position and overshoot. It will then find itself in an environment of higher density and it will accelerate in the opposite direction back towards its original position. The parcel will therefore oscillate about its equilibrium position with a frequency related to the density gradient. If the density gradient is large, then the buoyancy accelerations will be correspondingly increased and the frequency will also increase. This frequency is called the Brunt-Väisälä frequency or the buoyancy frequency (N).

Consider the buoyancy force on a parcel displaced a distance δz from its equilibrium density ρ_e. The buoyancy force is:

$$(\rho_e - \rho)g.$$

If $\rho = \rho_e + \left(\frac{\partial \rho}{\partial z}\right) \delta z$ then

$$g(\rho - \rho_e) = -g\left(\frac{\partial \rho}{\partial z}\right)\delta z$$

The vertical acceleration from the buoyancy force is:

$$\rho_e \frac{dw}{dt} = -g\left(\frac{\partial \rho}{\partial z}\right)\delta z$$

Since,

$$w = \frac{dz}{dt} \text{ then: } \frac{d^2}{dt^2}\delta z = -\left(-\frac{g}{\rho_e}\frac{\partial \rho}{\partial z}\right)\delta z$$

This equation is analogous to the simple harmonic motion equation:

$$\frac{d^2z}{dt^2} = -N^2z$$

where N is the frequency of the oscillation in units of radians/second. Hence,

$$N = \sqrt{\left(-\left(\frac{g}{\rho_e}\right)\left(\frac{\partial\rho}{\partial z}\right)\right)} \tag{2.10}$$

For the ocean the vertical density gradient will be negative for oscillations to occur.

The expression can be applied to the atmosphere by replacing density by the potential temperature.

$$N = \sqrt{\left(-\left(\frac{g}{\vartheta}\right)\left(\frac{\partial\vartheta}{\partial z}\right)\right)} \text{ where } \frac{\partial\vartheta}{\partial z} \geq 0 \tag{2.11}$$

The buoyancy frequency, N, provides a useful way of describing the stability of both the atmosphere and the ocean and is, therefore, a very important parameter in the advanced study of dynamical oceanography and meteorology. One of its most important uses is the determination of the maximum frequency of internal waves, which are continually produced by turbulent motions in both systems. In the ocean thermocline, N has a period of 10 min, whilst in the deep ocean, where the stability is weaker, it has a typical period of 50 hours. Internal waves in the thermocline, produced by surface storms, can have amplitudes of 10–50 m. In the troposphere the period of the waves is shorter than in the ocean, ranging from 30 s to 10 min. The more spectacular internal waves are produced by forced ascent over mountains, which in certain circumstances can produce wave trains (lee waves) extending for up to 100 km downstream of the mountain. The waves are most apparent when the wave crests are above the condensation level and lenticular clouds are seen.

3

Radiation, Temperature and Stability

3.1 Vertical variation of atmospheric constituents

The vertical variations of constituents in the atmosphere are important in understanding the way in which solar and planetary radiations are absorbed in the atmosphere, and the resulting effect on the temperature profile of the atmosphere. Here, the physical processes which control the distribution of atoms and molecules in the atmosphere will be considered.

First consider the major constituents of the atmosphere: oxygen, nitrogen, argon, water vapour and carbon dioxide. If all these constituents were initially mixed together and left for a long time in an undisturbed environment, the distribution of each constituent would vary with height according to its molecular or atomic weight. The lighter molecules would have a smaller variation with height than the heavier molecules and thus the lighter molecules would be more abundant in the upper atmosphere than the lower atmosphere compared with the heavier molecules.

The *equilibrium distribution* is given by the hydrostatic equation (see 2.4) for each constituent gas rather than for the mixture of gases. It can be seen that the rate of change of partial pressure (p_n) of each gas is directly proportional to the molecular weight (m_n) of the gas, i.e.

$$\frac{\partial p_n}{\partial z} = -\frac{m_n}{R^*}\frac{p_n}{T}g$$

However, the gases in the atmosphere are not in an undisturbed environment and do not show the distribution predicted by this simple model.

The turbulence in the atmosphere means that all of the gases, even carbon dioxide which is the heaviest of the major constituents, are well mixed to a height of 100 km and it is only above this level that evidence of gravitational settling is apparent.

The Atmosphere and Ocean: A Physical Introduction, Third Edition. Neil C. Wells.
© 2012 John Wiley & Sons, Ltd. Published 2012 by John Wiley & Sons, Ltd.

To understand this distribution, it is necessary to consider the rate at which diffusion occurs. The rate of gravitational settling of gases is determined by the molecular diffusion rate for each gas. The diffusion rate is defined by the product of the velocity of the molecules and their mean free path between collisions. The velocity of an atom or molecule increases with the temperature of the gas and decreases with molecular weight and pressure. Hence the diffusion rate is higher for light gases at high temperatures and low pressures than for heavier gases at low temperatures and high pressures. There is a rapid decrease of pressure with height in the atmosphere and, because of this, the most significant factor in the determination of diffusion rates is the mean free path between collisions. The diffusion rate is many orders of magnitude higher in the upper atmosphere, where the mean free path is very large, than in the lower atmosphere.

Typically, at a height of 120 km, it would take approximately one day for gravitational separation to occur but, for the atmosphere taken as a whole, the typical time scale is approximately 30 000 years. The atmosphere is always in a state of turbulent motion and therefore the molecules are being continually mixed by a variety of atmospheric motions. The rates of mixing are much larger than the corresponding rates of gravitational settling in the lower atmosphere and, therefore, the profiles of these constituents are well mixed. Only when the gravitational settling time becomes of the same order as the mixing time scale is diffusive separation observed. For molecular oxygen and nitrogen, this occurs typically at a height of 100 km. Above these levels the gases are distributed according to their molecular weights and hydrogen and helium, therefore, occur in increasing abundancies at high levels, whilst the abundancies of the heavier gases, such as oxygen and nitrogen, decrease with height.

It is only for the noble gases (e.g. argon) that the story finishes here. For all of the other constituents, their interaction with radiation has to be considered when discussing their vertical distribution in the atmosphere. These interactive processes include:

(i)   The dissociation of a molecule by incident radiation.
(ii) The combination of atoms by collisions.

The dissociation energy of a molecule is the energy required to break the molecular bond(s). The energy of one photon of radiation is given by $h2\pi/\lambda$, where h is Planck's constant and λ is the wavelength of the photon. It is therefore accepted practice to define the dissociation energy in terms of the wavelength of the radiation. For an oxygen molecule the dissociation energy corresponds to a wavelength of 0.24 µm and therefore all radiation having a smaller wavelength, and hence a higher energy, will dissociate the oxygen molecules into two atoms of oxygen, i.e.

$$O_2 + h\,2\pi/\lambda \rightarrow O + O, \quad \lambda \le 0.24\,\mu m$$

Nitrogen has a higher bond energy than oxygen and only photons having wavelengths less than $0.13\,\mu m$ are able to dissociate the nitrogen molecule.

Atomic oxygen becomes an increasingly abundant constituent above a height of 20 km, but it is only above 120 km that molecular oxygen is replaced by atomic oxygen as the dominant oxygen constituent. The reason for this distribution is that the rate of molecular dissociation is proportional to the product of the number of oxygen molecules and the number of photons having more than the required energy for dissociation. At very high elevations there are a large number of photons with the energy necessary to dissociate oxygen molecules but, due to gravitational settling and dissociation, there are few oxygen molecules and therefore the rate of production of atomic oxygen is low. However, in the lower atmosphere, though there is an abundance of molecular oxygen, there are only a few photons possessing the required energy to dissociate the molecules, because many of the photons have been absorbed in the upper levels of the atmosphere. Therefore, it is only in the middle levels, around 100 km, where there are sufficient molecules and photons to produce a large dissociation rate and hence to produce a high number of oxygen atoms. However, the predicted height of the maximum atomic oxygen production does not agree with the observed distribution. This is due to two additional processes in the lower atmosphere which have so far been neglected and which act to reduce the number of oxygen atoms.

The first process is the recombination of atomic oxygen into molecular oxygen by collision. This process is very effective at higher pressures and it is therefore a major sink of atomic oxygen at lower altitudes, between 70 and 100 km. The second process is the production of ozone, O_3, from a three-way particle collision between an atom of oxygen, a molecule of oxygen and another unspecified molecule, M:

$$O + O_2 + M \rightarrow O_3 + M^*$$

The unspecified molecule is necessary because it absorbs the energy released when a molecule of ozone is formed. The production of ozone depends on the relative numbers of oxygen molecules and atoms, and the number of three-way collisions which depends, in turn, on the total pressure of the gas. For a photochemical equilibrium, the rate of production of atomic oxygen must be equal to its rate of destruction. Given the absorption properties of the constituents, it is possible to obtain the vertical equilibrium distribution. This profile is known as the Chapman profile.

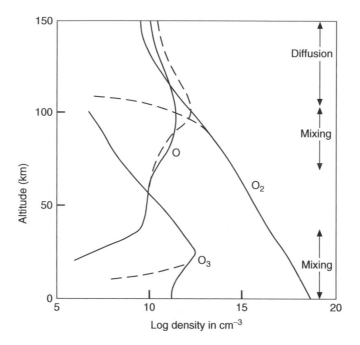

Figure 3.1   Vertical profiles of the densities of atomic oxygen (O), molecular oxygen (O₂) and ozone (O₃) in the Earth's atmosphere. Broken lines show the theoretical profile calculated on the assumption of photochemical equilibrium. Reproduced, with permission, from Goody, R.M. and Walker, J.C.G, 1972, Atmospheres, Prentice Hall: page 25, figure 2.2

Figure 3.1 shows the relative distribution of atomic oxygen, molecular oxygen and ozone in the atmosphere. It is noted that, because the ozone distribution is dependent on both the collision frequency and the amount of atomic oxygen, it tends to have its highest production rate between 30 and 100 km. Below 30 km, the infrequency of occurrence of atomic oxygen reduces the rate of ozone production to insignificant levels. Above 100 km, it is the low collision frequency which brings about the reduction of the rate of ozone formation.

However, there are three major processes which result in the destruction of ozone:

(i)   The photodissociation of ozone which occurs for photon wavelengths less than 11 μm and which therefore operates throughout the solar and planetary spectra.

$$O_3 + h\,2\pi/\lambda \rightarrow O_2 + O, \quad \lambda \leq 11\,\mu m$$

The strongest dissociation occurs for wavelengths between 0.2 and 0.3 μm and therefore ozone is a very effective absorber of ultraviolet radiation.

(ii)   The recombination of ozone with atomic oxygen

$$O_3 + O \rightarrow 2O_2$$

(iii) The removal of ozone at the Earth's surface during oxidation processes.

It can be seen that, at high levels in the stratosphere, ozone will be readily destroyed by processes 1 and 2, whilst process 3 will destroy ozone in the troposphere. In the troposphere, mixing rates are larger than in the stratosphere where most ozone is formed, and therefore ozone molecules entering the troposphere are brought to the surface in a matter of a few days and oxidation takes place. Hence, ozone tends to have a concentration maximum in the lower stratosphere where destructive processes are at a minimum. Here, individual ozone molecules may have lifetimes of the order of a few months. The detailed photochemistry of ozone is further complicated by other trace gases, such as nitrous oxide (which is related to biological production processes at the surface and vehicle exhausts), chlorofluoromethanes (present as aerosol propellants in the 20th century) and other man-made gases. All of these trace gases, in certain circumstances, can act to reduce the level of ozone in the atmosphere and hence reduce the ultraviolet shielding at present enjoyed by the Earth. The link between increasing levels of chlorine in the stratosphere and the decrease in ozone is well established, and efforts to reduce inputs into the atmosphere have been in place since the Montreal protocol agreed in 1987.

The interrelationship between atomic oxygen, molecular oxygen, ozone and radiation is an example of the complexity of photochemical processes which produce the observed distributions of the constituent gases of the atmosphere. For the other principal gases, water vapour and carbon dioxide, dissociation is only significant at altitudes greater than 100 km. Carbon dioxide is dissociated by ultraviolet radiation of wavelengths less than 0.17 μm and water molecules commence dissociation at 0.24 μm. The vertical distribution of water vapour in the atmosphere below 100 km is determined principally by the condensation process. The rapid decrease of pressure and temperature with height in the lowest 20 km of the atmosphere results in a reduction of the saturated mixing ratio (defined as the ratio of the mass of water vapour in a given volume of air, saturated with respect to a plane water surface, to a mass of dry air in the same volume) from 10^{-2} at the surface to 10^{-5} at 20 km. Therefore for every tonne of dry air in the stratosphere, there is only 10 g of water vapour. The temperature minimum at the tropopause, $\sim$10 km above the surface, acts as a cold trap for the movement of water vapour between the troposphere and the stratosphere. However, the water vapour in the upper atmosphere is still an important constituent when considering the absorption of solar and planetary radiation. At 80 km, where the temperature reaches a second minimum at the mesopause, occasionally there is sufficient water vapour to allow the formation of thin, noctilucent ice clouds. These clouds cannot be observed during the day, but are often seen just after sunset in high latitudes when the reflected sunlight from these clouds may be observed at the surface.

The reasons for the temperature minima at the tropopause and the mesopause will be discussed in Section 3.4.

3.2   The attenuation of solar radiation

The observation that the maximum in the solar-radiation spectrum occurs at $0.5\,\mu m$, whilst the planetary radiation peaks between 10 and $15\,\mu m$ (in the middle infra-red), allows separation of the radiation processes due to solar radiation absorption from those involved in planetary radiation absorption. Only 0.4% of the total solar radiation has a wavelength longer than $5\,\mu m$, whilst a similar percentage of planetary radiation from a surface at a temperature of $288\,K$ has a wavelength less than $5\,\mu m$. The propagation of solar radiation through the atmosphere is simpler to understand than that of planetary radiation because no re-emission of radiation occurs as the solar radiation travels through the atmosphere, and therefore it is only necessary to consider the attenuation factors. The situation with respect to the planetary-radiation spectrum is complicated by the re-emission of radiation at similar wavelengths to the absorbed radiation. This will be considered in Section 3.3.

The attenuation of solar radiation in a dust-free and pollution-free atmosphere with no clouds (Figure 3.2) would be the result of two processes:

(i)   Gaseous absorption.
(ii) Molecular scattering.

The primary gaseous absorption occurs in the ultraviolet region of the spectrum, at wavelengths less than $0.3\,\mu m$. As previously discussed, this absorption mainly results in the photodissociation of oxygen and nitrogen above $100\,km$, and the photochemical reactions involving ozone in the stratosphere. This primary gaseous absorption results in the nearly complete removal of lethal ultraviolet radiation by the upper atmosphere. However, a small fraction of the incident ultraviolet radiation does manage to reach the surface, particularly when the sky is clear and this can be damaging to human skin and, with excess exposure, it can cause skin cancers. Radiation at $0.26\,\mu m$ damages the DNA molecule which is present in all living organisms. In quantitative terms, the amount of ultraviolet radiation (wavelength less than $0.32\,\mu m$) absorbed is less than 2% of the total solar flux. Ozone has absorption bands in the visible region of the spectrum but these bands are generally weaker than the ultraviolet absorption bands. The only other

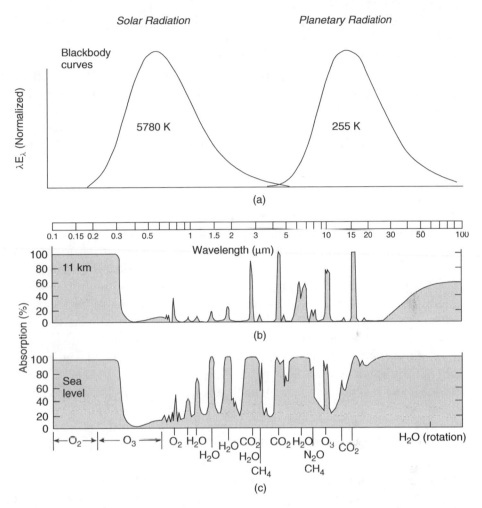

Figure 3.2 Attenuation of solar radiation. (a) Normalized blackbody curves for 5780 K (solar) and 255 K (planetary), plotted so that irradiance is proportional to the areas under the curves. (b) Atmospheric absorption in clean air for solar radiation with a zenith angle of 50° and for diffuse planetary radiation at 11 km height above the surface, near the middle latitude tropopause. (c) Same as for (b) but at the Earth's surface. Reproduced, with permission, from Goody, R.M., 1964, Atmospheric Radiation, Oxford University Press: Page 4, figure 1.1

significant absorber of solar radiation is water vapour, which has a number of absorption bands in the near infra-red part of the solar-radiation spectrum. The absorption bands of water vapour have a very significant effect on the amount of solar radiation actually reaching the surface, because of the large amount of solar radiation in this region of the spectrum. This effect is illustrated in Figure 3.2. The variability of the amount of water vapour present in the atmosphere means that the effect of its infra-red absorption will vary over the globe.

Molecular scattering is proportional to λ^{-4}, where λ is the wavelength of the incident photon. Thus, the shorter-wavelength light is scattered rather more than the longer-wavelength light. For example, blue light scattering is four times greater than red light scattering and hence the direct solar beam loses a significant fraction of its blue light. On a clear day, with a dust-free atmosphere, the shortest wavelength radiation is seen at the anti-solar point, i.e. the point at 180° to the direction of the Sun and at the same elevation as the Sun. This part of the sky often has a purple hue and ultraviolet radiation is at a maximum from this direction.

The presence of small suspended particles in the atmosphere, of diameter generally less than a few micrometres, has a significant influence on the radiation received at the surface. Such particles, called the aerosol, can both scatter and absorb radiation. The scattering of solar radiation tends to increase the path length of a photon through the atmosphere and therefore it leads, indirectly, to an increase in the probability of absorption. The scattering by the aerosol varies as λ^{-1}, i.e. it is not so strongly wavelength dependent as molecular scattering, and therefore it is significant throughout the solar spectrum. From Figure 3.3, it can be seen that the attenuation by the aerosol increases as the elevation of the Sun decreases. This is the result of the increased path length through the atmosphere and, therefore, the increased chance that a photon will be absorbed. Hence, in polar latitudes in summer, an aerosol layer will tend to produce enhanced absorption.

The aerosol scatters radiation back to space and therefore increases the global albedo (see 1.5) of the planet, which will influence the climate system.

There are many kinds of aerosol in the atmosphere, but a broad classification into two major types is possible. Considering the Earth as a whole, natural aerosol dominates over man-made aerosol although the latter may be dominant in industrial regions, for example North America, Europe and China.

The natural aerosol is produced by wind turbulence at the Earth's surface. Over land the particles are mainly silica and clay whilst over the ocean they are mainly salt. In addition, volcanic activity can inject large quantities of sulphur particles into the troposphere and the lower stratosphere. In the El Chichon eruption in 1982, it is estimated that 10^7 tonnes of material were injected into the stratosphere and, in Hawaii; the direct solar radiation was reduced by approximately 7% following the explosion. The aerosol produced by El Chichon was traced across the Pacific Ocean for many months. Other sources of the natural aerosol include forest fires, which can inject large quantities of carbon particles into the troposphere. These carbon particles are particularly effective in absorbing solar radiation. An important type of aerosol is formed from hygroscopic substances such as sodium chloride. These aerosols act as very effective cloud condensation nuclei and hence

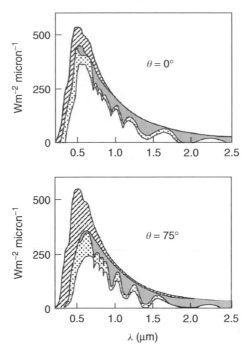

Figure 3.3 Molecular scattering, absorption and scattering by aerosols for a solar zenith angle of 0° (upper) and 75° (lower). The zenith angle is the angle between the vertical and the Sun. The upper curve shows the spectral distribution of the solar constant, while the lower curve represents the spectral distribution of the direct solar beam at the sea surface, for clear dry weather (equivalent to water thickness equal to 1cm). The areas of oblique lines, cross-hatching and dots represent, respectively, the effects of molecular scattering, absorption and scattering by aerosols. Reproduced, with permission, from Ivanoff, A., 1977 Ocean absorption of solar energy in Modelling and Prediction of the Upper Layers of the Ocean, ed. Kraus E.B.: (Oxford, Pergamon): page 49 figure 5.3

maritime salt haze becomes particularly effective in reducing solar radiation when the humidity reaches approximately 95%.

Human activities increase the input of sulphur into the troposphere in the form of sulphur dioxide and ammonium sulphates, and the effect of this input is important in both the formation of acidic rain and its influence on solar radiation. The carbon input from coal- or oil- based industries can also produce very significant reductions in the solar radiations. A dramatic example was the ignition of oil wells in Kuwait following the first Gulf war (1991). This produced a 5 km-deep plume of smoke which extended for many hundreds of kilometres downwind. In the centre of the plume, the solar radiation was reduced to zero and the surface temperature dropped by 10°C. Again, as with the salt haze, as humidity increases, condensation occurs on hygroscopic pollutants, such as ammonium sulphate, and this increases their influence on the solar radiation.

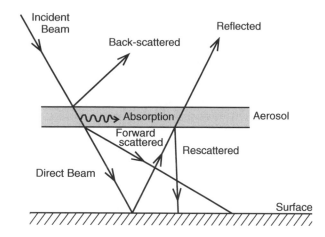

Figure 3.4   Schematic diagram of scattering and absorption by an atmospheric aerosol layer

Figure 3.4 shows how the aerosol may affect the path of a photon through the atmosphere. The back-scattering of incident photons from the aerosol will tend to enhance the albedo of the atmosphere. The photons which are not back-scattered into space, in the absence of gas absorption or clouds, will either be absorbed by the aerosol or reach the Earth's surface. Some of the photons will be reflected back into the atmosphere by the surface. A fraction of the reflected photons will be back-scattered to the surface by the aerosol and the remaining photons will be lost to space. The relative magnitudes of each fraction will depend on the absorption to back-scattering ratio of the aerosol and the surface albedo. Although these properties depend on the type of aerosol and the type of surface, some broad trends have been identified.

First, if the surface albedo is less than 30%, then the aerosol will tend to increase the planetary albedo. Surface albedos of such magnitudes are typical of all ocean surfaces and most land surfaces except for sand deserts. If the surface has a higher albedo, light trapping between the surface and the aerosol will tend to decrease the planetary albedo.

Second, the absorption to back-scattering ratio varies between 1 and 20% for maritime 'salt' particles and rises to 500% for carbon particles. Thus, maritime aerosols attenuate radiation primarily by scattering whilst carbon particles attenuate radiation by absorption. Figure 3.5 show the effect on solar radiation of a dust storm in the southern Sahara. It is noticed that the direct solar radiation is dramatically reduced by dust, although the global solar radiation, i.e. the radiation from all directions, which includes scattered radiation, is only reduced by approximately 15%. Hence the major influence of the dust is the scattering of the light rather than the absorption of radiation. Aerosols, both naturally occurring and man made, have an important influence on surface temperature from local scales to the global scale.

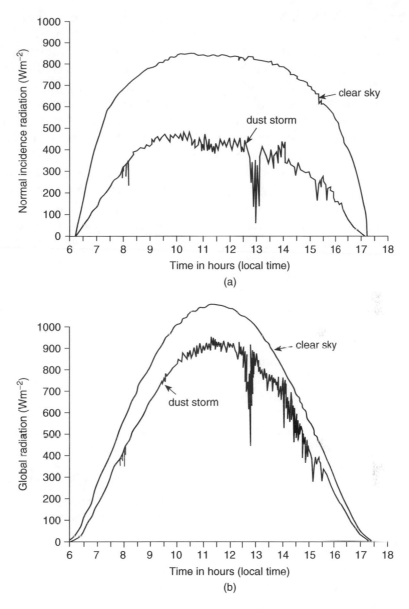

Figure 3.5 Solar radiation attenuation during a dust storm. (a) Normal incidence of radiation during the dust storm (13 November 1977) and clear sky (15 November 1977) as a function of local time. (b) Same as for (a) except for global radiation (direct + diffuse). Note the large depletion of direct solar radiation but the smaller depletion of global radiation. Reproduced, with permission, from Adetunji, J. *et al.*, 1979, Weather, 30: pages 434 and 435, figures 4 and 5

The last and most significant influence on solar radiation is that of clouds. Cloud droplets are typically larger than aerosol particles, usually varying between 1 and 20 μm in diameter. In this size range, the scattering is dominated by diffraction which tends to produce scattering at small angles from the angle of incidence. This diffraction-dominated scattering does not tend to be dependent on wavelength. The absorption to back-scattering ratio from clouds is very small and therefore the predominant influence on solar radiation is back reflection. Particularly spectacular is the reflection from cumulus clouds on an otherwise clear day. Of course, ice is an important scatterer of radiation as well, but it tends to be more effective at scattering radiation at larger angles from the angle of incidence than water droplets. Therefore, close to the Sun, ice clouds appear less bright than water clouds. However, the opposite is the case for clouds in the direction away from the Sun. In addition, because the ice crystals are hexagonal, plate and columnar prism optical processes are important, and these produce a fascinating series of optical phenomena. The more common phenomena include the appearance of a 22° halo around the Sun and the production of parhelia, or mock Suns.

The quantitative influence of clouds on solar radiation is dependent on:

(i)   The solar elevation.
(ii) The thickness of the cloud.
(iii) The size, distribution and amount of liquid water in the cloud.

Though absorption is small, typically less than 10% and probably often as low as 5%, it is enhanced by scattering, which increases the path length within the cloud and hence the probability of absorption. The transmission of the direct beam is also dependent on the type of cloud. For example, the direct beam is significantly attenuated by low-level clouds and therefore a stratocumulus cloud of 400 m thickness will be sufficient to obliterate the Sun's disc. As the thickness of the cloud increases, the transmission decreases and the albedo increases, approaching a value between 80 and 90%. A decrease in the elevation of the Sun will also result in a decrease in the transmission of the direct beam and an increase in back-scattering by clouds. A generalized quantitative analysis is not possible, however, because of the differences between clouds of the same general type.

Figure 3.6 shows a detailed breakdown of the downward and the upward components of solar radiation measured through the low-level stratocumulus cloud, which is particularly common over the ocean. The cloud is 200–400 m thick on average and, in common with other low-level clouds, it has a high liquid water content of between 0.1 and 0.5 g m^{-3}. The albedo of the cloud is 68% and, of the solar radiation not reflected, 25% is transmitted through to the surface and 7% is absorbed by the cloud. Though absorption is a relatively small contribution to the attenuation of the incident solar radiation

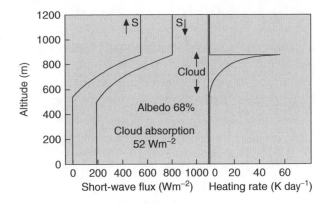

Figure 3.6 The observed short-wave (S) fluxes measured from three aircraft during the JASIN experiment (August 1978), corrected to a solar zenith angle of 43.7° for a stratocumulus cloud. The calculated heating-rate profile is shown in the right-hand diagram. Reproduced, with permission, from Schmetz et al., 1983, Results of the Royal Society Joint Air – Sea Interaction Project (JASIN), The Royal Society: page 380, figure 2

when compared with the amount reflected, it is, in this case, equivalent to $52\,\mathrm{W\,m^{-2}}$. This is sufficient to warm the upper part of the cloud by 20–40 $\mathrm{K\,day^{-1}}$ and offset the long-wave cooling from the cloud top. It is noted that only a small fraction of the downward solar radiation is reflected from the surface and this is consistent with the general observation of the low albedo of the ocean surface.

For a non-uniform broken cloud, it is more difficult to characterize and quantify the solar radiation. The solar radiation may vary by 50% of the average value between clear and cloudy air. In addition, side reflections from the cloud may enhance the radiation above that measured under clear-sky conditions. For these reasons, it is necessary to obtain statistical relationships between the fraction of cloud cover and the solar transmission. In this respect, satellite techniques are very useful for estimating the albedo for ensembles of clouds.

3.3   Absorption of planetary radiation

The predominant absorption of planetary radiation is by water vapour, carbon dioxide and ozone. All of these gases consist of triatomic molecules which have vibration-rotation absorption bands in the infra-red region of the spectrum, unlike diatomic molecules, such as oxygen and nitrogen, and the noble gases which exist as single atoms. The detailed properties of the absorption curves are complicated, particularly for water vapour, as illustrated by Figure 3.2, but some bands are more intense than others. These strong bands occur in the near infra-red at 1.4, 1.9, 2.7 and 6.3 µm and the continuum absorption band stretches from 13 to 1000 μm. The importance of this latter absorption band cannot be overemphasized, since more than 50% of the planetary radiation occurs at wavelengths greater than 13 μm. Carbon dioxide has its principal absorption bands at 2.7 and 4.3 μm, with a further strong band at 14.7 μm. Ozone has bands at 4.7, 9.6 and 14.1 μm. The 9.6 μm absorption band coincides with the maximum of the planetary radiation emitted from the surface. One of the broad features of the absorption band typical of the atmosphere as a whole is the absence of absorption bands between 8 and 13 μm except for the 9.6 μm absorption band of ozone. It is noted that this region between 8 and 13 μm is within the peak emission region of the planetary radiation spectrum and some 32% of the total planetary emission resides within it. The region is known as the 'atmospheric window' because, in the absence of clouds, planetary radiation should pass through the atmosphere almost unimpeded. However, recent work has shown that, with high water vapour concentrations typical of the tropical latitudes, additional absorption peaks appear in the window region as the result of the formation of water-vapour dimer, $(H_2O)_2$. The water dimer is formed by the joining together of two water molecules by hydrogen bonding. Although it only exists in low concentrations when compared with water vapour, it has significant rotation – vibration absorption bands in the atmospheric window. With very high water-vapour concentrations, typically greater than 20 mb in tropical environments, the dimer has an important influence in reducing the long-wave radiation lost to space. The presence of this dimer may lead to doubling of the absorption at 10 μm (under conditions favourable to its formation). Other minor gases, such as nitrogen oxides, methane and carbon monoxide, also have absorption peaks in the infra-red spectrum and therefore contribute to the absorption of planetary radiation. These gases are collectively known as 'greenhouse' gases, though carbon dioxide is the most effective absorber of planetary radiation.

In addition to gaseous absorption, the continuous absorption throughout the infra-red spectrum of liquid water droplets is also of prime importance in the absorption of planetary radiation. At a wavelength of 10 μm, a liquid water film only 60 μm thick is sufficient to absorb 95% of the incident radiation, although care has to be exerted to distinguish between the absorption by plane water surfaces and by spherical droplets. In the latter case, scattering becomes an important process. As an example, a cloud 200 m thick with a liquid water concentration of $0.1 \, g \, m^{-3}$ would absorb about 90% of the incident planetary radiation. Therefore, a reasonable thickness of unbroken cloud would be sufficient to absorb all the incident planetary radiation. Ice has a similar infra-red absorption to liquid water, but the total water content in ice clouds is generally much less than that found in mixed and water droplet clouds lower in the troposphere because of the low temperatures that prevail. Hence, ice clouds will not be as effective in absorbing infra-red radiation.

Typically cirrus cloud has an infra-red absorptivity of 35% compared with an absorptivity of 100% for stratocumulus.

Additional absorption by aerosols occurs, but this effect is not as marked in the infra-red as it is at visible wavelengths because of the λ^{-1} dependence. However, the importance of hygroscopic nuclei when considering infra-red absorption cannot be overstressed, since the particles will grow by condensation and absorption will increase because of the water which becomes attached to the nuclei. It should therefore be expected that a significant increase in absorption of planetary radiation will occur as the humidity rises and such an increase could well be anticipated in the marine atmosphere and over industrial areas where hygroscopic nuclei may be produced in large concentrations.

3.4   Vertical temperature profile and its relation to radiation

The mean vertical temperature profile for the atmosphere is shown in Figure 3.7. The most important features are the three turning points, known

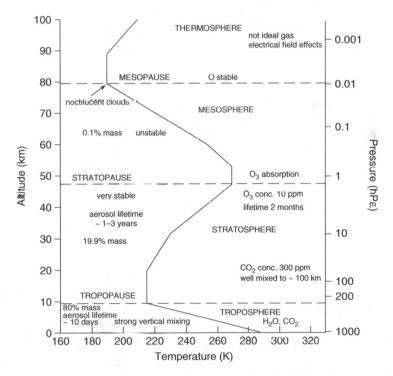

Figure 3.7   Global vertical temperature profile, delineating the four major regions of the atmosphere: the troposphere, stratosphere, mesosphere and thermosphere. Reproduced, with permission, from Wallace, J.M. and Hobbs, P.V., 1977, Atmospheric Science, Academic Press: page 23, figure 1.8

as the tropopause, the stratopause and the mesopause, which allow a quantitative division of the atmosphere into four distinct layers; the troposphere, the stratosphere, the mesosphere and the thermosphere.

The troposphere accounts for 80% of the mass of the atmosphere and it is a region of strong vertical mixing. It is associated with our experience of 'weather' and is most important for the condensation of water vapour, the average decrease of temperature with height being $6.5\,K\,km^{-1}$. The stratosphere contains about 19.9% of the mass of the atmosphere and it is a region of very pronounced stability. The temperature increases with height at a rate of approximately $2\,K\,km^{-1}$, compared with the decrease of $9.8\,K\,km^{-1}$ expected for a dry neutrally buoyant atmosphere. The stability of the stratosphere is most marked between 30 and 50 km, where the absorption of solar radiation by ozone reaches a maximum. It is noted that the maximum ozone concentration occurs in the lower stratosphere. Although the tropopause forms a pronounced stability barrier between the stratosphere and the troposphere, it is highly variable in position. Occasionally, the troposphere may extend down to 6 km and, in these conditions, stratospheric air may be drawn into the troposphere, most commonly along frontal boundaries. There is a high ozone concentration in this stratospheric air and it is therefore possible to trace, from ozone concentrations, its incursion into the troposphere. In the tropical atmosphere the tropopause may extend to over 15 km in height, particularly in regions of intense thunderstorm activity.

The inherent stability of the stratosphere inhibits the vertical mixing of aerosols and they therefore have a long residence time, typically of the order of one to three years compared with some 10 days in the troposphere. This fact has important consequences for the climatic effects of volcanic aerosols. The El Chichon explosion of 1982, though of smaller magnitude than the Mount St Helens eruption of 1980, injected considerably more volcanic aerosol into the stratosphere. The dust cloud, composed mainly of sulphur particles and sulphuric acid droplets, was tracked across the Pacific Ocean for many months following the explosion and it produced significant reductions in solar radiation in Hawaii, over 6000 km from the volcano in Mexico, as has been previously stated. The Pinatubo eruption in June 1991 injected about three times as much sulphur dioxide into the stratosphere as did the El Chichon eruption. This resulted in a global surface cooling of 0.5°C about 12 months after the explosion. Man-made pollutants will have similar long residence times in the stratosphere and it is in this region, therefore, that their climatic effects may be first observed. The production of chlorofluorocarbons (CFC's) in the 20th century has caused the reduction in stratospheric ozone observed in the last 50 years. Ozone levels will continue to decrease for many years to come, despite the significant decrease in the production of the CFC's since Montreal protocol in 1987, because of the long residence time of this pollutant in the stratosphere.

The stratosphere interacts strongly dynamically with the troposphere below and is important on time scales of weeks to climate scales.

The mesosphere contains 0.1% of the mass of the atmosphere and it is less stable than the stratosphere, with temperatures decreasing with height to reach a minimum of 190 K at 80 km. The mesosphere is dynamically active, and is associated with atmospheric tides and waves (Section 8.6). The temperatures are sometimes sufficiently low to produce freezing of the low concentration of water vapour present and to produce noctilucent clouds. These clouds are normally translucent to the direct solar radiation incident upon them but they can sometimes be seen in summer. The mesosphere and stratosphere together are known as the middle atmosphere.

Above the mesopause is the thermosphere, where the temperature increases steadily with height. In this region, the molecules become very well separated and collisions become so infrequent that we can no longer treat the atmosphere as an ideal gas. Furthermore, the thermosphere does not behave as a simple fluid because the electrically charged particles present will move differently to neutral charge entities.

Consideration of the processes which produce the observed temperature distribution in the lower 100 km of the atmosphere is now in order. The detailed processes of absorption of solar radiation and long-wave planetary radiation in the atmosphere have already been discussed and this information may now be used to obtain a model of the vertical distribution of radiation and temperature. Consider a horizontal slab of the atmosphere as depicted in Figure 3.8. The net rate of heating (or cooling) of the slab will depend on:

(i)   The absorption of solar radiation, ΔE.
(ii)   The absorption of long-wave radiation from the atmosphere above, ΔL_d, and below the slab, ΔL_u.
(iii) The re-emission of long-wave radiation at the slab temperature, L_e.

Hence, following the above notation, the energy budget of the slab is given by:

$$mC_p \frac{dT}{dt} = \Delta E + \Delta L_u + \Delta L_d - 2L_e \qquad (3.1)$$

Where m is the mass of the slab and C_p is the specific heat of air at constant pressure.

In radiative equilibrium, $dT/dt = 0$, i.e. the rate of change of temperature with time is zero. Therefore the radiation absorption is balanced by the

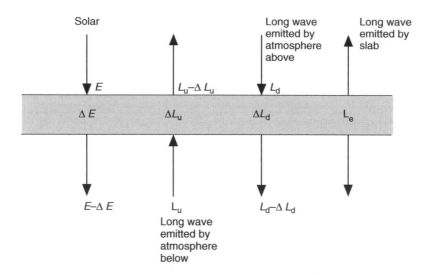

Figure 3.8 Radiation budget of a thin layer of the atmosphere (see text for nomenclature)

radiation re-emission. As has been noted earlier, the absorption in a clear, cloud-free atmosphere depends on:

(i)   The mass of the absorbing gas.
(ii) The path length through the gas.
(iii) The absorption coefficient for the gas for each wavelength.

Ideally, the absorption coefficients would be calculated for each radiation wavelength between 0.1 and 100 μm but this is very time consuming even with high-speed computers, and therefore approximations to the absorption spectra are made. The simplest approximation is to associate a single absorption coefficient with the solar radiation band and another with the long-wave planetary radiation band. Similarly, for emission, an effective emissivity, ε, can be defined which depends on the mass of carbon dioxide, water vapour and ozone present. Hence, if the vertical distribution of these gases is known, then the atmosphere can be divided up into a number of levels and equations similar to equation 3.1 may be written down for each level. The long-wave emission term, L_e, where $L_e = \varepsilon \sigma T^4$, can be adjusted by iteration until it balances the other terms in the equation.

Figure 3.9 shows the results of this type of calculation for a cloud-free atmosphere composed of carbon dioxide, water vapour and ozone. The results of the same calculation for

(i)   Carbon dioxide only.
(ii) Carbon dioxide and water vapour only.
(iii) Water vapour only.

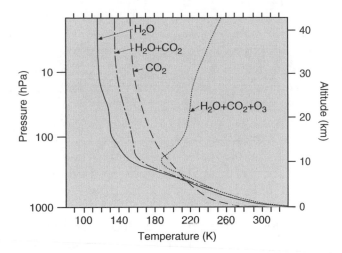

Figure 3.9 Radiative equilibrium temperature profile for various atmospheric absorbers in the absence of clouds. The absorbers are water vapour (H_2O), carbon dioxide (CO_2) and ozone (O_3). The effects of both long-wave radiation and solar radiation are included. Reproduced, with permission, from Manabe, S. and Strickler, R.F., 1964, Journal of Atmospheric Science, 21: page 371, figure 6a

are also shown in Figure 3.9. It can be seen that, without ozone absorption, the temperature would decrease with height up to 100 km and the stable stratosphere would not exist. This profile is similar to the atmospheric profiles of Venus and Mars, where ozone occurs in insufficient quantities to produce a stratosphere. Another interesting feature is the role of water vapour which is very important in the absorption in the lower troposphere. In a water-free atmosphere, the troposphere and the surface temperature would be considerably cooler than present observations show. It is also interesting to note the shielding effect of tropospheric water vapour on the stratospheric temperature. In the absence of water vapour, long-wave radiation is transmitted from the surface into the stratosphere and hence produces an increase in stratospheric temperature. The carbon dioxide does contribute significantly to the warming in the atmosphere but, because it is well mixed throughout the atmosphere, it tends to warm the stratosphere as much as the troposphere. It is noted that in the realistic model, as illustrated by Figure 3.9, the calculated surface temperature is 330 K, whilst the observed surface temperature is approximately 288 K. The reason for this discrepancy is that the vertical temperature gradient in the lowest 10 km is unstable (see Section 2.6) for the radiative equilibrium profile. If the model is adjusted to a stable temperature profile, then the surface temperature approaches the observed value and the temperature in the upper troposphere is increased. The assumption inherent in this temperature profile stabilization is that convection will continually readjust the vertical temperature profile by transferring heat from the surface to the upper troposphere. Hence the radiation will always be endeavouring to produce an unstable gradient near the surface but, because of the efficiency of convection in the vertical transfer of heat, the surface will not come into radiative equilibrium. However, very close to the surface, below a height of about 1 m, unstable temperature gradients may be maintained. This occurs when the solar heating of the surface is large and small-scale convection near the surface is not sufficiently vigorous to completely adjust the vertical temperature gradient.

These model results will be complicated by clouds, aerosols and horizontal variations in the distribution of water vapour and ozone in the atmosphere. It is model calculations such as these that have been used to obtain estimates of the increase in surface temperature as a result of increasing levels of carbon dioxide in the atmosphere.

The fluxes of radiation for the observed global atmosphere are indicated in Figure 3.10, which assumes a solar flux of 100 units at the top of the atmosphere but, because of the variability of the absorption and reflection processes, the diagram can only act as a guide to the magnitude of the processes involved. In common with the approach in Sections 3.2 and 3.3, the solar and planetary radiation will be divided into two separate streams. Of the 100 units of solar radiation incident on the top of the atmosphere, 49 units are absorbed by the Earth's surface, 31 units are reflected back into space and 20 units are absorbed by the atmosphere. The major reflectors of

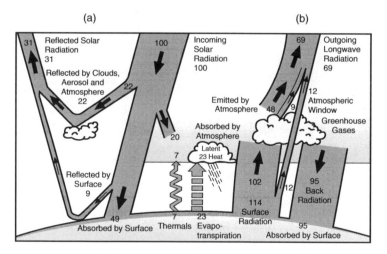

Figure 3.10   The globally and annually averaged components of the solar (a) and planetary (b) radiation streams. The proportion of the radiation intercepted by various components of the atmosphere and the Earth's surface is normalized to the incident radiation at the top of the atmosphere (100 units). 100 units equivalent to 342 W m^{-2}. Reproduced, with permission, from Trenberth, K.E., 1997, Bulletin of the American Meteorological Society, 78: page 206, figure 7. See plate section for a colour version of this image

solar radiation are the clouds and aerosols, whilst the main absorbers are the atmospheric aerosol; water vapour; ozone and carbon dioxide. It is noted that, globally, the Earth's surface reflects only nine units back to space because of the low albedo of the oceans. The Earth's surface and atmosphere, to maintain thermal equilibrium, must lose heat at the same rate as it is absorbed from solar radiation.

The planetary radiation budget for the surface and atmosphere is rather more complicated than for solar radiation because of the re-emission of radiation by the atmosphere, known as the 'greenhouse effect'. For instance, 114 units of planetary or long-wave radiation are emitted from the Earth's surface, of which 102 units are absorbed by the atmosphere and the remaining 12 units are lost to space through the atmospheric window. The atmosphere re-emits 95 units back to the Earth's surface and therefore the net long-wave flux from the surface is only 19 units. Furthermore, the surface gains 49 units by solar radiation and therefore has a net surplus of 30 units of total radiation (i.e. solar + planetary). To maintain the thermal equilibrium of the surface, it is necessary to transfer this surplus heat to the atmosphere by convection and by the evaporation of water.

Globally, about 75% of the surplus heat is used to evaporate water, though this fraction will vary between land and ocean. Over land about 50% of the energy is used for evaporation whilst over the ocean about 90% of the available energy is used. The atmosphere absorbs 102 units of long-wave radiation emitted from the surface but re-emits 95 units back to the surface and 57 units to space. Therefore the atmosphere has a net deficit of long-wave radiation of 50 units. Furthermore, solar absorption by the atmosphere accounts for only 20 units and therefore there is a net total radiation deficit of 30 units for the atmosphere. The global atmosphere will always be continually cooling by radiation processes at the rate of 30% of the incident solar radiation $\sim 100\,\mathrm{W\,m^{-2}}$. The mass of the atmosphere per unit area is $\sim 10^4\,\mathrm{kg\,m^{-2}}$ and the specific heat at constant pressure is $1000\,\mathrm{J\,kg^{-1}\,K^{-1}}$. Therefore the cooling rate is $10^2/(10^3 \times 10^4)\,\mathrm{K\,s^{-1}}$ or $1\,\mathrm{K\,day^{-1}}$. Hence the atmosphere will continue to cool by radiation at $\sim 1\,\mathrm{K\,day^{-1}}$. Therefore to maintain the thermal equilibrium of the atmosphere, it is necessary to transfer heat from the Earth's surface by turbulent heat conduction and by the latent heat of evaporation of water. Upon condensation the heat is released to the atmosphere.

Figure 3.11 shows the net heating rates observed for the atmosphere. The continual convection of heat, both horizontally and vertically, implies that the atmosphere will never be in radiative equilibrium. It is noted that, in general, the tropical stratosphere is always heating up, mainly as a result of ozone and carbon dioxide absorption, whilst the troposphere is cooling at rates between 0.5 and $1.8\,\mathrm{K\,day^{-1}}$, except over the poles where some heating is observed

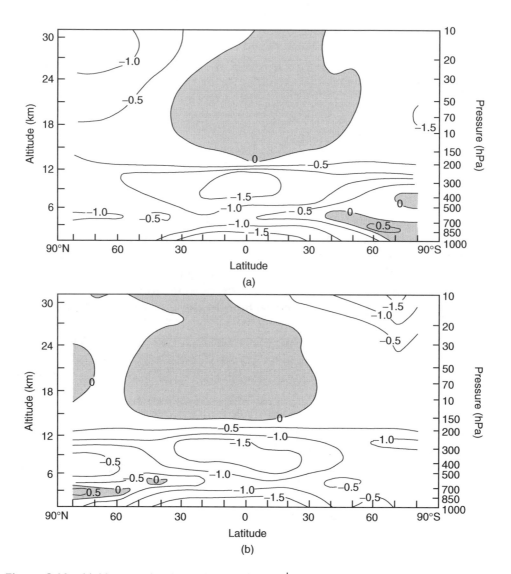

Figure 3.11 (a) Mean total radiative heating (K day^{-1}) for December–February. (b) Mean total radiative heating (K day^{-1}) for June–August. Areas of heating are shaded. Reproduced, with permission, from Dopplick, T.G., 1972, Journal of Atmospheric Science, 29: pages 1291 and 1292, figures 21 and 23

in the summer hemisphere. The largest cooling rates occur over the surface and at the top of the equatorial troposphere, where high temperatures and high humidities result in a large loss of long-wave radiation. To maintain this cooling rate, heat must be continually convected from the surface into the atmosphere.

3.5   The absorption of solar radiation in the ocean

The ocean absorbs the largest fraction of solar radiation incident on the Earth's surface because of its great area (70.8% of the Earth's surface) and its low surface albedo. The manner in which this radiation is absorbed is of importance in understanding both the heating distribution in the ocean and the stability of the ocean. Furthermore, the biological activity of the ocean is primarily dependent on the direct absorption of the incident solar radiation by microscopic phytoplankton. The rate of phytoplankton growth, or primary production, is a major control on the development of food chains which ultimately determine the productivity of fisheries and the population of the larger sea mammals at the ends of these chains.

It can be shown that the attenuation of solar radiation by pure sea water can be represented by an exponential law of the following form:

$$E_\lambda(z) = E_{\lambda 0} \exp(-k_\lambda z) \tag{3.2}$$

where k_λ is the attenuation coefficient at a wavelength λ and $E_\lambda(z)$ is the solar radiation intercepted at a depth z below the ocean surface. $E_{\lambda 0}$ represents the solar radiation incident on the sea surface at a wavelength λ. The variation of k_λ with wavelength is depicted in Figure 3.12a for the solar spectrum.

The main feature is that the coefficient varies by a factor of 10^7, with a minimum value at $0.46\,\mu m$, in the blue region of the spectrum and maximum values in the ultraviolet and near infra-red regions of the spectrum. The inverse of the absorption coefficient, $1/\lambda$, gives a measure of the depth of penetration of the radiation, $z_p(\lambda)$. Substituting in equation 3.2:

$$E_\lambda(z_p) = E_{\lambda 0}\, e^{-1} \quad \text{or} \quad 0.37\, E_{\lambda 0}$$

Thus the penetration depth, $z_p(\lambda)$, is the depth to which 37% of the incident radiation penetrates. At $0.46\,\mu m$, this depth is $55\,m$, whilst at $0.3\,\mu m$, in the ultraviolet, it is $10\,cm$ and at $1\,\mu m$, in the infra-red, it is $1\,cm$. The strong absorption of ultraviolet radiation was important in shielding early organisms from high doses of ultraviolet radiation at a time when the ozone layer in the atmosphere was non-existent, as previously discussed in Section 2.1. Less than 1% of the most penetrating blue radiation will reach $500\,m$ and even the human eye, which has a truly remarkable ability to see objects at radiation levels as low as $50 \times 10^{-6}\,W\,m^{-2}$, would not be able to detect any solar radiation below a depth of $800\,m$. For these reasons the majority of the ocean depths are dark and do not directly benefit from solar radiation absorption.

The absorption of the solar spectrum in pure sea water, depicted in Figure 3.12b, shows the majority of infra-red radiation is absorbed in the

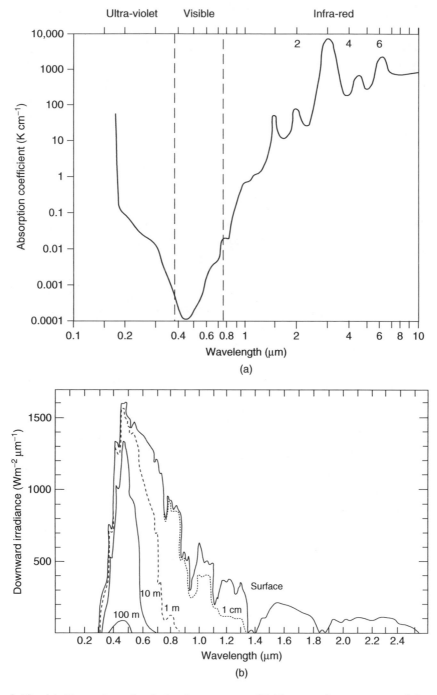

Figure 3.12 (a) Absorption of radiation by sea water. (b) The complete spectrum of downward solar irradiance in the sea between the surface and 100 m. Reproduced, with permission, from Jerlov, N.G., 1976, Marine Optics, Elsevier: page 140, figure 7.4

first 1 cm of the ocean and that all of it is absorbed within 1 m of the surface. Thus over 55% of the incident solar radiation is absorbed in the first metre of sea water but, once the infra-red and ultraviolet components have been eliminated, the rate of absorption with depth decreases, progressively removing the red and orange components of the solar spectrum and leaving only the most penetrating blue component. The sea water, therefore, acts as a monochromo ter, removing all colours except blue below a depth of 10 m. The blue colour of sea water is predominately the result of the absorption process rather than the Rayleigh scattering from sea water molecules. A person looking downwards into the ocean sees light that has been back scattered from successive layers of water. Since all of the other colours except blue have been absorbed very rapidly, a large proportion of the back-scattered light comes from layers where only blue light has penetrated. The blue colour of the ocean surface is enhanced by the reflection of 'sky light'.

Thus far only the absorption of solar radiation by pure sea water has been considered. Only in regions of very low biological productivity situated at a considerable distance from estuarine environments does the pure sea water absorption approach reality. The centres of the large sub-tropical gyres, such as the Sargasso Sea, are such regions of very clear ocean water. In regions of significant primary productivity, the colour of the sea water changes from dark blue to blue-green. The minimum of absorption no longer occurs at 0.46 μm but is shifted towards the green and occurs at approximately 0.48 μm. The shift is primarily the result of the presence of chlorophyll which absorbs strongly in the purple and the blue and, less significantly, in the red part of the spectrum. In addition to the colour change, the presence of chlorophyll also reduces the overall penetration depth of solar radiation. In very productive waters, such as the equatorial upwelling region, the penetration depth for blue light at 0.46 μm is ~10 m, compared with a penetration depth in pure sea water of 55 m. However, the most significant effects are found in coastal waters over the continental shelves. Here, strong tidal mixing and large inputs of suspended solids from rivers increase the absorption coefficient dramatically and penetration depths of ~1 m for blue light are observed. In addition, chlorophyll pigment and other biologically-derived pigments produce variations in the perceived colour of sea water. For example, a yellow pigment, called gelbstoff, which is prevalent in European coastal waters, has a very strong absorption in the blue part of the spectrum. Thus the most penetrating wavelengths occur around 0.55 μm in the yellow part of the visible spectrum. The presence of large quantities of suspended mineral sediment in some coastal waters tends to impart a grey colour because of scattering by the mineral particles. Occasionally, a red pigment, produced by a particular species of phytoplankton, can cause a phenomenon known as 'the red tide'. This pigment can be remarkably toxic to many other organisms

Table 3.1   Penetration depth (metres) of solar radiation (0.3 − 2.4 µm) in typical ocean and coastal wafer masses

Water type	Example of water type	Depth for attenuation of 90% of incident solar radiation	Depth for attenuation of 99% of incident solar radiation
Ocean I	Sargasso Sea	33	85
Ocean II	North-eastern Atlantic	14	45
	Equatorial East	8	25
Ocean III	Pacific (upwelling)		
Coastal 1	Continental shelf-break	7	20
Coastal 9	Baltic Sea	2	5

and it is, therefore, usually associated with a temporary absence of fish and other forms of marine life.

Table 3.1 shows the typical penetration depths of solar radiation for different types of water masses. The most important point to notice is that, because of the presence of biologically derived pigments and mineral and calcareous particles in the ocean, the average penetration depth is much less than that of pure sea water. All of the incident solar radiation is effectively absorbed within the uppermost 100 m of the ocean. Thus the ocean is warmed from above and the formation of a warm surface layer enhances the stability of the ocean, in the same way that the stability of the stratosphere is enhanced by ozone absorption. This is in marked contrast to the troposphere, where radiative heating of the surface maintains hydrostatic instability and convection.

3.6   Diurnal and seasonal temperature cycles in the ocean

We will now consider the temperature in the surface layers of the ocean in response to the diurnal and seasonal cycles of solar radiation. Absorption of solar radiation, in the absence of vertical mixing, may produce temperature changes in the upper 10 cm of the ocean ranging from 0.1°C day^{-1} in winter to 4°C day^{-1} during summer. This is because of the rapid absorption of the solar near infra-red radiation (>0.8 µm) shown in Figure 3.12b.

The observed temperature changes over a diurnal cycle (Figure 3.13) are, however, typically less than 1°C in the upper 1 m over the day. The weak vertical mixing caused by the surface wind stress and waves during the day mix the warm surface layers with underlying cooler layers. Even on a perfectly calm and clear day, there is a cooling of the surface 1 mm of the ocean, known as the surface skin, caused by latent heat of evaporation and long-wave radiation from the ocean surface to the overlying atmosphere. On calm days, the maximum temperature occurs in the afternoon, about two hours after the maximum of solar heating. The diurnal thermocline is

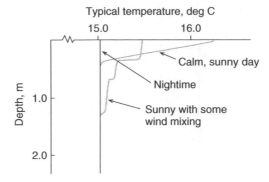

Figure 3.13 Diurnal change in temperature in the surface of the ocean. Reproduced, with permission, from Robinson, I.S. 1995, Satellite Oceanography, Wiley-Praxis Series in Remote Sensing: page 210, figure 7.5

removed during the night due to surface cooling, which produces vertical convection and mixing of this layer. The diurnal thermocline only occurs when the wind generated vertical mixing is weak and the solar heating is large, and therefore is most commonly found in the tropical oceans.

The seasonal temperature changes in the surface layer shown in Figure 3.14a, b, c are similarly governed by the processes of the absorption of solar radiation, cooling at the sea surface and vertical mixing. The cooling at the sea surface is due to heat lost from the surface to the atmosphere by latent heat of evaporation, long wave radiation and heat conduction. In late winter (March in Northern Hemisphere, Figure 3.14a), the surface water column is homogeneous down to depths of ~100–200 m, due to a combination of winter cooling and wind mixing. Warming of the surface layers starts to occur as the solar radiation increases in spring and a seasonal thermocline becomes established at the boundary of the warm surface layer and cooler deeper water. As more solar radiation is absorbed in the surface layer, the buoyancy increases and the layer becomes hydrostatically stable. Increased stability reduces the vertical mixing. Once the seasonal thermocline is established, it is remarkably resilient to vertical mixing by transient storms. The surface layer continues to increase in temperature for about two months after the summer solstice, when the maximum solar radiation occurs. The sea surface temperature reaches its maximum value in late summer (Figure 3.14). In autumn, surface cooling by evaporation, heat conduction and long-wave radiation exceeds the solar heating and this results in surface cooling of the surface layer. The surface cooling will produce a hydrostatically unstable surface layer. Vertical convection will transport the colder surface water downwards and warmer deeper water upwards. This cooling, in conjunction with increased wind stress, increases the vertical mixing of the warm surface layer which, in turn, increases the

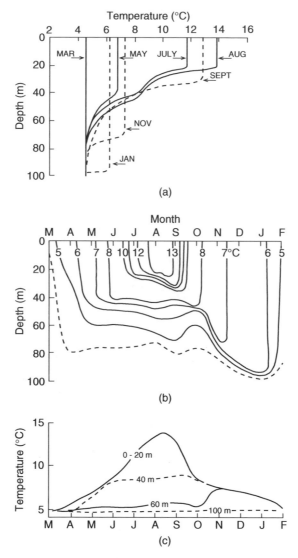

Figure 3.14 The annual cycle of temperature in the upper layers of ocean in the eastern North Pacific. (a) Vertical temperature profiles for different months. (b) Isotherms (°C) of temperature over the seasonal cycle. (c) The seasonal temperature cycle for different depths. The seasonal thermocline is the region of maximum vertical temperature gradient which occurs at depths of 30 m in summer and 100 m in winter. The temperature at 100 m depth shows very little seasonal variation. The N. Pacific halocline (see Section 3.6) inhibits vertical mixing below 100 m, and therefore the maximum depth of the seasonal temperature cycle. Reproduced, with permission, from Pickard G.L. and Emery, W.J. Descriptive Physical Oceanography (5th Edition), Pergamon Press,: page 45, figure 4.6, reproduced by permission of Butterworth Press

cooling of the surface mixed layer. The heat added to the surface layer during summer is removed during the autumn and winter, and is transferred to the overlying atmosphere. As the heat is depleted, vertical mixing is able to penetrate to deeper levels in the ocean. In the North Atlantic, the maximum depth of surface mixing is typically between 100 and 200 m, though in the Labrador and Irminger Seas, the mixing may extend to depths of 1000 m or more. In the North Pacific, the presence of a halocline limits the depth of mixing to about 100 m. The halocline is the boundary between the surface fresher layer and the more saline lower layer, and this enhances the vertical stability of the North Pacific Ocean (see Section 2.6).

Thus it has been shown that the diurnal and seasonal cycles of temperature in the upper ocean are caused by the absorption of solar radiation, mediated by the heat exchange between the ocean and atmosphere, and wind mixing.

4

Water in the Atmosphere

4.1 Introduction

The atmosphere is the smallest reservoir of water on the planet Earth, containing only 0.001% of the total mass of water present. In comparison, the ocean contains 96.8%, the ice caps of Antarctica and Greenland contain 2.0% and the freshwater reservoirs (groundwater, lakes, and rivers) contribute the remaining percentage. However, the effect of the water present in the atmosphere is by no means small. It has already been shown that water vapour in the atmosphere is an important absorber of both solar and long-wave planetary radiation and that, upon condensation, liquid water droplets can both reflect solar radiation back into space and intercept long-wave radiation from the surface and the atmosphere.

When water vapour condenses there is a large release of latent heat. For each kilogram of condensed water produced, 2.4 MJ of energy are released. The input of energy into the atmosphere, from condensation of water vapour, is equivalent to 23% of the solar flux incident on the Earth or $78\,W\,m^{-2}$ on average. It is, therefore, a major source of energy for driving the circulation of the atmosphere. For example, trade-wind circulations of both hemispheres are enhanced by the release of latent heat in the inter-tropical convergence zones (ITCZ) which straddle the thermal equator. The latent heat of condensation of water vapour may be released many thousands of kilometres away from where the evaporation of the water molecules took place and the wind circulation in the lower troposphere is important in determining the location of the energy release. Globally, the precipitation rate must balance the evaporation rate but relatively small variations in the wind circulation can drastically influence the spatial distribution of precipitation and produce devastating effects on agriculturally-based economies.

The Atmosphere and Ocean: A Physical Introduction, Third Edition. Neil C. Wells.
© 2012 John Wiley & Sons, Ltd. Published 2012 by John Wiley & Sons, Ltd.

4.2   The moist atmosphere

Figure 4.1 shows the phase diagram for water within the range of atmospheric temperatures. The curve AB is the saturated vapour pressure (SVP) over a plane ice surface and the curve BC corresponds to the SVP over a plane liquid water surface. The point B represents the point where water vapour, liquid water and ice can co-exist in equilibrium. It is known as the triple point.

Let us first consider the range of vapour pressure that can exist in the atmosphere. At temperatures of −50°C, typical of winter-time at the surface in the polar regions, the SVP over ice is only 0.04 hPa, whilst at 30°C, typical of the equatorial regions, the SVP approaches 43 hPa. Between 0°C and 30°C, the SVP approximately doubles for each 10°C temperature rise. Though there is a large range of water-vapour concentration in the atmosphere, the maximum value of the SVP is only 4.3% of the surface pressure. This is important because it allows water vapour to be included in the equation of state as an ideal gas. At higher concentrations of water vapour this assumption would not be valid.

In the atmosphere, liquid water can exist below the freezing point and this is known as a supercooled state. The SVP over a supercooled water surface is higher than the SVP over an ice surface; thus, a vapour in equilibrium with a supercooled water surface would be supersaturated with respect to an ice surface and water vapour would condense preferentially onto the ice surface (Figure 4.9). This is an important process in the formation of precipitation in cold clouds and will be discussed in Section 4.5.

Water vapour is a gas with a molecular weight of 18 a.m.u., compared with a mean molecular weight of 29 a.m.u. for dry air. Hence a mixture of water vapour and dry air will have a lower mass than the equivalent volume of dry

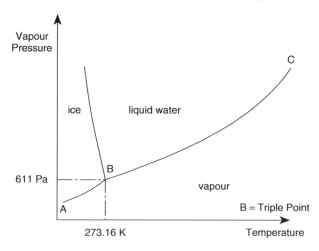

Figure 4.1   Phase diagram for water in the vapour, liquid and solid phases showing the triple point

air, assuming that temperature and pressure remain constant. The equation of state for moist air therefore requires another variable, namely the mass of water vapour, to determine the relationship between pressure, temperature and density. The equation of state for moist air gives the mean molecular weight, $\overline{m}$, of a mixture of dry air of mass M_d and of water vapour of mass M_v in a given volume as:

$$\frac{1}{\overline{m}} = \frac{M_d}{m_d(M_d + M_v)}\left[1 + \frac{M_v/M_d}{m_v/m_d}\right] \tag{4.1}$$

where m_d and m_v are the molecular weights of dry air and water vapour respectively.

The concentration of water in the atmosphere is defined by a mixing ratio, q, where

$$q = M_v/M_d \tag{4.2}$$

The mixing ratio is dimensionless, but meteorologists often express it in terms of grams of water vapour per kilogram of dry air. Since, for a given volume, masses are proportional to densities, the above equation may be rewritten as:

$$q = \rho_v/\rho_d \tag{4.3}$$

where ρ_v and ρ_d are the densities of water vapour and dry air respectively. The ideal gas equations for water vapour and dry air yield the following:

$$\rho_v = \frac{e}{(R^*/m_v)T} \tag{4.4}$$

$$\rho_d = \frac{p - e}{(R^*/m_d)T} \tag{4.5}$$

where p is the total pressure of the dry air and water vapour and e is the vapour pressure of water. Water vapour is not in general an ideal gas but at low concentrations, such as found in the atmosphere, the approximation is a valid one.

Substituting equations 4.4 and 4.5 into 4.3 yields:

$$q = \frac{e}{p - e}\left(\frac{m_v}{m_d}\right) \tag{4.6}$$

Substituting for M_v/M_d in equation 4.1 gives:

$$\frac{1}{\overline{m}} = \frac{1}{m_d}\left[\frac{1 + q(m_d/m_v)}{1 + q}\right] \tag{4.7}$$

Substituting equation 4.7 into the equation for a mixture of ideal gases, the ideal gas equation for a moist atmosphere is obtained, namely:

$$p\alpha = \frac{R^*}{m_d}\left[\frac{1 + q(m_d/m_v)}{1 + q}\right]T \qquad (4.8)$$

This equation is similar in form to the ideal gas equation for dry air except for the addition of the factor involving the mixing ratio, q. An example of the use of this equation is now appropriate. If the vapour pressure of water in the atmosphere is 10 hPa and the total pressure is 1000 hPa, then using equation 4.6, q is found to be 6.22×10^{-3} or 6.2 g of water vapour per kilogram of dry air.

At this stage it is useful to define a virtual temperature, T_v, where

$$T_v = T\left[\frac{1 + q(m_d/m_v)}{1 + q}\right] \qquad (4.9)$$

This virtual temperature is always higher than the actual temperature but, in practice, the difference is less than 5 K. In the middle latitudes the difference is typically only 1 K. In physical terms, the virtual temperature is the temperature a volume of dry air would have if it had the same density as an equal volume of moist air. Because water vapour has a lower molecular weight than dry air, this temperature will always be higher than the actual temperature.

Another useful parameter is the relative humidity, r, given by

$$r = \frac{q(T)}{q_s(T)} \times 100\%$$

where q is the actual mixing ratio of the sample of air and q_s is the saturation mixing ratio at the same temperature. It gives a useful measure of the humidity of the moist air mass relative to its saturation value. Hence a 100% relative humidity corresponds to a saturated air mass.

4.3   Measurement and observation of water vapour

First consider the measurements of water vapour in the laboratory. The most direct method is to pass air over a flat silver plate which is cooled. The temperature at which dew first forms, i.e. the dew-point temperature, T_d, is measured. The water vapour content of the air is unaltered during the cooling, as shown in Figure 4.2, and hence the saturated mixing ratio at the dew point temperature is equal to the mixing ratio of the air at its original temperature, T. The saturated mixing ratio is a known function of dew-point

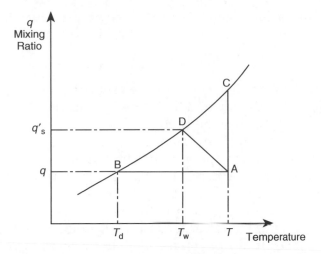

Figure 4.2   Dew-point temperature (T_d) and wet-bulb temperature (T_w) for a parcel of moist air at temperature T and mixing ratio q (point A). The saturation mixing ratio is given by the curve BDC. The saturation mixing ratio for the wet bulb temperature is q'_s at the point D. The mixing ratio at the dew point temperature has the same value as the original parcel (point A). The mixing ratio at the wet bulb temperature (point D) is larger than the original parcel (point A), because of the evaporation of water into it. Therefore, the wet bulb temperature is lower than the parcel temperature T, but because of the higher mixing ratio, it reaches saturation at a higher temperature than the dew point

temperature and it can, therefore, be easily obtained from tables. It is noted that the cooling process must take place at constant atmospheric pressure to maintain a constant mixing ratio, as shown in equation 4.6. At lower temperatures, frost rather than dew is more likely to form and, in this case, the saturated mixing ratio with respect to ice must be used instead of that with respect to water.

A more practical method for surface measurements of humidity uses an aspirated psychrometer. In this technique, air is drawn past two thermometers. One thermometer has its bulb covered with a clean muslin cloth which is kept moist with distilled water. The wet-bulb temperature always lies between the dry-bulb temperature and the dew-point temperature, except at saturation when all three temperatures coincide, as shown in Figure 4.2. The wet bulb evaporates water into the surrounding air, which:

(i)   Decreases the temperature of the wet bulb due to the latent energy input required to produce evaporation.
(ii)   Increases the mixing ratio of the surrounding air.

Hence the saturated air surrounding the wet bulb has a mixing ratio q_s' and is represented by point D in Figure 4.2. The difference between the saturation mixing ratio at the wet-bulb temperature, q_s', and the mixing ratio of the environmental air, q, can be shown to be proportional to the difference between the wet- and dry-bulb temperatures, i.e.

$$T - T_w = A[q'_s - q]$$

where A is a constant which depends on the ventilation rate of the instrument. q'_s can be obtained from tables and hence q is obtained.

Other modern devices use the change in electrical resistance or electrical capacitance to measure humidity changes.

The distribution of water vapour in the atmosphere is shown in Figure 4.3. The mixing ratio reaches a maximum in the equatorial zone and rapidly decreases with both altitude and latitude. This is a general reflection of the variation of saturated mixing ratio with temperature. The majority of the water vapour occurs in the lower troposphere, although the small amounts present in the upper troposphere and the stratosphere have already been shown to be important for the radiation balance in these regions. If all the water vapour were to be condensed out of the atmosphere, it would result in an average global precipitation of 2.5 cm, ranging from 4 cm in the equatorial regions to less than 0.5 cm over the poles. This can be compared with the observed global precipitation rate of about $100 \, \text{cm year}^{-1}$. The ratio of the water content of the atmosphere to the observed precipitation rate gives the time scale for the replacement of atmospheric water vapour by evaporation from the surface. Thus the average lifetime of a water molecule from evaporation at the surface to precipitation is approximately nine days and this is sufficient for the water molecule to be carried around the globe by the tropospheric wind system before it is precipitated. The relative humidity distribution shown in Figure 4.3 gives an indication of the efficiency of the water-holding capacity of the atmosphere. In the equatorial zone, between 0 and $10°$, and at latitudes greater than $50°$, the surface layers of the atmosphere have relative humidity above 80%. However in the sub-tropical zone of the mid-troposphere the relative humidity values are low, attaining values of only 30% as a result of the descent of drier air from the upper troposphere. This air will warm by adiabatic compression as it moves downwards through the troposphere but it will retain its low water content. Hence the relative humidity will decrease as air moves from the upper to the lower troposphere, although close to the surface evaporation into the air mass will act to increase the relative humidity.

4.4   Stability in a moist atmosphere

It has already been shown that water vapour will change the thermodynamic properties of a sample of air by:

(i)   Decreasing the mean molecular weight of the moist sample in comparison with a sample of dry air.
(ii) The release of latent heat of condensation when a moist sample reaches its saturation vapour pressure.

The change in the first property is directly proportional to the mass of water vapour in a given sample of air, whilst the second property is proportional to the mass of water vapour that condenses to form liquid water. In order to obtain a qualitative feel for the relative effects of the two processes, Figure 4.4 illustrates the temperature history of a parcel of moist air typical of the middle latitudes, as it is lifted through the atmosphere. At the surface, it is assumed that the air has a water-vapour content of 3.5 g per kilogram of dry air, which corresponds to a virtual temperature 1 K higher than the actual temperature. From the surface to a height of 1 km, the parcel of air cools at the dry adiabatic lapse rate of $9.8\,K\,km^{-1}$. A moist unsaturated sample of air will cool very close to the dry adiabatic lapse rate, because of the very small mass of water vapour present. At the point at which the condensation level is reached, release of latent heat slows down the rate of cooling of the parcel. The amount of latent heat released between 1 and 3 km depends solely on the difference between the saturation mixing ratios of the air at the two levels, which is approximately 7 g per kilogram of dry air in this case. This is equivalent to the release of 17 500 J for each kilogram of dry air. The temperature difference between a parcel of moist air at 3 km and a dry parcel of air at the same height is approximately 8 K, due solely to the release of the latent heat of condensation. It is noted that the difference between the actual and the virtual temperature is generally smaller, though not negligible, compared with the difference between dry air parcels and moist air parcels after condensation.

Now consider the change in temperature with height of a parcel of air at the level of condensation. If the parcel rises an infinitesimally small distance, dz, then the difference in the saturation mixing ratio, brought about by adiabatic cooling, will be $-dq_s$. The negative sign denotes a decrease in the saturation mixing ratio with increasing height. The latent heat released for each kilogram of dry air is given by $-Ldq_s$, where L is the latent heat of condensation. This heat is, therefore, an additional source of energy, equivalent to the term dQ in equation 2.6. Thus:

$$-Ldq_s = C_p dT - \alpha dp \qquad (4.10)$$

And from the hydrostatic equation (equation 2.4):

$$\alpha dp = -gdz \qquad (4.11)$$

Hence, by substitution in equation 4.11:

$$-Ldq_s = C_p dT + gdz \qquad (4.12)$$

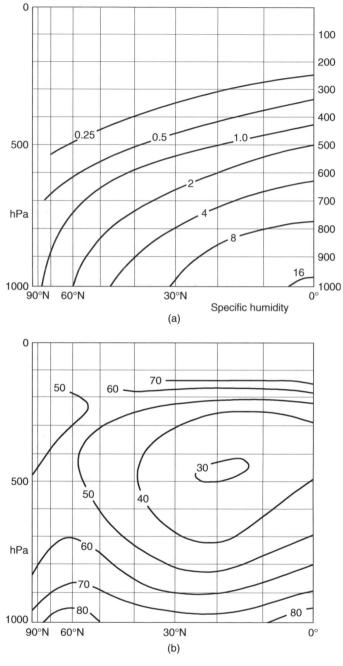

Figure 4.3 (a) The longitude-averaged specific humidity (q) in northern winter (October – March). Values are in g kg^{-1}; (b) The longitude-averaged relative humidity (q/q$_s$) in northern winter. Values are percentages. Reproduced, with permission, from Lorenz, E.N., 1967, The Nature and Theory of the General Circulation of the Atmosphere, World Meteorological Organization: 218 TP 115, pages 42 and 43, figures 14 and 16

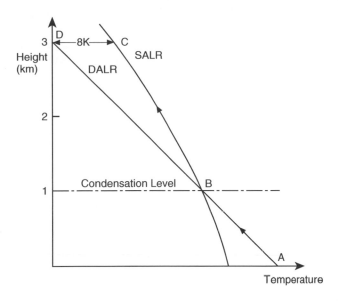

Figure 4.4 Typical ascent of a moist air parcel (ABC) and a dry air parcel (ABD). See text for details

Rearranging and dividing by dz yields:

$$\frac{dT}{dz} = -\frac{L}{C_p}\frac{dq_s}{dz} - \frac{g}{C_p}$$

Since $dq_s/dz = (dq_s/dT)/(dT/dz)$

$$\frac{dT}{dz} = \frac{-g}{C_p}\left[\frac{1}{1 + (L/C_p)(dq_s/dT)}\right] = \Gamma_s$$

where Γ_s is known as the saturated adiabatic lapse rate (SALR).

It is noted that as dq_s/dT is always positive, the SALR will always be less than the dry adiabatic lapse rate, given by $\Gamma_d = -g/C_p$. At a temperature of 20°C and at a pressure of 10^3 hPa, the SALR is approximately 0.41 Γ_d or $4\,\mathrm{K\,km^{-1}}$, whilst at 10°C it is 0.59 Γ_d or $5.8\,\mathrm{K\,km^{-1}}$. At high levels in the troposphere, the saturation mixing ratio will become very small and the SALR will approach the dry adiabatic lapse rate asymptotically.

The addition of latent heat of condensation also increases the potential temperature of the air and hence, in any region of the atmosphere where condensation occurs, the potential temperature will no longer be conserved. The stability of the atmosphere will therefore be significantly changed by the condensation process. Consider a parcel of dry air displaced from its equilibrium position in a stable environment where the temperature gradient, $d\theta/dz$, is positive (Figure 4.5). As discussed in Section 2.6, the parcel will be subjected to a negative buoyancy force as it rises and this force will act to return

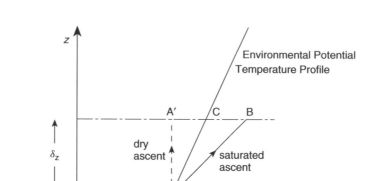

Figure 4.5 Conditional instability of a moist atmosphere. The parcel of air at A is in equilibrium with its environment. If the parcel is unsaturated and is perturbed upwards, it will follow AA' and it will become cooler than its environment. Negative buoyancy forces will return it to its initial position and therefore the parcel is stable with respect to its environment. If the parcel is saturated, it will follow path AB and will become warmer than its environment. Buoyancy forces will accelerate the parcel upwards, away from its initial position, and it will now become unstable

it to the equilibrium position. Now consider a parcel of air which is saturated at the equilibrium position, z_e. On rising a small distance, δ_z, it will condense some of its water vapour and this will warm the parcel by an amount $\delta\theta_1 > \delta\theta$ (where $\delta\theta$ is the temperature change of the environmental air over a small distance δ_z), then the moist parcel will be warmer than its environment (path AB in Figure 4.5) and it will accelerate upwards from its equilibrium position. Therefore the result of the condensation process is to make the previously stable environment profile unstable. Clearly, if the warming is insufficient, i.e. $\delta\theta_1 < \delta\theta$, the profile will still be stable. This form of instability is known as 'conditional instability' and is produced by the condensation process. However, it is not possible to determine the conditions for instability from the potential temperature alone and an equivalent potential temperature, θ_e, must be defined.

$$\theta_e = \theta \exp\left(\frac{Lq_s}{C_p T}\right)$$

This equivalent potential temperature is conserved in saturated conditions. It is defined as the potential temperature a parcel of air would have if all the water vapour in the parcel were to be condensed and all the latent heat of condensation converted into internal energy. It has the property of remaining constant during a saturated adiabatic process. Thus, the stability condition for saturated air is determined from the vertical profile of the equivalent potential temperature rather than from the potential temperature.

If, in the case considered above, $d\theta_e/dz < 0$, then the vertical profile would be unstable for saturated air. Parcels of air would have positive buoyancy when they rose and they would accelerate away from their initial surroundings. A profile with $d\theta_e/dz > 0$ would be stable for both dry and saturated air. It is therefore possible to compile a simple stability table:

	$d\theta_e/dz < 0$	$d\theta_e/dz > 0$
$d\theta/dz < 0$	Always unstable	Not possible
$d\theta/dz > 0$	Conditionally unstable	Always stable

It is noted that the stability of an environment temperature profile can be determined either from the equivalent potential and the potential temperature profiles or by comparison of the dry adiabatic and saturated adiabatic lapse rates.

An additional form of moist instability is termed the potential instability. This occurs when the profile has a negative equivalent potential temperature gradient but the air is not saturated. In this case, the lifting of the air mass is required to cool the air to its condensation point and to release the instability. This type of instability is often associated with the advection of an upper-level dry air mass over a very moist, low-level air mass. The advection produces a large negative equivalent potential temperature gradient between the two air masses. Usually uplift by mountains or uplift ahead of a depression may be sufficient to release the instability, which will often be associated with vigorous convection and occasionally produces severe thunderstorms with hail.

It is noted that a vertical gradient of humidity may produce instability in unsaturated air. Consider a neutral temperature profile onto which is superimposed a humidity gradient where the humidity decreases with height. In this case, the virtual temperature would be higher nearer to the surface than above and parcels of air would have positive buoyancy if they were displaced upwards and would, therefore, accelerate upwards. This type of instability is probably the most important in the surface boundary layer over the tropical ocean where the virtual temperature can be several degrees higher than the actual temperature. It is also possible to introduce instability by the evaporation of precipitation in unsaturated air. Precipitation falling from a cloud may evaporate and cool the air beneath to the wet-bulb temperature. The air will become negatively buoyant and accelerate downwards. This process is responsible for the sudden drop in temperature experienced in the downdraughts of thunderstorms and for sudden gusts of wind, up to 20 m s^{-1}, felt at the surface. Re-evaporation of precipitation is particularly common in the tropics, where cloud bases are generally higher than those found in the middle latitudes.

4.5 Processes of precipitation and evaporation: The formation of clouds

In the previous discussion it was implicitly assumed that condensation would occur when the relative humidity reached 100%, i.e. when the vapour pressure was equal to the saturation vapour pressure. However, the saturation vapour pressure described in Section 4.2 was defined with respect to a plane liquid surface and not a curved surface, such as a rain droplet. In a sample of pure moist air, where all other particles have been removed, condensation will not occur at 100% relative humidity but will only commence spontaneously at a relative humidity of 800%, equivalent to a supersaturation of 700%. The reason for this phenomenon is that the saturated vapour pressure over a curved surface is higher than over a plane surface. The relationship between the saturated vapour pressure over a curved surface, e_r, and that over a plane water surface, e_∞, is given by:

$$\ln \frac{e_r}{e_\infty} = \left(\frac{2\gamma m_w}{\rho_w R^* T} \right) \frac{1}{r} \tag{4.13}$$

where γ, m_w and ρ_w are the surface tension, molecular weight of water and density of water respectively, r is the radius of curvature and R^* is the universal gas constant. At a temperature of 273 K, the term in the bracket is 1.19×10^{-9} m.

If the value of $e_r/e_\infty = 8$ is substituted into equation 4.14, a value of $r \sim 0.6 \times 10^{-9}$ m is obtained and this is typical of a small group of water molecules. Therefore condensation in pure air samples will occur on groups of water molecules randomly brought together by Brownian motion. However, if a sample of air is taken directly from the atmosphere, condensation will occur at much smaller supersaturations, typically less than 1%. This is because of the presence of particles in the atmosphere which provide surfaces for condensation. These particles are known as condensation nuclei. The size of the nuclei, estimated by substitution into equation 4.14, is approximately $0.1 \, \mu\text{m}$ for a supersaturation of 1%. Basically there are two types of condensation nuclei:

(i)   Hygroscopic nuclei, usually ionic salts such as sodium chloride and ammonium sulphate. These nuclei have the property that they are able to initiate condensation at relative humidities as low as 80%.

(ii) Hydrophobic particles, not soluble in water, and requiring supersaturation conditions for condensation to occur.

Equation 4.14 has to be modified for hygroscopic nuclei. Figure 4.6 shows SVP curves for pure water droplets and droplets containing dissolved hygroscopic nuclei. It is noted that the difference between the two SVP curves is most marked for small droplets having radii less than 0.1 μm. For larger water droplets, the hygroscopic nuclei will become well diluted and the effect becomes less pronounced. In addition, the radius of curvature also becomes large enough to approximate to a plane surface and, for larger water droplets bigger then 1 μm, the difference between the two types of nuclei is no longer important. One interesting aspect of the behaviour of hygroscopic nuclei can be seen by comparing the growth of two small droplets. In an environment where the relative humidity is less than 100%, only hygroscopic nuclei will present possible condensation sites but, as the water droplet grows, the salt

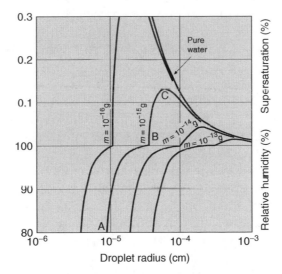

Figure 4.6   The equilibrium relative humidity (or supersaturation) as a function of the droplet radius for solution droplets containing the indicated mass (m) of sodium chloride. The supersaturation is $\left(\frac{q}{q_S} - 1\right)$ 100%. A small droplet (0.1 μm) containing 10^{-15} g. of sodium chloride can remain in air with a relative humidity of about 85% (A). This is a typical haze droplet. A larger droplet (B) will require a RH of 100% to remain in the air. A slightly larger droplet (C) will need supersaturated air (∼0.1%) to remain in the air. A pure water droplet will always need a higher supersaturated airmass than the droplet containing sodium chloride to remain in equilibrium. Reproduced, with permission, from Mason, B.J., 1975, Clouds, Rain and Rainmaking, Cambridge University Press: page 21, figure 4

becomes more dilute and a higher relative humidity is required for further growth. This process is represented by path AB in Figure 4.6.

Once the droplet has reached the size dictated by the prevailing relative humidity, its growth will cease. Only if there is an increase in relative humidity of the environmental air, or if supersaturation is achieved, will there be further growth. An example of this process is the development of haze. A maritime haze layer will develop a few hundred metres above the surface of the ocean where salt nuclei are readily available because of bubble bursting at the sea surface. Haze over land may form as the result of the introduction of hygroscopic pollutants from industrial processes or by photochemical reactions of gaseous car exhaust emissions.

In an environment where there are very few hygroscopic nuclei, supersaturation of the air of the order of 1% will be required to start the condensation process. However, once a droplet starts to grow, its radius will increase. The SVP (with respect to the curved surface) will decrease and the droplet will find itself in a continually saturated environment and will carry on growing. Thus, once a critical supersaturation has been reached, represented by the point C in Figure 4.6, droplets will grow but, as the supersaturation curve approaches 100% relative humidity, the rate of growth will decrease. Water droplets larger than 10 μm in radius grow only very slowly by this process.

The availability of condensation nuclei depends on the rate of supply of suitable particles and on their rate of removal, either by gravitational settling or by precipitation, known as 'washout'. Figure 4.7 shows the typical sources and sinks of atmospheric aerosols and the size distributions for the various types of aerosols.

The small Aitken nuclei, less than 0.1 μm in radius, tend to be produced by smokes and gases, and are plentiful in industrial areas. They tend to be too small to be efficient as condensation nuclei on their own, but coagulation can produce larger particles that are very efficient condensation nuclei. The majority of the aerosol is removed by 'washout' whilst the remainder is removed by dry deposition or gravitational settling. A particle 1 μm in diameter would take about one year to fall 1 km, whilst a 10 μm particle would take only 3.5 days. Thus gravitational settling is an important process only for the deposition of giant aerosols. For smaller Aitken nuclei, scavenging by precipitation will be the principal removal process.

The number of nuclei reduces from 10^{12} m^{-3} over industrial areas to about 10^{8} m^{-3} over the ocean, where the dominant nuclei are salt particles. These are principally produced by bursting bubbles in foam patches which are produced, in their turn, by breaking waves. The number of particles entering the marine atmosphere is dependent on the area of foam over the ocean and this is determined by the sea state and the surface wind speed. The number of particles can vary from 10^{6} m^{-3} in low wind conditions to 10^{9} m^{-3} in storm-force conditions. Many of the larger particles will re-enter the

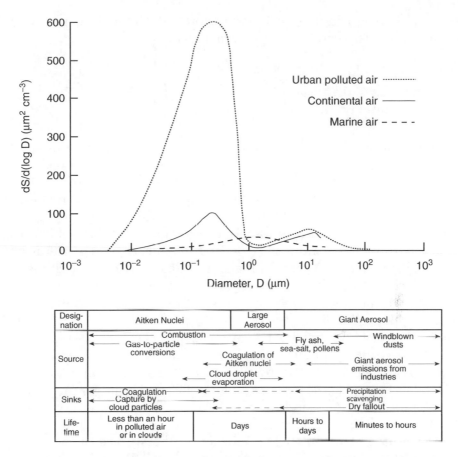

Figure 4.7 Schematic curves of aerosol surface area distributions for urban polluted air, continental air and marine air. Shown below the curves are the principal sources and sinks of atmospheric aerosols and estimates of their mean lifetimes in the troposphere. Reproduced, with permission, from Slinn, N. and George, W., 1975, Atmospheric Environment, 9: page 763, figure 1

ocean but a small proportion will be taken aloft by atmospheric turbulence. These salt particles can have a very large range and they may be found in clouds many hundreds of kilometres away from the ocean. As a general rule, there are fewer condensation nuclei over the ocean than over the continents (Figure 4.7). The giant aerosols are produced mainly by 'mechanical' wind action at the ocean or land surface under storm conditions. Wind-blown sand can be transported considerable distances. Red Saharan dust is occasionally reported in the British Isles and in the Caribbean, over 3000 km from the Sahara. In addition large industrial plants produce considerable numbers of giant nuclei.

The typical size distribution of water droplets in clouds is now considered. Water droplets may be obtained by exposing an oil-covered slide from an aircraft and measuring droplet population with a microscope. A typical distribution is described by three parameters:

(i)   The mean radius.
(ii)   The median radius.
(iii) The maximum radius.

Table 4.1 shows the statistics of drop distribution in various cloud types. In continental clouds there are much larger numbers of droplets than in ocean clouds, because there is an increased population of large condensation nuclei over land. The amount of water vapour available for condensation is determined by large-scale processes, such as uplift, thus ensuring that the liquid water contents of maritime and continental clouds are similar. Hence, the sizes of droplets in continental clouds are smaller than in maritime clouds. It is noted from Table 4.1 that the maximum droplet size in maritime clouds can be twice as large as those in continental clouds. This observation is of critical importance in the determination of the time scale for the growth of droplets in clouds.

Table 4.1 also shows that the maximum droplet size is related to the type of cloud. Weak vertical motions in the stratocumulus result in a reduction in the maximum size, whilst the strong vertical motions found in cumulonimbus result in large droplets. The water content of cumulonimbus clouds is much larger than in cumulus clouds because of the different surface-volume ratio of the two types of clouds. In cumulus cloud, usually 100 m to 1 km thick, this ratio is large and mixing dry air across the surface of the cloud will significantly modify the water content. In cumulonimbus, the ratio will be smaller because of the size of the cloud, typically 5–10 km thick, which implies that mixing across the surface of the cloud will have less effect in reducing the water content.

Table 4.1   Cloud drop distributions and liquid water content in typical maritime and continental clouds

Cloud type	Droplet number density (cm^{-3})	Range of radii of droplets (µm)	Most frequent radius of droplet (µm)	Liquid water content (g m^{-3})
Continental cumulus	420	3–10	6	0.4
Maritime cumulus	75	2–20	11	0.5
Cumulonimbus	72	2–100	5	2.5
Altostratus	450	1–13	4	0.6
Continental stratus	260	1–22	4	–
Maritime stratus	24	2–45	13	0.35

A qualitative discussion of the growth of a small water droplet in a super-saturated environment has already been given. This growth is dependent on:

(i) The supersaturation of the environment.
(ii) The size of the drop.

It is also related to the rate of diffusion of water vapour onto the droplet. As the droplet grows the curvature effect on the SVP decreases and therefore the rate of growth decreases. The growth of a small droplet, 1 µm in diameter, is depicted in Figure 4.8(a). It takes approximately 30 min for the droplet to reach 10 µm, a further 1.5 hours to reach 20 µm and 10 hours to grow to 40 µm. Even a 40 µm droplet is a long way short of the 200 µm of a respectable drizzle droplet. Clearly the diffusion process cannot account for the growth of raindrops on the time scale of a shower cloud, which is typically 0.5–2 hours. The additional process is the growth of raindrops by collision and coalescence. A droplet may collide with another drop, coalesce and increase its mass. Its vertical velocity will increase and it will sweep out more droplets at a faster rate, soon reaching precipitation size. The process is non-linear in that initial growth will be slow because of the low terminal velocity of the small droplet but the growth rate will become larger and larger as the droplet grows. This is shown in Figure 4.8b.

The rate of increase in mass of a droplet (M) with time is given by:

$$\frac{dM}{dt} = \pi r_1^2 \left(v_1 - v_2 \right) q_L E_c \qquad (4.14)$$

Where r_1 is the radius of the droplet and $v_1 - v_2$ is the difference between the terminal velocities of the growing droplet and the surrounding cloud droplets respectively. q_L is the liquid water content and E_c is the collision efficiency. Except in the initial stage of growth, $v_1 \gg v_2$ and substituting

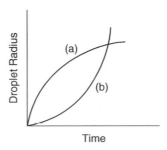

Figure 4.8 Schematic curves of droplet growth by (a) condensation from the vapour phase, and (b) coalescence of droplets. Reproduced, with permission, from Wallace, J.M. and Hobbs, P.V., 1977, Atmospheric Science, Academic Press: page 171, figure 4.16

$M = (4/3)\pi r_1^3 \rho_L$ into equation 4.15 yields:

$$\frac{dr_1}{dt} = \frac{v_1 q_L E_c}{4\rho_L} \tag{4.15}$$

The terminal velocity, v_1, for different droplet sizes is shown in Table 4.2.

A 10 μm water droplet has a terminal velocity of 10^{-2} m s^{-1}, whilst a drizzle droplet has a terminal velocity of about 0.5 m s^{-1} and a raindrop has a maximum terminal velocity of 9 m s^{-1}. Thus the rate of growth, which is proportional to the terminal velocity, will accelerate with droplet size. The growth rate is also dependent on the liquid water content of the cloud and the collision efficiency. As already noted, cumulus and stratocumulus tend to have lower water content than the larger cumulonimbus and therefore the precipitation process will be more effective in the latter cloud. The collision efficiency increases with droplet size and is low for droplets less than 20 μm in radius. When the cloud droplets are small, they tend to follow streamlines around the collector droplets and do not get captured. For larger droplets, of 70 μm radius or more, turbulent wakes develop behind the droplet, drawing in more droplets than were originally in the volume swept out.

The production of precipitation from a water-droplet cloud is a non-linear process. In a given distribution of cloud droplets, only a few droplets will be bigger than 20 μm and thus be sufficient in size to grow efficiently by collision and coalescence. At first, the process will be slow but, because of the non-linearity of the process (Figure 4.8(b)), the droplet will soon grow quickly to the size of a raindrop. In a maritime cumulus, with droplet concentrations of 100 cm^{-3}, droplets will grow from 25 to 1000 μm within 10 min. However, only a small proportion at the extreme end of the droplet population will be involved in the precipitation process. The differences in the size distributions in maritime and continental clouds have already been noted, as well as the fact that in many cases r_{max} is less in continental than

Table 4.2 Terminal velocities of water droplets in still air at surface atmospheric pressure and temperature of 20°C

	Radius of droplet(μm)	Terminal velocity in still air(m s^{-1})
Fog	1	0.0001
Cloud droplet	5	0.003
Large Cloud droplet	25	0.072
Drizzle	50	0.256
Drizzle	250	2.04
Drizzle	500	4.03
Rain	1000	6.49
Rain	2500	9.09

in maritime clouds. A continental cloud will therefore take longer to produce rain than a corresponding maritime cloud and it has been calculated that a maritime cloud 1.5 km deep will produce rain within 30 min of its formation whereas a continental cloud will take more than an hour. This illustrates the dramatic influence that the concentration of condensation nuclei has on the rate of growth of water droplets and hence on the formation of precipitation.

The growth of a shower cloud is dependent on:

(i)   The cloud droplet size distribution.
(ii)   The liquid water content of the cloud.
(iii) The thickness of the cloud.
(iv)   The updraft within the cloud which is essential for the growth of cloud droplets into raindrops.

Small droplets have small terminal velocities and will be maintained in the cloudy environment for longer than larger droplets, provided that they are not ejected through the top of the cloud. As the updraft increases, the size of the final raindrop increases up to a maximum size of 5 mm and, at this stage, the raindrop will break up into smaller droplets. A maritime cumulus will have to be at least 1 km thick to produce light precipitation. However, drizzle may be produced by stratus cloud whose thickness is less than 1 km because the weak vertical motion in such clouds allows the droplets to be maintained in the cloudy air for a sufficient time to reach drizzle drop size.

The growth of precipitation in clouds where the temperatures are above 0°C has been discussed. These are known as warm clouds and include tropical cumulus clouds and low-level clouds in middle latitudes. However, below 0°C additional physical processes come into play as the result of the introduction of the ice phase. Consider a cloud of liquid water droplets with a mean size of 10 μm. As the droplets are cooled below 0°C, only a few droplets will freeze and the remainder will be in a supercooled liquid state. As the temperature is reduced still further, the number of ice crystals will increase, but only when the temperature reaches −39°C will all of the water droplets have frozen. At this temperature, spontaneous nucleation will occur, caused by the chance aggregation of water molecules into an ice embryo. This process is analogous to the condensation of water vapour in a cloud chamber in the absence of ions or other particles. However, the observation that ice crystals are formed at temperatures higher than −39°C indicates the existence of ice nuclei. These ice nuclei are different from condensation nuclei in two important respects. First, the most successful ice nuclei tend to have hexagonal molecular structures similar to water ice. The most common types of ice nuclei are clay minerals, obtained form the soil, and organic materials, particularly decayed plant matter and derivatives of phytoplankton from the ocean. Second, the ice nuclei have a threshold temperature at which

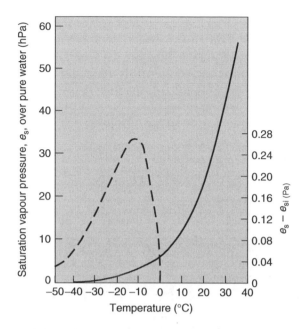

Figure 4.9 Variations of the saturation vapour pressure (e_s) with temperature over a plane surface of pure water (_____) and the difference between (e_s) and the saturation vapour pressure over a plane surface of ice e_{si}(- - -). Reproduced, with permission, from Wallace, J.M. and Hobbs, P.V., 1977, Atmospheric Science, Academic Press: page 73, figure 2.7

they become efficient and below which they act as preferential sites for ice nucleation. For clay particles, this temperature is −15°C whilst organic nuclei have threshold temperatures as high as −4°C. Thus the presence of ice nuclei only has a significant effect below −4°C and this explains the observation that clouds between 0 and −4°C exist in the supercooled liquid state.

The processes which lead to the growth of ice crystals in a supercooled cloud will now be considered. Figure 4.9 shows the SVP over a plane water and a plane ice surface. It is noted that the SVP over ice is always lower than over liquid water except at 0°C. The water vapour in equilibrium with a water surface will therefore always be supersaturated with respect to an ice surface below 0°C. The difference in SVP with respect to ice and water reaches a maximum at −12°C. In a saturated environment with respect to supercooled liquid water, the vapour would have a supersaturation of 10% with respect to ice. In a mixed cloud, consisting of supercooled liquid water droplets and ice crystals, the ice crystals will grow at the expense of the water droplets. This process is known as the Bergeron process, after the Norwegian cloud physicist who first noted this behaviour. It is most effective in clouds where the temperature is between −10 and −30°C (i.e. where the difference in SVP is greatest). Between 0 and −10°C there are too few nuclei present for the mechanism of ice crystal formation to be effective, whilst below −30°C there are too many ice crystals competing for too few liquid water droplets for the process to be significant. The growth rates of ice crystals in mixed clouds are much greater than the growth rates of water droplets in warm clouds because of the higher supersaturations occurring in mixed clouds compared with those in warm clouds, typically 10% supersaturation compared with less than 1% supersaturation. For example, an ice crystal can grow to 1000 μm within 30 min by the Bergeron process.

An ice crystal can grow by two additional processes other than the Bergeron process. First it can grow by aggregation of ice crystals, because supercooled water films attached to an ice crystal can act as binding for other ice crystals as a result of surface tension between the water and the ice. The thin film of supercooled water then freezes and the crystal grows. The process is more effective between 0 and −10°C, where there are a large number of supercooled water droplets and freezing does not occur immediately on impact. A snowflake may grow from 1 mm to 1 cm in about 30 min by aggregation and the largest snowflakes occur at temperatures close to 0°C. At lower temperatures, supercooled water droplets may freeze on impact with ice crystals to produce a rimed snowflake. This process is known as accretion and it is responsible for the growth of ice crystals between −10 and −30°C. It is most effective in clouds such as cumulonimbus where the liquid water content is high. In contrast, the Bergeron process is most effective in small cumulus and stratocumulus clouds where the water content is low. Accretion and riming of ice particles can produce soft hail and ice pellets and, more exceptionally, hailstones. The accretion process has many similarities with the collision-coalescence mechanism in that it is non-linear due to the increase in terminal velocity with size. For crystals formed by aggregation rather than accretion, the terminal velocity does not tend to increase with the size of the crystal because of the crystal's large surface-to-mass ratio. The drag force on a crystal increases in proportion to its mass and hence an unrimed crystal 2 mm in diameter may have a terminal velocity of 0.5 m s^{-1}, whilst a rimed ice pellet of similar diameter would have a terminal velocity of 1.8 m s^{-1}. Therefore an ice pellet falling through a cloud of high water content could grow very quickly by accretion.

Riming, or accretion, can also cause a rapid build up of ice on solid surfaces, such as aircraft wings and television transmitters. In such cases, impact freezing will generally occur at all temperatures below 0°C and it is especially important between 0 and −4°C where large numbers of supercooled water droplets exist. In March 1969 a television transmitter in Yorkshire, England, collapsed as a result of excessive loading of rimed ice on its structure.

Figure 4.10 shows a simplified diagram of the precipitation processes occurring in layered clouds associated with middle-latitude (extra-tropical)

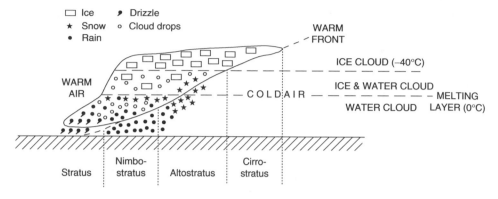

Figure 4.10 The Norwegian model of the microphysical structure of a warm front. The horizontal scale is typically 100 to 200 km, the vertical scale is about 10 km and the uplift of the warm air over the cold air is about 0.1 ms^{-1}. Reproduced, with permission, from Wallace, J.M. and Hobbs, P.V., 1977, Atmospheric Science, Academic Press: page 255, figure 5.32

cyclones. The high cirrus and cirrostratus clouds are composed entirely of ice crystals, whose growth is dependent on the number of ice nuclei and the amount of water vapour available. The freezing process in cirrus clouds produces ice splinters, thereby increasing the number of ice nuclei. These ice splinters have fall speeds of approximately 0.5 m s^{-1} and they can, in turn, seed the supercooled clouds below. Ice crystals do not evaporate as readily as water droplets and they may fall through layers of clear air up to 500 m thick before finally evaporating. This can occasionally be observed in cirrus, where fall streaks give a characteristic hook shape to the cloud. In the middle-level clouds, such as altostratus, the clouds are mixed and both Bergeron and accretion processes are important. At distances of 1.5 km or less above the freezing level, aggregation becomes the dominant process and, depending on the temperature of the lower layer of the atmosphere, precipitation will fall as either rain or snow.

The growth of hailstones is an extreme example of the accretion process and it occurs in deep convective clouds which have both a high water content and a very strong updraught. Under exceptional circumstances, hailstones as large as 13 cm in diameter (about the same size as a grapefruit) have been recorded, although most hailstones are less than 1 cm across. Even in the United Kingdom, hailstones as large as 7 cm have been recorded in summer storms. Hailstones are formed in only a small region of a cumulonimbus cloud on the forward side of the updraught region as shown in Figure 4.11. When a thin-section of a hailstone is viewed in transmitted light, a banded appearance is usually seen with alternating layers of clear and opaque ice. This banded structure is the result of the impact of supercooled water droplets at different heights within the cloud. In the upper levels of the cloud, where

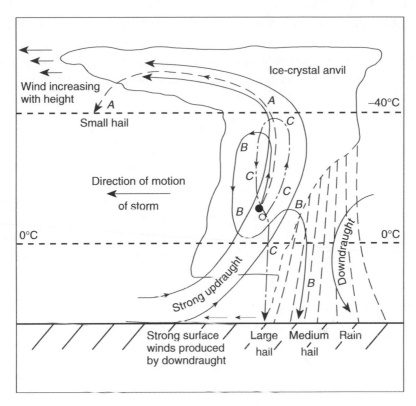

Figure 4.11 The structure of a hailstorm. The horizontal and vertical scales are typically 10 km. though the anvil may spread many times this distance. Large hail occurs in the region of cloud where recirculation in the updraught is possible. Reproduced, with permission, from Mason, B.J., 1975, Clouds, Rain and Rainmaking, Cambridge University Press: page 94, figure 24

the temperature range is from −20 to −40°C, water droplets are not as numerous and freezing occurs on impact. Air is trapped between the layers thereby giving a milky appearance. At lower levels in the cloud, where the temperature range is −20 to 0°C, there are more water droplets and the hailstone collects water droplets at too fast a rate to be able to dissipate the latent heat of fusion by conduction to the air, as is required if spontaneous freezing is to occur. The surface of the hailstone becomes wet and it eventually freezes to form a layer of clear ice. A hailstone will often require a number of cycles to grow to a reasonable size and, for this to occur, the updraught must exceed $10\,\mathrm{m\,s^{-1}}$. As a general rule, the maximum updraught in the upper troposphere is equivalent to the final terminal velocity of the hailstone. A hailstone 1 cm in diameter has a terminal fall speed of $25\,\mathrm{m\,s^{-1}}$, whilst a 10 cm hailstone has a fall speed of $65\,\mathrm{m\,s^{-1}}$. The latter size hailstone would have a mass of about 0.5 kg and would be able to shatter glass, break roof tiles and indent cars and caravans.

4.6   Macroscopic processes in cloud formation

One advantage the meteorologist has over the oceanographer is the ability to see the formation of clouds and to use his or her direct observations to make deductions about atmospheric motions and processes. From a microphysical point of view, it is possible to distinguish three types of cloud. At temperatures below −39°C all clouds are composed of ice crystals. These clouds are known as 'cirrus' and they tend to have fuzzy edges because of the slow evaporation of ice crystals ejected by the cloud, as discussed in Section 4.5. These cirrus clouds are optically thin and they allow solar radiation to penetrate to lower levels in the atmosphere. The alignment of these clouds gives information on the direction of the wind and the fall streaks allow an estimate of the vertical variation of the horizontal wind to be made. The hexagonal structure of the ice crystals gives rise to preferred directions of scattering and thus to a variety of optical phenomena. The most common of these phenomena is a white halo around of the moon or sun at an angle of 22.5°.

At temperatures between −39 and 0°C, supercooled water droplets and ice crystals co-exist and, in this temperature range, the clouds are termed 'mixed'. At temperatures colder than −30°C, the water droplets are outnumbered by the ice crystals and so mixed clouds tend to behave optically like pure ice clouds. Between −10 and 0°C, ice crystals are few in number and therefore these clouds behave optically as if they were water clouds. Typical mixed clouds are altocumulus and altostratus, which form at mid-levels in the troposphere. In thin mixed clouds with small water droplets, coloured diffraction rings can appear around the Sun or Moon, forming a corona.

At temperatures above 0°C, clouds are composed solely of water droplets. Such clouds include small cumulus, stratus and stratocumulus and they exist at low levels in the troposphere. Water-droplet clouds tend to have distinct edges due to the rapid evaporation of ejected water droplets in an unsaturated environment. Such sharp delineation is often seen in growing cumulus.

However, the microscopic classification of clouds has its limitations in ascertaining cloud processes since, for example, it does not distinguish between cumulus- and stratus-type clouds. These two types of clouds, which are the basis of the descriptive classification of clouds formulated in the 19th century by Luke Howard, give information about the stability of the atmosphere. Layered, or stratus, clouds form in a locally stable environment whilst lumpy, or cumulus, clouds are indicative of vertical hydrostatic instabilities (see 2.6).

Low-level stratus-type clouds can be formed by:

(i)   Advection of warm moist air over a cooler surface. This may cause advection fog, particularly over cold ocean currents and snow covered surfaces.

(ii) Radiation cooling of a moist air mass during its transition from low to high latitudes.

(iii) Gentle uplift of a stable air mass, either by a topographic feature or in a large frontal region associated with an extra-tropical cyclone.

On the other hand, cumulus-type clouds are formed in a locally unstable environment produced by the advection of cool, drier air over a warm (moist) surface. In summer, a stable air mass may be rendered unstable in the lower 1 km by surface heating over land. This gives rise to small cumulus, known as fair weather cumulus, providing that the convection reaches the condensation level. Cumulus clouds may form over hills and mountains. Surface heating of mountain slopes may generate upslope winds which can lift the surface air mass above the condensation point to form a cumulus cloud over the mountain peak. Over the ocean, polar air masses moving equatorwards over warmer surface waters can give rise to convection. The depth of the convection depends upon the dynamical motions maintaining the stability of the troposphere. For example, in an atmosphere of weak vertical stability, such as occurs in the rear of an extra-tropical cyclone or in the ITCZ, convection may extend from the sea surface to the top of the troposphere. In such cases, deep convective cumulonimbus clouds will form. These clouds will consist of water droplets in the lower troposphere and predominantly ice crystals in the cirrus anvil at the top of the troposphere. In the trade-wind region or in a middle-latitude anticyclone, where the air has strong vertical stability, the convection will be limited and cumulus will form in the moist boundary layer. However, progressive warming of the trade-wind air mass as it crosses the ocean may be sufficient to produce convection 2–3 km deep, which can produce showers. A less commonly observed convective cloud is termed 'arctic smoke' and it is formed by the movement of a very cold air mass over a warmer sea surface. The shallow layer of air adjacent to the sea is strongly unstable and transfers heat and, more importantly, water vapour, into the air mass. This water vapour condenses on mixing with its cold environment and this process quickly warms the shallow arctic air mass.

Instability may be induced not only by surface heating but also by the advection of colder air over a warm air mass. This type of instability often occurs in summer when a cooler maritime air mass is advected over a very warm surface air mass of continental origin. A notable feature of this type of instability is the appearance of lines of convective cloud in the middle troposphere, which are known as castellanus clouds because of their 'castle-like' appearance.

Unstable and stable air masses may co-exist and produce a variety of clouds. As noted earlier, a stable air mass may be rendered unstable by surface heating which, in turn, leads to the development of cumulus. If there is sufficient moisture present and if the surface heating is strong enough, the cumulus may develop to a depth of 1 km, which may be enough for the development of an isolated light shower. However, the stability of the air above the cloud may be strong enough to limit further upward development of the cloud which will, therefore, spread laterally. This cloud, known as stratocumulus, will often produce a complete layer of cloud cover which will persist until the surface heating decreases or until sufficient dry air from above the inversion is mixed into the clouds to promote evaporation. This type of cloud often has a diurnal cycle, with maximum thickness occurring in the late afternoon and usually clearing at dusk, when the surface heating is reduced. Alternatively a layered cloud may break up into convective type cloud because of solar radiational heating. As discussed in Chapter 2, solar radiation penetrates a cloud to a considerably deeper extent than long-wave planetary radiation. Thus, the top of a cloud will cool by long-wave radiative emission into space whilst the bulk of the cloud mass will heat up by solar radiation absorption. This process results in an increase the vertical temperature gradient within the cloud which, in turn, leads to instability. This phenomenon is more often observed in thin clouds like altostratus or cirrostratus which are transformed into altocumulus and cirrocumulus.

The mixing of air masses is another important macroscopic cloud process. The importance of the mixing of cloudy air with dry air to dissipate cloud has already been mentioned but if two air masses are close to the saturation vapour pressure, clouds may form. The reason for this may be understood by referring to Figure 4.12.

If air mass A and air mass B are mixed, then the resultant air mass will lie along the line AB and its exact location will depend on the relative properties of the two air masses. The saturation mixing ratio does not vary linearly with temperature and so the points lying along the line CD will represent saturated

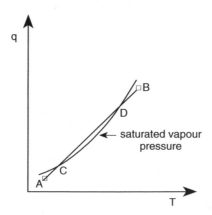

Figure 4.12   Mixing of two unsaturated air masses (A, B). A mixture lying on the line CD will become saturated

or supersaturated air and thus condensation would be expected. Only for mixtures lying on the lines AC and BD will the air remain unsaturated. A common example of this phenomenon is the emission of hot, moist exhaust gases from a car or an aeroplane into a cold moist environment. In this case, because of the low environmental temperature, only a small amount of water vapour is required for condensation.

Vertical mixing of air masses may also result in cloud formation. One particular example is the formation of stratocumulus at the base of an inversion in an anticyclonic system. In this case, mixing of cool, moist surface air with warm dry air aloft will result in a decrease in temperature but an increase in the mixing ratio at the top of the layer. The converse will occur in the lower layer, thus resulting in condensation in the upper layer. This process is often responsible for the development of stratocumulus over the sea, when the sea temperature is lower than the temperature of the air above it. In a winter anticyclone, over a moist land surface, a thick layer of stratocumulus may persist for many days, giving rise to a condition known as 'anticyclonic gloom'. Clearly, the formation of this low-level cloud is dependent on a number of processes. The upward vertical flux of water vapour from the surface and vertical mixing will tend to enhance cloud formation whilst the downward mixing of warmer, drier air from above the inversion will tend to dry out the boundary layer and dissipate the cloud.

Vertical mixing is also important in the development of surface fogs. These may develop by long-wave radiational cooling of the land at night which, if sufficient, will lower the ground temperature to the dew point of the overlying air. In light winds, only the surface layers of air will reach the dew point and a shallow layer of fog will form. However, with stronger vertical mixing, the depth of air cooled to the dew point will increase and hence a thicker layer of fog will develop. Another example is the formation of advection fog, formed by the flow of warm, moist air over a cold surface, such as a cold ocean current or a snow-covered landscape. In this case, vertical mixing will tend to enhance the depth of the fog.

An important contribution to the formation of clouds is adiabatic cooling. A spectacular example is the formation of funnel clouds in a tornado vortex. The low pressure of such a vortex is often sufficient to adiabatically cool the air to the condensation point. For example, a fall of 50 hPa in the pressure will cause an adiabatic cooling of 4°C at the surface.

By far the most common cause of condensation is adiabatic cooling by vertical uplift. This vertical motion may be caused by the forced lifting of an air mass over a topographical feature by convection, as previously discussed, or by the large-scale ascent associated with frontal or synoptic scale features. The lifting of air masses over topography may give rise to relatively strong vertical motions of $1 \, \mathrm{m \, s^{-1}}$ and to the development of 'local' clouds over the topographic feature. In a stable environment, wave clouds may form over the mountain and downstream of the mountain for up to a horizontal distance of 100 km. Clouds formed by large scale ascents in fronts tend to be associated with weaker vertical motions of approximately $0.1\,\mathrm{m\,s}^{-1}$ and the clouds therefore tend to be layered. However, regions of stronger uplift, associated with rain bands in a frontal system, may produce more convective cloud.

5

Global Budgets of Heat, Water and Salt

5.1 The measurement of heat budgets at the surface

In Chapter 3 the heat budget of the atmosphere and the Earth's surface was discussed. At the top of the atmosphere the absorbed solar radiation is exactly balanced by the outgoing planetary radiation. By contrast at the Earth's surface the fluxes due to latent heat of evaporation and sensible heat flux are necessary to include in the budget, to enable a surface equilibrium temperature to be maintained. In this chapter these fluxes are considered over land and ocean, in particular the relation of these fluxes to the heat, fresh water, and salt budget of the ocean.

The net heat flux, Q_T (positive downwards) across a horizontal surface is given by

$$Q_T = Q_{SW} - Q_{LW} - Q_E - Q_H \qquad (5.1)$$

where Q_{SW} is the net downward solar radiation flux, Q_{LW} is the net upward long-wave or planetary radiation flux, Q_E is the heat lost by the surface due to the latent heat of evaporation and Q_H is the upward sensible heat flux. Sensible heat flux is sometimes referred to as heat conduction. The net radiation flux, Q_R is given by $Q_R = Q_{SW} - Q_{LW}$. The fluxes have units of $W\,m^{-2}$. The four separate quantities must be determined in order to calculate the total heat flux across an ocean or land surface, and the method used over land and ocean to make the measurements over limited areas will now be discussed.

The solar radiation flux, Qs, can be determined using a solarimeter, which measures both the direct and diffuse radiation between 0.35 and 2.8 μm, i.e. over most of the solar spectrum. The solarimeter consists of two elements, one blackened to both solar and long-wave radiation, and the other covered with a white paint that has a high short-wave reflectivity. The temperature

The Atmosphere and Ocean: A Physical Introduction, Third Edition. Neil C. Wells.
© 2012 John Wiley & Sons, Ltd. Published 2012 by John Wiley & Sons, Ltd.

difference between the two elements, measured by a series of thermocouples, is proportional to the incident short-wave radiation. A glass dome covering the sensor elements gives the instrument protection from the weather, and it can be mounted on a ship and give reliable readings for long periods of time. The accuracy of the instrument is about 2%.

The net long-wave flux, Q_{LW} can be determined from measurements of the total net radiation and the solar radiation. A net radiometer consists of two blackened thermopiles, one pointing upwards and the other pointing downwards. The voltage difference between the two thermopiles is a measure of the net radiation across the surface, which may thus be determined with an accuracy of about 2.5%. The thermopiles are enclosed in a polythene dome which transmits both long- and short-wave radiation. The dome is ventilated with dry nitrogen gas to avoid water-vapour absorption and condensation. Measurements of net long-wave radiation are obtained routinely at many meteorological stations but measurements at sea are more difficult. It is necessary to avoid contamination by reflections from the ship and thus a boom is required to obtain uncontaminated fluxes. However, providing that the sea surface temperature is known, the upward flux of long-wave radiation can be calculated from Stefan's law, $E = \varepsilon\sigma T^4$, with an emissivity, ε, of 0.97, and thus only the downward long-wave flux from the sky needs to be measured. This can be done using an upward-looking radiometer.

One of the most accurate determinations of the latent heat flux, Q_E, and the upward sensible heat flux, Q_H, is by the eddy correlation method. Over both land and sea, the upward flux of heat (and water vapour) is caused by turbulent motions close to the surface. Figure 5.1 shows the fluctuations of

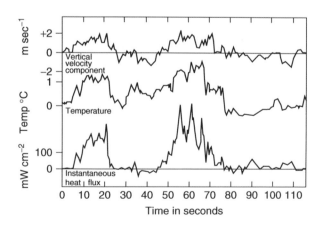

Figure 5.1   Temperature fluctuations, θ, vertical velocity, w, and sensible heat flux $\rho_a C_p \overline{w'\theta'}$ at 2 m above the Earth's surface. Reproduced, with permission, from Priestley, C.H.B., 1959, Turbulent Transfer in the Lower Atmosphere, University of Chicago Press: page 67, figure 18

vertical velocity and temperature for a 2 min record at a height of 23 m above a land surface. Despite the irregularity of the fluctuations, it can be seen that higher temperatures are generally associated with the rising motion, whilst lower temperatures are associated with the descending motion. In this situation, a mean vertical gradient of temperature exists with highest temperature close to the surface. Fluctuations in vertical velocity, associated with convection and surface friction (see Box 5.1), cause an exchange of parcels of air across the temperature gradient and this, in turn, gives rise to an upward flux of heat. The correlation coefficient between the vertical velocity and temperature gives an accurate measurement of the heat flux at the level of instrumentation. Providing that there is no loss of heat between the surface and the instrument, the correlation coefficient will give a measurement of the surface sensible heat flux. In a similar manner, the correlation coefficient between the vertical velocity and humidity fluctuations allows a determination of the latent heat flux.

Box 5.1 Atmospheric Boundary Layer

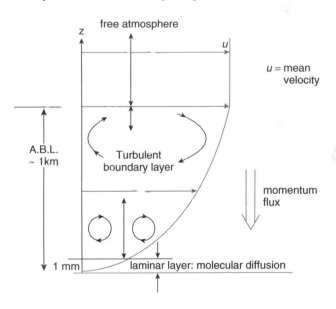

Atmospheric flow is strongly influenced by the underlying surface, due to frictional forces. The above figure shows the vertical variation in horizontal flow over a flat, uniform surface. The frictional forces acting on the atmospheric flow are turbulent forces, and therefore are characterised by eddies. These eddies will transfer high velocity air downwards and lower velocity air upwards, and produce a tangential stress on the atmospheric layer. This stress is measured by the correlation of vertical velocity and horizontal velocity. Provided there is mean vertical gradient of horizontal velocity there will be tangential stress. This frictional layer is known as the atmospheric boundary layer. These turbulent eddies will also transfer heat and water vapour between the surface and atmosphere above the boundary layer as shown in Figure B5.1.

The mean wind profile in the boundary layer measured over any surface has a characteristic form.

$$U = \frac{U_*}{k} \ln \left(\frac{z}{z_0} \right) \qquad \text{(B5.1)}$$

where U_* is the friction velocity and k is the von Karman constant which has a value of 0.4. z_0 is the roughness coefficient. The friction velocity is defined by:

$$\tau = \rho U_*^2$$

Thus, from a vertical profile of U, the gradient will determine the friction velocity, U_* and thence the stress, τ. The roughness coefficient, z_0, can also be determined from the velocity profile. Typical values of z_0 vary from 0.01 cm over smooth sand to about 100 cm over scrub country.

The equation for the mean wind profile is shown here for conditions where the vertical temperature and humidity profiles are uniform but modifications to this equation can be made to include these variations.

Over a sea surface the roughness coefficient also depends on the friction velocity and is given by:

$$z_0 = \frac{\alpha U_*^2}{g}$$

where g is the acceleration due to gravity $= 9.81 \, \text{ms}^{-2}$ and α is the Charnock constant $= 0.016$. The waves are generally driven by the wind stress, and represented by the friction velocity.

The sensible heat flux is given by:

$$Q_H = \rho_a C_p \overline{w'\theta'} \qquad \text{(5.2)}$$

and the latent heat flux by:

$$Q_E = L\rho_a \overline{w'q'} \qquad \text{(5.3)}$$

where the overbar denotes a time average of the instantaneous values of the vertical velocity, w', the mixing ratio, q', and the potential temperature, θ', over a period of 10 minutes. ρ_a is the density of the air, C_p is the specific heat at constant pressure of the air, and L is the latent heat of evaporation.

This method requires sophisticated and accurate instrumentation in order to measure the high-frequency variations in w, q and θ. In one 10 min average, about 10 000 observations are required to obtain reliable fluxes. Over the sea, there are more difficulties because of the ship's movements and the contamination of the instruments by salt spray. For these reasons, the above technique is only practicable for the determination of latent heat and sensible heat fluxes from a research ship for limited periods and in light to moderate wind conditions. Another direct method, which can be applied from ships in strong wind conditions, is the dissipation method. This uses the high-frequency wind observations to determine the dissipation of turbulent kinetic energy. This measurement can be used directly to obtain the surface wind stress or it may be combined with additional high-frequency temperature and humidity measurements to determine sensible heat and latent heat fluxes. These methods are very useful for the calibration of other simpler formulae which are more widely used.

From turbulence measurements, the surface wind stress can be measured directly (see Box 5.1). However, these sophisticated measurements can only be made by research ships. To use the data from voluntary observing ships a simpler but accurate method is needed. The bulk aerodynamic method relates the surface wind stress τ_0 to the wind speed at 10 m above the sea surface

$$\tau_0 = \rho_a C_D (U_{10} - U_s)^2 \tag{5.4}$$

where U_{10} is the mean wind speed 10 m above the surface and U_S is the surface current speed. C_D is the drag coefficient and it also has a dependence on wind speed given by $C_D = (0.61 + 0.063\ U_{10}) \times 10^{-3}$ which is valid for neutral stability conditions. This coefficient includes the effects of surface roughness caused by the waves and their interaction with the overlying air. The surface current speed, U_s, is roughly about 2% of the wind speed U_{10}, however it may be much larger in regions where strong ocean currents exist. Therefore from mean winds measured at 10 m above the ocean surface the surface stress can be obtained for winds up to 26 m s^{-1}. At higher wind speeds the drag coefficient, C_D, is not well determined, but it is believed that the dependency on wind speed weakens and C_D becomes constant.

A similar approach can be used to calculate latent heat fluxes, Q_E, and sensible heat fluxes, Q_H, over the sea surface.

$$Q_E = \rho_A L C_E U_{10}(q(T_s) - q_a) \tag{5.5}$$
$$Q_H = \rho_A C_p C_H U_{10}(T_S - T_A) \tag{5.6}$$

At the sea surface, $q(T_s) = 0.98 \times q_s(T_s)$, where q_s is the saturation mixing ratio. These relationships are only applicable for the computation of fluxes over the sea surface. They require observations of the wind speed (U_{10}), the mixing ratio, q_a, and the air temperature, T_A, at 10 m height above the sea surface and sea surface temperature, T_s . All these quantities can be obtained from ships' reports as a matter of routine.

The exchange coefficients, C_H and C_E, can be determined from the direct turbulence measurements discussed earlier.

Over the land, the aerodynamic equations 5.5 and 5.6 are not as useful because the drag coefficient will vary with the type of surface. However, over land the profile method (See Box 5.1) and the eddy correlation method can be more easily used.

Apart from the use of the aerodynamic method, all of the previous methods rely on instrumentation which is only available on research ships and some automatic ocean buoys. If it is assumed that there are, at most, 100 specially equipped research vessels at sea at any one time, and this is probably a considerable overestimate, then, provided that they were evenly distributed, one observation would be made for each 2000 km square of ocean. This is an unsatisfactory data set for the consideration of surface fluxes on a global scale. The global maps of surface fluxes in the scientific literature are compiled from Voluntary Observing Ships (VOS) rather than those from research ships. As noted previously, the bulk aerodynamical method is suitable for estimating the sensible heat flux and the latent heat flux from standard weather reports, but even here there are problems. The height of the wind observations will vary from ship to ship and wet-bulb thermometers can be contaminated by salt spray and may, therefore, give a systematic error in humidity if they are not regularly maintained. In addition, the water temperature is measured at the engine intake point, which may be a few metres below the sea surface, and the recorded temperature may be systematically different from the sea surface temperature, particularly in low wind conditions when vertical mixing is reduced. The observations from VOS have to be subjected to careful error checking before they are used to obtain the surface fluxes.

The problems posed by the global measurement of solar radiation and long-wave radiation are also not straightforward because of the variable effect of cloud and atmospheric aerosol. In clear sky conditions, provided that the aerosol is low, both the solar radiation and the long-wave radiation can be determined reasonably accurately. The solar radiation impinging on the top of the atmosphere can be calculated for a given latitude and time of year, as demonstrated in Chapter 3. Allowance can be made for the depletion of the solar radiation by atmospheric gases and a 'standard' maritime aerosol to obtain a value for the solar radiation incident at the surface. However, recent research has shown that horizontal variations of aerosol can cause variations of as much as 5% in the solar radiation.

Reed (1977) showed that an accurate formula for estimating daily mean values of solar radiation is given by:

$$Q_{SW} = (1 - \alpha)Q_C(1 - 0.62n + 0.0019\theta_n) \tag{5.7}$$

Here α is the surface albedo and therefore $(1 - \alpha)$ is the fraction of solar radiation absorbed at the sea surface. Q_c is the clear sky solar radiation, which depends on the elevation angle of the sun and the atmospheric transmission in the absence of clouds. The second term in brackets accounts for cloud, where n is the cloud fraction and θ_n is the solar elevation at local noon.

Net long-wave radiation can be estimated from a formula proposed by Clarke (1974):

$$Q_{LW} = \varepsilon_o T_s^4(0.39 - 0.05\sqrt{e_a})(1 - \lambda n^2) + 4\varepsilon_o T_s^3(T_s - T_a) \tag{5.8}$$

where ε is the emissivity of the sea surface ($\varepsilon = 0.97$), T_S is the surface temperature measured in Kelvin, σ is the Stefan-Boltzmann constant, T_a is the temperature and e_a is the vapour pressure of the atmosphere at 10 m. The dependence on humidity is due to the re-emission of the long-wave radiation back to the surface by atmospheric water vapour. The majority of the re-emission comes from the first 100 m above the surface and the humidity at 10 m can, therefore, be used to estimate the downward long-wave radiation. To account for variations in cloud there is an extra term in equation 5.8 which depends on the fraction of the sky covered by cloud, n, and a latitudinal coefficient, λ.

There are two points to note about formula 5.8. First, over the observed range of sea surface temperatures, the net long-wave flux, assuming saturated air and clear skies, only varies from $70\,W\,m^2$ at 271 K to $23\,W\,m^2$ at 300 K, which is much smaller than the latitudinal variation of solar radiation. Second, the net long-wave radiation decreases with increasing temperature because the downward long-wave radiation, which is proportional to the vapour pressure, increases more quickly with surface temperature than the upward radiation emitted by the surface.

These methods can provide surface heat fluxes over the ocean to an accuracy of about $10\,W\,m^{-2}$. However, global estimates are still subject to biases due to the lack of observations over remote areas of the Southern and South Pacific Oceans.

5.2   Observations of surface heat fluxes and budgets

5.2.1   *Seasonal cycle of surface heating and cooling*

The global seasonal cycle of the atmosphere-ocean system is responsible for the largest changes in temperature and precipitation on the planet. In

Chapter 1.4 it was explained that the seasonal cycle of solar radiation, in conjunction with the thermal inertia of the atmosphere and upper ocean, was the reason for these large changes in temperature and precipitation. In Section 3.6 it is shown that the main changes in the seasonal ocean temperature occur in the upper ocean typically in the 0–100 m layer of ocean. Solar radiation at the top of the atmosphere is the main driver for these changes but how do the surface fluxes vary over the cycle and what are the differences between the land surface and ocean surface?

Figure 5.2 shows the variation of the four surface fluxes (equation 5.1) for equatorial and middle latitudes in the North Atlantic. In middle latitudes the solar cycle, Q_{SW}, has a maximum radiation in June and a minimum in December, similar to that seen at the top of the atmosphere. The latent heat of evaporation, Q_E, is very much larger than the longwave cooling, Q_{LW}, and the sensible heat flux, Q_H. Furthermore, the latter two terms contribute little to the seasonal variability. The net heat flux, Q_T, shows a very strong seasonal cycle dominated by the variation in solar heating. From April until September, the ocean is warming, whilst from October until March the ocean cools. This cycle is similar over the middle latitudes of the North Pacific and North Atlantic. Within the equatorial latitudes the sun is overhead at the equinoxes and therefore the maximum surface solar radiation tends to occur at those times. However, this is mediated by seasonal variations of cloud

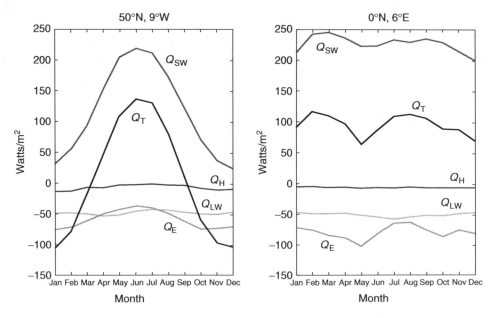

Figure 5.2 Annual cycle of the vertical heat fluxes for equatorial and mid-latitudes in the Atlantic Ocean. Q_{sw} = solar radiation, Q_{LN} = longwave or planetary radiation, Q_E = latent heat of evaporation, Q_H = sensible heat, and Q_T = net heat flux

associated with monsoonal circulations. The heat lost from the ocean surface is dominated by the latent heat of evaporation, which shows a strong seasonal variation, whilst the other two fluxes show little variation and are smaller. There is a net heat gain by the equatorial ocean throughout the year.

Table 5.1 shows the annually averaged components of the heat budget for land and ocean. Over the annual cycle, the net heat flux into the land will be very close to zero because soils and other surfaces are generally poor conductors of heat, and the seasonal temperature cycle does not penetrate beyond 5–10 m. Therefore the relatively small amount of heat gained by the surface layer in the summer period will be lost to the atmosphere during the winter. Over the ocean, there is a net gain of heat between 10°S and 10°N, and at higher latitudes there is a net heat loss. These heat deficits and surpluses must be compensated for by a horizontal transfer of heat from the equatorial zones to the middle and high latitudes via ocean currents.

The annual radiation budgets at lower latitudes generally show more heat going into the ocean than into the land. This observation is related to two factors. First, the albedo of the sea surface is much lower on average than that of the land at corresponding latitudes, especially over the sub-tropical deserts, and therefore more solar radiation is absorbed by the ocean. Second, the surface temperatures of the tropical ocean are lower than those observed over land and therefore the net long-wave radiation losses are less. Over the ocean, the largest net loss of heat is by evaporation; the sensible heat conduction accounting for only 10% on average of the net heat loss. In the equatorial zone, Q_H contributes only 5% to the total heat loss, whilst at higher latitudes it contributes about 20%. Over land, both latent heat and heat conduction contribute equally to the overall net heat loss. In the sub-tropical deserts, heat conduction dominates over evaporation, whilst the opposite is true in the equatorial zone and over the wetter middle latitudes.

Table 5.1 Annual surface heat budgets ($W\,m^{-2}$) over ocean and land. Q_R is the net radiation budget, Q_E is the latent heat of evaporation, Q_H is the upward conduction of heat and Q_T is the net heat flux into the ocean

	Ocean				Land		
Latitude zone	Q_R	Q_E	Q_H	Q_T	Q_R	Q_E	Q_H
50–70°N	40	63	16	−39	33	22	11
30–50°N	98	107	11	−20	70	31	39
10–30°N	128	130	9	−11	94	33	61
10°S–10°N	154	128	6	20	96	66	30
30–10°S	122	128	8	−14	96	46	50
50–30°S	67	73	6	−12	69	33	36

To gain an impression of the variations in surface heat fluxes over the ocean, the annual average of the net heat flux, Q_T, is shown in Figure 5.3 for the global ocean, calculated from three decades of observations by voluntary observing ships. Large heat losses of over $100\,W\,m^{-2}$ are seen over the Kuroshio, Gulf Stream, North Atlantic Current and Norwegian Current. The latter warm current extends into the Arctic Ocean. The large heat losses in the western North Atlantic are caused by a combination of the high sea surface temperatures and the frequent outbreaks of cold, dry air from the North American continent in winter. A similar explanation can be applied for the large heat loss over the Kuroshio where very cold air from the Asian continent spills out over the Western Pacific Ocean in winter.

In the Southern Hemisphere, the major area of heat loss is to the south of Africa, where the warm Agulhas Current extends into the Southern Ocean. Other areas of cooling occur both to the east and west of Australia, and over the Southern Ocean. Strong cooling would also be expected around the Antarctic Continent, where very cold air from the continent will spread over the adjacent ocean.

Heat gain by the ocean occurs generally throughout the equatorial zone and along continental margins. These regions are associated with upwelling of cooler waters from the thermocline and the lower sea surface temperatures reduce the heat loss by evaporation. The eastern equatorial Pacific Ocean

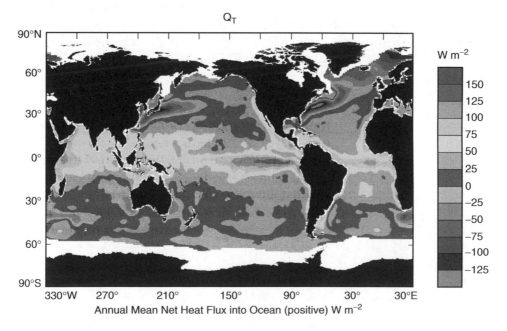

Figure 5.3 Annual mean net heat flux over the global ocean $W\,m^{-2}$. Reproduced, with permission, from Ocean circulation and climate Seidler, G., Church, J. and Gould, J. Ocean Heat Transport by H.L. Bryden and S. Imawaki Plate 6.1.4 715pp. See plate section for a colour version of this image

has the most intense net heating in the global ocean. These cold upwelling zones are generally found along the western and eastern African coasts, and western American coasts.

In summary, the long term net heat flux into and out of the ocean is mainly controlled by the latitudinal variation of the solar radiation and the heat loss by evaporation. The latter process is strongly influenced by ocean currents and the upwelling of cooler water to the surface.

It will now be shown how the distribution of the net heat flux across the ocean surface can be related to the transport of heat in the ocean.

First, consider the heat budget for a slab of ocean extending across an ocean basin and over the whole depth. For the conservation of heat (Figure 5.4)

$$S_o + T_{i+1} - T_i = Q_T \tag{5.9}$$

where S_o is the change of storage of heat in the slab which is proportional to the rate of change of mean temperature. T_i and T_{i+1} are the net horizontal transports of heat into and out of the slab. Their difference $T_{i+1} - T_i$ is known as the heat flux divergence. Equation 5.9 shows that a net heat flux, Q_T, into the ocean must be accompanied by either a change in the heat storage, S_o, or by the net transport of heat out of the slab. To obtain the heat transport across the boundary of each slab, it is assumed that the northward flux across the strip closest to the Antarctica continent is zero. Hence, by summing consecutive strips northwards, T_{i+1} can be obtained:

$$T_{i+1} = Q_T - S_o + T_i \tag{5.10}$$

The small contributions of frictional dissipation by tides and currents, by precipitation and by radioactive decay in the Earth's interior have all been neglected. All of these factors contribute less than $1\,W\,m^{-2}$ to the net heat flux (see Section 9.1). In the ocean, and in marked contrast to the land surface, the storage component, S_o, can be as large as the net heat flux, Q_T, during the

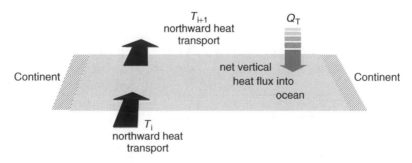

Figure 5.4 Change in heat storage S_O for an ocean basin caused by difference in horizontal heat transports, $T_{i+1} - T_i$, between two latitudes and the net heat flux across the ocean surface, Q_T

seasonal cycle because of the important role played by the ocean increasing the thermal inertia of the atmosphere – ocean system (see Section 1.4).

Figure 5.5 shows a map of S_o, $T_{i+1} - T_i$ and Q_T for the Northern Hemisphere during the seasonal cycle. Comparison of S_o and Q_T shows the large accumulation of heat in the middle latitudes during the summer, and the release of that stored heat during the winter (see Section 3.6) However, it is noted that the middle-latitude heat loss in winter is greater than the heat accumulated in summer, and therefore there must be a poleward flux of heat to maintain the balance. This flux is, on average, about 2–3 petawatts (1 petawatt $= 10^{15}$ watts) at 25°N.

The seasonal change of heat storage occurs mainly over the upper 100–200 m of the ocean, except in high latitudes of the Atlantic Ocean where it occasionally occurs to depths of 1000 m. It is also noted that, averaged over the seasonal cycle, the storage term tends to cancel out, and it is therefore reasonable to assume that, over a long period of observations, say three decades, the change in storage will become small and a balance between the transport of heat and the net heat flux across the surface will occur (equation 5.10). This is a steady-state assumption which implies that, in the long term, the mean ocean temperature is a constant. Measurements of the change in global ocean heat storage from 1950 to 2009 in upper 0–700 m have shown an increase equivalent to an additional heat flux of 0.3 W m^{-2}. This warming is associated with the increase in the greenhouse gases. When this is compared with the net heat divergence shown in Figure 5.5, the steady-state assumption yields an error which is generally less than 1%.

Figure 5.6 shows the long-term mean meridional transport of heat in the ocean, estimated by applying equation 5.10 to each ocean basin. Of particular interest, is the difference between the Atlantic and Pacific Oceans. In the Atlantic Ocean there is a northward heat flux throughout the whole ocean which is maintained by a net heat flux from the Indian and South Pacific Oceans. True heat transfers assume that mass is conserved, which is not the case for the South Indian and South Pacific Oceans because of the Indonesian Throughflow (IT). The total poleward heat transport across 30°S is 0.6 PW, whilst at 30°N it is 1.8 PW and therefore there is a large difference in ocean heat transfer between the two hemispheres of the globe. It is also noted that the Arctic Ocean has a substantial heat flux from the North Atlantic Ocean, but gains no significant heat from the Pacific Ocean through the Bering Straits.

An overall conservation equation for the entire atmosphere-ocean can be determined in a similar way to equation 5.9. A box extending from the base of the ocean to the top of the atmosphere is shown in Figure 5.7. The net heat flux, R, into the top of the atmosphere is simply the sum of the net solar radiation and net outgoing planetary radiation, which can be determined from satellite observations. The divergence of atmospheric energy, i.e. internal energy, potential energy and latent heat flux, is represented by T_A ; the atmospheric

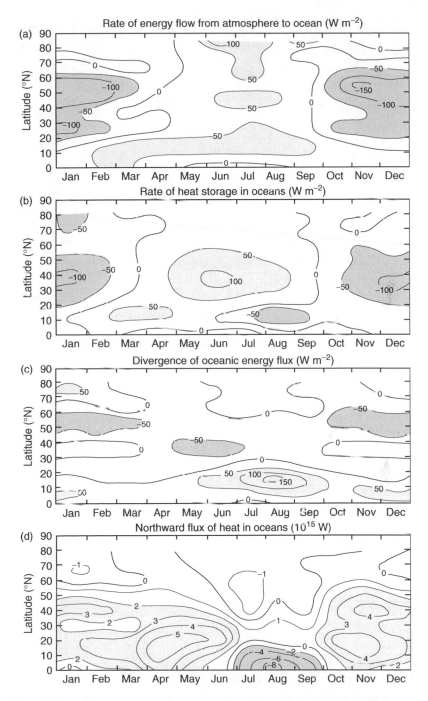

Figure 5.5 (a) Vertical heat flux into the ocean, Q_T; (b) Storage, S_O; (c) Heat divergence; (d) Heat transport, T_i, in the Northern Hemisphere. Reproduced, with permission, from Oort, A.H. and von der Haar, T.H., 1976, Journal of Physical Oceanography, 6: pages 789–795, figures 5, 8, 9 and 13

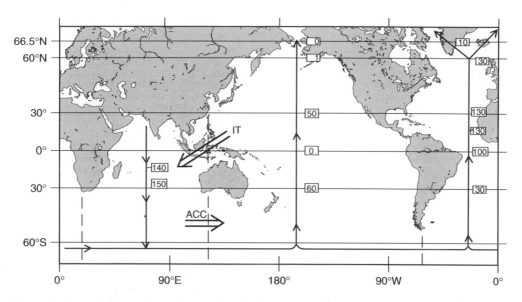

Figure 5.6 Estimates of annual mean meridional heat transport within the ocean, in units of 10^{13} W. The latitude and magnitude of the maximum meridional transport is shown for each ocean basin. Broken lines show meridians where mass is exchanged between ocean basins. The Antarctic Circumpolar Current (ACC) connects all three ocean basins (Atlantic, Indian and Pacific Oceans), and the Indonesian Throughflow (IT) connects the Pacific and the Indian Oceans. True heat transports require no net mass exchange between ocean basins

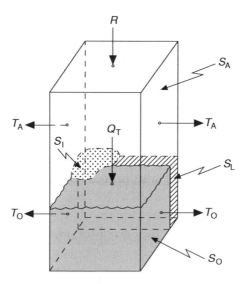

Figure 5.7 Schematic diagram of the different terms in the Earth's energy budget. R and Q_T are the net radiation fluxes at the top and bottom of the atmosphere, respectively. S_A, S_I, S_L and S_O are the rates of energy stored in the atmosphere, snow and ice, land, and ocean, respectively. T_A and T_O are the horizontal transports of energy in the atmosphere and the ocean. Reproduced, with permission, from Oort, A.H. and von der Haar, T.H., 1976, Journal of Physical Oceanography, 6: page 783, figure 1

storage term by S_A; the land storage and ice storage by S_L and S_I respectively; the ocean storage by S_O and the divergence of the ocean by T_O, then:

$$R = T_A + T_O + S_A + S_O + S_L + S_I \qquad (5.11)$$

If, over a long period of time, the storage terms are assumed to be negligible, then:

$$R = T_A + T_O \qquad (5.12)$$

The term $T_A + T_O$ is the total energy divergence of the atmosphere- ocean system. In a similar process to the determination of the poleward transport of heat in the ocean, equation 5.12 can be used to determine the total poleward transport by atmosphere and ocean. Figure 1.7 shows that poleward of 40° there is a net deficit of radiation, because the outgoing long-wave radiation exceeds the incoming solar radiation, and so a poleward transport of energy is required to maintain the balance. The atmospheric energy transport can be obtained by subtracting the ocean transport from the total transport as illustrated in Figure 5.8. These studies have shown that the atmosphere and ocean carry similar quantities of heat from low latitudes to high latitudes over an annual period. The ocean heat transport is a maximum at 20° latitude, whilst the atmosphere heat transport reaches a maximum at about 40° latitude.

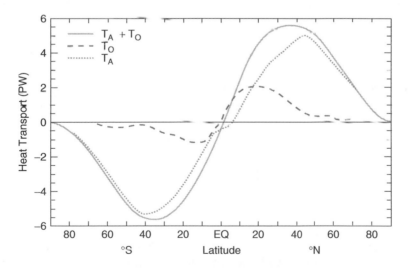

Figure 5.8 Poleward transfer of heat by: (a) ocean and atmosphere together (T_A and T_O), (b) ocean alone (T_O) (c) atmosphere alone (T_A). The total heat transfer (a) is derived from satellite measurements at the top of the atmosphere, that of the ocean (b) is obtained from direct measurements of the ocean, and (c) the atmosphere transfer is calculated as the difference between (a) and (b) (1 petawatt = 1 PW = 10^{15} W). Reproduced, with permission, from Trenberth, Kevin E., Julie M. Caron, 2001: Estimates of Meridional Atmosphere and Ocean Heat Transports. *J. Climate*, 14, 3433–3443, figure 7

5.3   The measurement of the water budget

The vertical flux of water, W, across a horizontal surface is given by:

$$W = P - E \qquad (5.13)$$

where P is the precipitation, including dewfall, and E is the evaporation. Both quantities are usually expressed in units of millimetres per year or centimetres per year. The precipitation is measured over land using rain gauges. At a standard climatological station, when care is used in siting the rain gauge, the precipitation can be determined to an accuracy of approximately 1–2%. However, over the oceans, precipitation measurements using rain gauges are neither as frequent nor as accurate as over the land. The reasons for this are related to the problem of wind flow around the ship carrying the gauge and, to a lesser extent, the problem of spray. Comparisons between rain gauges mounted on different parts of a ship have shown that differences of 50–100% can be expected, and a number of indirect techniques have, therefore, been employed to measure the rainfall over the ocean.

One technique uses the visual observations of rain intensity, which are recorded regularly by voluntary observing ships as part of their meteorological reports, and converts them into numerical values of rainfall intensity. As an example, light continuous rain would be assigned a rainfall intensity of $0.6\,\text{mm hour}^{-1}$, whilst moderate continuous rain would be given a value of $1.9\,\text{mm hour}^{-1}$. By this method, annual mean distributions of rainfall can be obtained to an accuracy of about 10%. This technique is probably best suited to higher latitudes where the rainfall is less intense, but more frequent, than in the tropics. Another indirect method is to use radar to measure the back-scattered radiation from large water droplets. The radar can be calibrated by comparing simultaneous measurements of back-scattered radiation with rainfall rates determined by a conventional, land-based, rain-gauge network. This radar technique is not always accurate because the back-scattered radiation is dependent on the size distribution of the droplets. It will not be reliable when there is a large component of frozen precipitation. It is most useful for the detection of intense mesoscale rainfall systems and is limited to a range of about 100 km.

In recent decades there has been considerable interest in the use of satellites to measure precipitation. One technique measures the temperature of the cloud tops using infra-red sensors. This can be used to obtain the depth of a cloud which can, in turn, be empirically related to the intensity of precipitation from the cloud. This method is most successful for deep convective clouds and is, therefore, best suited to determine the rainfall distribution over the tropical ocean. However, as with all satellite techniques, its principal value lies in the determination of the spatial distribution of precipitation rather than for obtaining accurate point values. A second satellite method uses a microwave radiometer to measure the small amounts of microwave radiation emitted by the suspended liquid water in the atmosphere. Changes in the liquid water content can be related to the precipitation rate and thus an instantaneous picture of the precipitation pattern may be obtained. Both satellite techniques are limited by the frequency of satellite passes over a particular region; perhaps only one to three passes per day. Rainfall is highly variable in both time and space, and therefore these techniques require further development if they are to be of use in obtaining quantitative climatological precipitation over the globe.

Another way of measuring rainfall at sea is by acoustic methods. Rainfall on a sea surface will create an acoustic signal which can be measured by underwater hydrophones.

The direct measurement of evaporation from a ship is no less difficult than that of precipitation. Because of these difficulties the indirect techniques, such as the bulk aerodynamic method discussed in Section 5.1, are found to more reliable with systematic errors, of less than 10%. These indirect techniques can be used to make routine calculations from ships' reports. It is noted that an evaporative heat flux of $100 \, \mathrm{W \, m^{-2}}$, such as observed at the equator, is equivalent to an evaporation rate of $126 \, \mathrm{cm \, year^{-1}}$.

5.4   Observations of the water budget

For the reasons discussed above, the distribution of precipitation is only accurately known over land areas. In the Pacific Ocean, the large number of islands in the western and central regions allows a good estimate of the rainfall over the ocean to be obtained, although it is important to realize that islands, especially if they have hills or mountains, will tend to be preferential sites for tropical convection and precipitation. In the tropical South Atlantic, the paucity of island sites results in a large uncertainty in the rainfall distribution. Between the equator and 40°S, there are only seven islands, not all of which record rainfall. However, despite some uncertainty in quantitative measurements over the ocean, the general global distribution is now well-established. In particular the previously described satellite techniques in conjunction with conventional observations have enabled maps of the major precipitation regions to be obtained.

The most significant feature of the precipitation distribution in tropical latitudes is the relationship between the sea surface temperature and the rainfall belts. High rainfall in both the Pacific and Atlantic Oceans occurs along the Inter-Tropical Convergence Zone (ITCZ), which corresponds closely to the maximum sea surface temperatures, known as the thermal equator.

Of particular note is the very high rainfall over the equatorial regions of South America and Africa, and also the maritime continent of Indonesia, which exceeds 300 cm year^{-1}. By contrast, in parts of the eastern tropical and sub-tropical ocean, the rainfall is less than 10 cm year^{-1}. This dramatic decrease is particularly noticeable in the south equatorial Pacific, where the Galapagos Islands have a desert climate, whilst only 300 km northwards, in the ITCZ, the rainfall may exceed 300 cm year^{-1}. The oceanic desert regions are associated with lower sea surface temperatures which, in turn, are related to the upwelling of cold sub-surface water and the advection of colder water by the sub-tropical anticyclonic gyres from higher latitudes.

With the use of climatological cloud distributions derived from satellites, measuring in the visible and infra-red, convergence zones and regions of higher precipitation have been located in all three ocean basins. One of the most significant is the South Pacific convergence zone which extends south-eastwards from Indonesia into the central South Pacific. Similar cloud zones have been observed over the South Atlantic extending outwards from the Amazon basin and from East Africa into the southern Indian Ocean. In higher latitudes, the precipitation rates are much lower due to the lower water-holding capacity of the cooler atmosphere.

The evaporation rate, which is related to both the vertical gradient of humidity and to the wind speed (equation 5.5), is large over the tropical trade-wind zones where both the wind speed and the vertical gradients of humidity are large. In these regions, evaporation reaches 150–200 cm year^{-1}. In the middle latitudes, large evaporative fluxes of up to 250 cm year^{-1} are seen over the western North Pacific and western North Atlantic due to advection of dry continental air over a warm ocean surface, often associated with extra-tropical storms.

Over the whole globe, evaporation must be in balance with precipitation. The global average precipitation required to balance evaporation is about 100 cm year^{-1} and a horizontal transfer of water must occur to balance regions of water deficit with regions of water surplus. The horizontal transfer over land is by rivers; ground water; glaciers and ice sheets, whilst in the ocean it is the freshwater water component of the ocean currents which needs to be determined (see Section 5.5). Pack ice and icebergs will also contribute to ocean freshwater water transport from polar latitudes.

Figure 5.9 shows the variation of the components of the water budget over the ocean. A net surplus of water (i.e. P > E) exists over the equatorial region, associated with the ITCZ, and polewards of 40° latitude, associated with the extra-tropical cyclones. In the tropical and sub-tropical zones there is a net deficit (i.e. E > P) as a result of the low precipitation over the ocean, and the high evaporation in the trade-wind zones. To maintain this latitudinal variation of P − E it is necessary to transfer fresher water in the ocean from the higher latitudes and the equatorial zone into the sub-tropical regions. In the

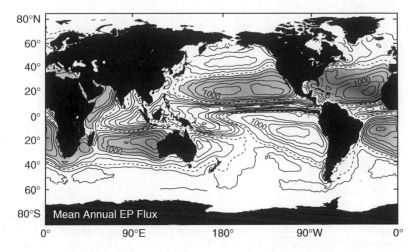

Figure 5.9   Evaporation-Precipitation (E-P) over the ocean. Units: mm/year. Shaded areas indicate E > P, whilst clear areas indicate P > E. The dotted lines are where E = P. Reproduced, with permission, from Wijffels, SE., 2001, In Ocean Circulation and Climate, Ed: Seidler, G., Church J., and Gould J., Academic Press, International Geophysics Series Volume 77: page 478, figure 6.2.1(a)

atmosphere, a transfer of water vapour in the opposite direction is required to maintain the regions where precipitation exceeds evaporation. Most of the water vapour is contained in the lower troposphere and therefore these water-vapour transfers also occur at these levels. The trade winds, which occupy the lowest 2–3 km of the atmosphere, are responsible for the transfer of water vapour from the sub-tropical ocean into the ITCZ. During the Indian monsoon, the low-level south-west monsoon winds transfer water vapour from the southern Indian Ocean and the Arabian Sea onto the Indian sub-continent. At higher latitudes, the warm moist poleward winds, associated with the warm sectors of extra-tropical cyclones, transfer water vapour into regions of precipitation surplus.

Over the continents, the precipitation generally exceeds the evaporation by about 27 cm year^{-1}, whilst over the oceans, evaporation exceeds precipitation by 11 cm year^{-1}. The continental surplus is transferred by runoff (rivers; ground water; glaciers and ice sheets), into the ocean to balance the ocean freshwater deficit (Figure 5.10). These continental surpluses must be balanced by a net flux of atmospheric water vapour from the ocean to the continents. Although evaporation exceeds precipitation over the global ocean, there are significant differences between individual ocean basins. Over large areas of the Atlantic and Indian Oceans evaporation exceeds precipitation, whilst in the Pacific and Arctic Oceans precipitation is larger than evaporation. For the balance to be maintained, a net transfer of fresh water from the Pacific and Arctic Oceans into the Atlantic and Indian Oceans is required as shown in Figure 5.11.

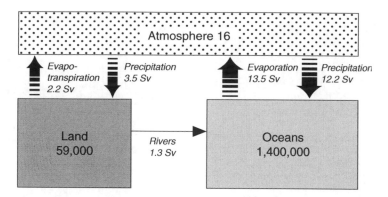

Figure 5.10   Estimates of the fresh water storage in the ocean, atmosphere, and on land. Units: 10^3 km³. The arrows are transfers of freshwater between the ocean, atmosphere and land. Units: 1 Sv $= 10^6$ m³s⁻¹. Reproduced, with permission, from Charnock H.1996, In Oceanography – An Illustrated Guide, Ed. Summerhayes, C.P. and Thorpe, S.A.T. Manson Publishing: page 28, figure 2.2

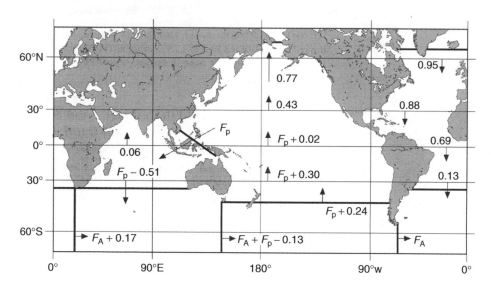

Figure 5.11   An estimate of the transfer of fresh water ($\times 10^6$ m³s⁻¹) in the world's ocean. In polar and equatorial regions,precipitation and river runoff exceed evaporation, and hence there is an excess of fresh water, while in subtropical regions there is a water deficit. A horizontal transfer of fresh water is therefore required from regions of surplus to regions of deficit. F_P and F_A refer to the freshwater transports of the Pacific-Indian Throughflow and of the Antarctic Circumpolar Current in the Drake Passage, respectively. True freshwater water transports require no net mass transports between ocean basins (see equation 5.26). Reproduced, with permission, from Wijffels, SE., Schmitt, R., Bryden, H. and Stigebrandt, A., 1992, In Journal of Physical Oceanography, 22: page 77, figure 2

In summary, it has been shown that the surface water budget allows a picture of the global transfers of atmospheric water vapour and of fresh water in the ocean to be ascertained. However these estimates are still prone to errors because of the inaccuracies of the measurement of precipitation and evaporation over the ocean.

5.5   The salt budget of the ocean

The total mass of the ocean is 1.4×10^{21} kg and, if a global average salinity of 34.7 is assumed, then the total mass of dissolved salt is 4.8×10^{19} kg. It has been estimated that the rate of salt input into the ocean is about 3×10^{12} kg year^{-1} and therefore the replacement time for the ocean salt is approximately 16 million years. This calculation illustrates the point that the sources and sinks of salt in the ocean are only of concern on a geological time scale. For ocean circulation, a time scale of about 2000 years is applicable (see Section 5.7), and therefore it is a good approximation to assume that the total salt content of the ocean is constant. This has the important consequence that observed salinity variations in the ocean are the result of dilution and concentration brought about by imbalances in the surface water budget discussed in Section 5.4.

Over the ocean basins, away from major river discharges, the salinity distribution is well correlated to the net surface water flux, W. The principal features are that the largest surface salinities occur in the subtropical zone where there is a net deficit of water, and that the lowest surface salinities occur in high latitudes where precipitation exceeds evaporation. A minimum is also apparent in the region of the ITCZ.

As might have been anticipated from the discussion in the previous section, there are considerable differences between the various ocean basins (Figure 5.12). The North Pacific Ocean has an average surface salinity of about 34.2, whilst the North Atlantic Ocean has a salinity of 35.5. These observations are related to larger input of fresh water into the Pacific Ocean than into the Atlantic Ocean. The input of freshwater causes a low salinity surface layer to form in the higher latitudes of the North Pacific. This maintains a stable density stratification, known as a halocline, which inhibits the formation of deep water. Similarly in the Arctic Ocean the excess of precipitation over evaporation and the influx of fresh water from runoff, particularly by Canadian and Siberian rivers, causes the formation of a stable halocline. A concern has been that a reduction of runoff caused by a redirection of Siberian rivers from the Arctic Ocean to the Caspian Sea may affect the stability of the Arctic halocline. It has been argued that weakening of the halocline would enable deep convection to occur which would, in turn, bring warmer water to the surface and inhibit the development of sea ice.

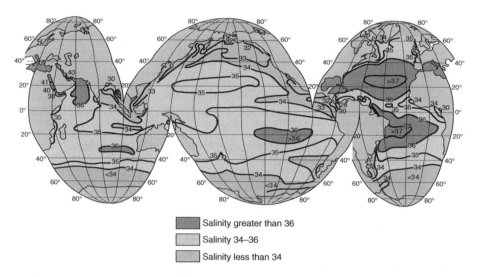

Salinity greater than 36

Salinity 34–36

Salinity less than 34

Figure 5.12 Surface salinity in the Northern Hemisphere summer. Reproduced, with permission, from Charnock H.1996, In Oceanography – An Illustrated Guide, ed. Summerhayes, C.P. and Thorpe, S.A.T. Manson Publishing: page 30, figure 2.5. See plate section for a colour version of this image

Although this is only a possibility, it does demonstrate the point that small variations in the water balance may produce changes in the polar sea ice distribution and subsequently the Arctic climate.

A useful example of the application of the principle of the conservation of salt may be found in the Red Sea. There is a surface flow of low-salinity water into the Red Sea through the Strait of Bab el Mandeb and an outflow of high-salinity water over a sill at a depth of 110 m. This two layer model is a simplification of the actual flow which is often a three layer flow. If F_I is the volume flux of water into the Red Sea, with a salinity of S_I, and F_O is the volume flux of water out of the Red Sea, with a salinity S_O, then for the conservation of salt across the Straits of Bab el Mandeb:

$$\rho_I F_I S_I = \rho_O F_O S_O \tag{5.14}$$

To a good approximation $\rho_I = \rho_O$ and for the mass conservation of water:

$$F_I = F_O - W - R_O \tag{5.15}$$

where W is the net water flux through the surface of the Red Sea and R_O is the runoff. Thus, eliminating F_O:

$$F_I S_I = (F_I + W + R_O)S_O$$

Therefore:

$$F_I = \left[\frac{S_O - S_I}{S_O} \right] = -[W + R_O] \tag{5.16}$$

The salinity, S_O, is 40 and S_I is 37. Measurements across the Strait of Bab el Mandeb suggest $F_I = 0.5 \times 10^6\,\mathrm{m^3s^{-1}}$, and therefore the net surface flux of water is approximately $-3.75 \times 10^4\,\mathrm{m^3s^{-1}}$. The negative sign indicates an excess of evaporation over precipitation and runoff. The surface area of the Red Sea is $438 \times 10^9\,\mathrm{m^2}$, and therefore the excess evaporation over precipitation and runoff is $2.70\,\mathrm{m\ year^{-1}}$. Since precipitation is generally less than $20\,\mathrm{cm\ year^{-1}}$ and the runoff term is negligible, the evaporation rate is close to this value.

Equation 5.16 is generally applicable to any semi-enclosed sea or estuary where the vertical circulation can be represented by a two-layer exchange with the ocean as shown in Figure 5.13. Some conclusions can be drawn from Equation 5.16. When evaporation exceeds precipitation and runoff, $(W + R_O) < 0$ and $S_O > S_I$ and therefore the outflow will be more saline than the inflow. If the temperature changes are much smaller than the salinity changes, then the more-saline water will be more dense, will sink and flow out underneath the less-saline inflow. This situation is typical of the Red Sea, the Persian Gulf and the Mediterranean Sea, all of which produce saline bottom waters. In the northern extremities of the Red Sea and in the Persian Gulf, salinities as high as 42 occur, whilst in the Mediterranean Sea, salinities of 38 are typical.

In higher latitudes, precipitation and runoff exceed evaporation, i.e. $W + R_O > 0$, and therefore $S_O < S_I$. The outflow will thus be less saline, and, provided the temperature differences between inflow and outflow are small, the low salinity water will flow out in the surface layer. The more-saline ocean water will flow into the semi-enclosed sea or estuary at depth to replace the low salinity water. This situation is representative of the Baltic Sea, where surface salinities as low as 10 can occur.

From equation 5.16 it can also be seen that the salinity difference, $S_O - S_I$, is proportional to the ratio of $(W + R_O)/F_I$ and, if the volume flux into the basin is considerably larger than the net water flux, $W + R_O$, then the salinity

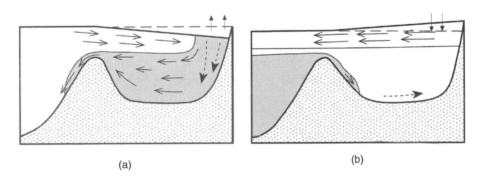

(a)

(b)

Figure 5.13 The two principal types of sill basin. (a) Evaporation > precipitation + runoff. (b) Evaporation < precipitation + runoff. Reproduced, with permission, from Tchemia, P., 1980, Descriptive Regional Oceanography, Pergamon Marine Series 3: page 38, figure 3.5

difference between the inflow and outflow will be small. On the other hand, large salinity changes will occur if the volume flux is small relative to the net water flux. The Baltic Sea, for instance, has a very shallow sill depth of 18 m, the volume flux is very small compared with the freshwater input, and hence the surface water has a low salinity.

The value of equation 5.16 lies in obtaining either a consistency check on the salt budget of an estuary or semi-enclosed sea, or in determining the value of one of the terms in the equation such as precipitation which, as previously discussed, is difficult to measure directly. The principle can also be applied to the salt balances of ocean basins such as the Arctic Ocean. In this case, the situation is complicated by the large number of inflows and outflows, and by the fact that the vertical circulation may be rather more complicated than that assumed for the Red Sea. In ocean basins such as the Atlantic Ocean direct measurements of the volume fluxes are much more difficult. In the World Ocean Circulation Experiment (1990–1997) sections across ocean basins were measured from which estimates of the large-scale transports of both freshwater and heat have been made. This is based on the equations for the conservation of salt, mass and heat, and a mathematical method known as inverse analysis.

5.6   Temperature and salinity relationships in the ocean

In the previous section it was shown how the surface water flux, runoff and circulation determine the observed salinity and it is reasonable to acknowledge that the surface heat budget, in conjunction with circulation, can determine the observed temperature of the ocean. It is clear that the relationship between temperature and salinity will give useful information about the sources of water masses and about the circulation and mixing rates in the ocean. Close to the surface, the temperature and salinity characteristics will be constantly changing in response to seasonal variations in heat fluxes and water fluxes between the atmosphere and ocean. However, below the depth of seasonal influence, on average about 100–200 m, the temperature and salinity of a water parcel will only change by mixing with the surrounding water masses, and therefore it can be said that the temperature-salinity (θ-S) properties of the parcel will be conserved. In the ocean, potential temperature, θ, is used rather than the actual temperature, T, because of the adiabatic compression.

Figure 5.14 shows the temperature and salinity profiles taken at two oceanographic stations in the Gulf Stream about 200 km apart. At first glance there is no obvious similarity but, when the information is plotted on a θ-S diagram, also shown in Figure 5.14, it can be seen that they have a similar characteristic shape. The values of two profiles only diverge from each other at the higher temperatures and salinities typical of the surface layer.

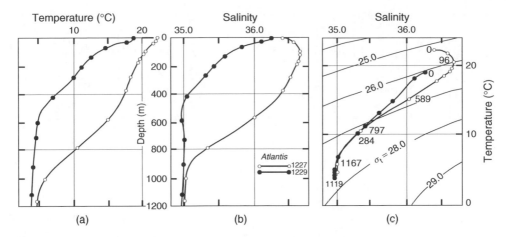

Figure 5.14   Relation between temperature and salinity for two neighbouring *Atlantis* stations 1227 and 1229 in the Gulf Stream east of Cape Hatteras. (a) Vertical profile of temperature (°C). (b) Vertical profile of salinity. (c) Temperature-Salinity diagram with lines of constant density (σ_t). Reproduced, with permission, from Dietrich, G. *et al.*, 1980, General Oceanography: An Introduction, Wiley Interscience: page 197, figure 78

It can be seen that the water mass at one station has been vertically displaced by 600 m relative to the other station. This example shows that a θ-S relationship will allow a particular water mass to be labelled and to be followed in the ocean. In oceanography, a point on a θ-S diagram is referred to as a water type, whilst a straight-line relationship is known as a water mass. Thus, the water type refers to the original θ-S during formation at the surface, whilst the water mass evolves by mixing between different water types.

Figure 5.15a shows the progressive mixing of two water types, A and B. In order to conserve temperature and salinity, mixtures of the two water types must lie on the line AB and the position on the straight line is determined by the relative concentration of each water type. Hence, a succession of different water types is formed. The ensemble of water types is known as a water mass.

Figure 5.15b shows the mixing of three water types, not dissimilar to those in the Atlantic Ocean. The mixed water will be located within the triangle ABC and its actual position will depend on the relative proportions of the three water types. If the rate of mixing between the water mass above and the water mass below is equal, then the salinity minimum will be located along the line BD. An example of the progressive mixing of water types is the distribution of the Antarctic Intermediate Water in the Atlantic Ocean, as shown in Figure 5.16. This water type is found in the Antarctic convergence zone at about 50°S, where it has a temperature of 2.2°C and a salinity of 33.8. It sinks to a depth between 700 and 1500 m, and then it progresses northwards

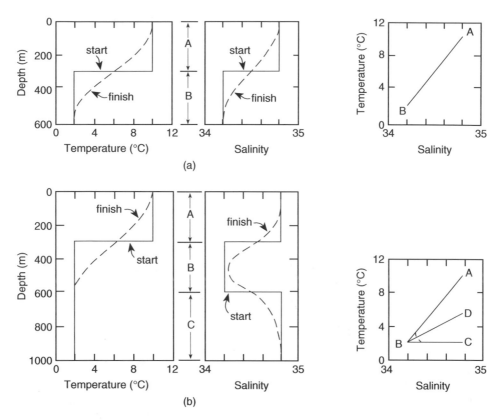

Figure 5.15 Temperature-salinity relationships resulting in the mixing of water masses. (a) Mixing of two water types. On the Temperature (T)- Salinity (S) diagram, mixing of the two water types is shown by the straight line connecting the T-S points characteristic of the two original water types. (b) Mixing of three water types. On the T - S diagram, the mixing process is shown by two straight lines, indicating that water types A and B mix, and that water types B and C mix. The lines meet at a point representing water type B, now altered by mixing with water above and below so that its original identity has been lost and it can no longer be represented by a single point on the T-S diagram. For equal mixing above and below, the mixture will lie along the line BD. Reproduced, with permission, from Grant-Cross, M., 1987, Oceanography: A View of the Earth, Prentice-Hall, page 449, figure 7.15

as a tongue of low-salinity water. It can be identified on the θ-S diagram of the Atlantic Ocean, shown in Figure 5.16 as a salinity minimum. Successive θ-S diagrams show that the water mass becomes more saline as it mixes with the Central Atlantic water above and the North Atlantic Deep Water below. From the θ-S diagrams, the rate of mixing of the Antarctic Intermediate Water can be deduced, and thus an estimate of the volume of the original water type can be made.

Figure 5.17 shows a three-dimensional diagram representing a census of the world's water masses by Worthington (1981). The vertical coordinate is the volume of water having a mean temperature in the range $\pm 0.05°C$ and a

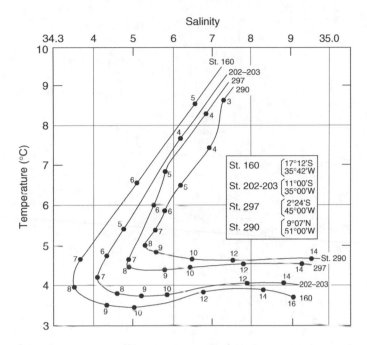

Figure 5.16 Variations with latitude of the salinity minimum characterizing the Antarctic Intermediate Water (AAIW) between 17°S and 9°N in the Atlantic. The AAIW is sandwiched between warmer, saline Central waters above and saline North Atlantic Deep Water below. These water masses mix with the AAIW (salinity minimum) as it progresses northward and it becomes more saline. Reproduced, with permission, from Tchernia, P., 1980, Descriptive Regional Oceanography, Pergamon Marine Series 3: page 157, figure 5.34

mean salinity in the range +0.005. Thus, the peak in Figure 5.17 corresponds to a potential temperature between 1.1 and 1.2°C and a salinity between 34.68 and 34.69. It has a volume of 26×10^{15} m^3. This water mass has a larger volume than all of the water above 19°C and is also larger than the total volume of all the continental ice caps. The census shows that the major water masses of the ocean have a relatively small range of temperature and salinity, distributed about a global mean temperature of 3.51°C and a mean salinity of 34.72. The much warmer surface waters and the saline deep waters of the Mediterranean and Red Seas are insignificant compared with the deep polar waters that fill all of the ocean basins. Thus, the major deep ocean water masses are probably not significantly different to those occurring during the Ice Ages and the amelioration of the climate by the ocean is limited to the effect of the rather shallow, warm surface layer above the main thermocline.

The most uniform water masses are found in the North Pacific Ocean where there are no sources of bottom water. All these deep-water masses have their origin in the Atlantic and Antarctic Oceans, at a considerable distance from the Pacific Ocean, and therefore the mixing process has had sufficient time to

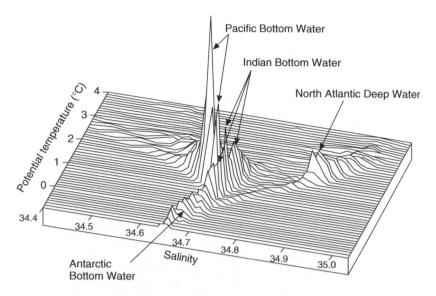

Figure 5.17 Simulated three-dimensional T-S diagram of the water masses of the world ocean. The potential temperature (θ) is used rather than in situ temperature (T), in order to remove the adiabatic effect. Reproduced, with permission, from Worthington, L.V., 1981, In Evolution of Physical Oceanography, ed. Warren, B.A. and Wunsch, C., MIT Press: chapter 2, page 47, figure 2.2

homogenize the deep-water masses here. The Indian Ocean tends to have a relatively uniform salinity of 34.70, but a larger range of temperatures. The Atlantic Ocean is quite different from either the Indian or the Pacific Oceans. It has a large number of water masses ranging from −1 to + 4°C, and most are relatively saline compared with the Indian and Pacific Oceans.

Of note is the formation of North Atlantic Deep Water in the Labrador and Greenland Seas during late winter, and the formation of Weddell Sea Bottom Water. The former is the largest of the deep water masses by volume in the global ocean and has a temperature of 2°C and a salinity of 34.9. However, the Weddell Sea Bottom Water is the densest water mass in the ocean, having a temperature of −0.5°C and salinity of 34.66.

The conductivity, temperature and depth recording system (CTD) is the main method for determining the θ-S properties of the world's oceans. The characteristics of the CTD are high precision and accuracy. CTD systems consist of three sensors whose resistances vary with the temperature, electrical conductivity and pressure of their surroundings. The resistances of the three sensors are measured by an audio-oscillator whose frequency is determined by the resistance. The signals are sent up a cable to be processed on board the research ship. The measured electrical conductivity of the sea water is dependent on both temperature and salinity. Thus, a temperature correction to the conductivity signal is required to obtain accurate salinity values. Usually, two temperature sensors are employed. One is a thermistor, which has a very fast response time, and the other is a platinum resistance thermometer, which is very accurate. The thermistor measures small variations in temperature, and the platinum resistance thermometer is used in the absolute calibration of the thermistor and the conductivity sensor. By this method, *in situ* temperatures can be measured to $+0.001°C$, salinity to $+0.005$ and pressure to an accuracy of 0.5 db or 0.5 m depth. Thus, from one cast of a CTD system, a high-resolution vertical distribution of temperature and salinity can be obtained from the surface to the ocean bottom.

The conductivity cell is susceptible to contamination by biological particles and therefore samples of sea water are also taken to provide an absolute calibration of the sensor.

These measurements have shown that the θ-S relationships for water masses are not as uniform as shown by classical methods. Distinct layers of uniform water have been found to be separated by rapid changes in temperature and salinity. These uniform layers range from as little as 1 m in depth up to 100 m. The small-scale variations in θ-S relationships are known as the fine structure, and it is now known that these fine structures are direct evidence for water-mass transformation in the deep ocean. One process which can cause this fine structure at the boundaries of water masses is double-diffusive convection. One example of this process is salt fingering, which depends on the much larger diffusivity of heat compared with the molecular diffusivity of salt. Consider two layers of water, the upper one being salty and warm, whilst the lower layer is cold and fresh. If it is assumed that temperature dominates the density of each layer, the upper layer will be less dense than the lower layer and the two fluids will be hydrostatically stable. Consider now a parcel of fluid moved from the upper layer to the lower layer where, because of the larger heat diffusivity, it will lose heat more rapidly than it will lose salt. The parcel will lose its buoyancy, will become statically unstable and sink. A parcel moving from the lower layer to the upper layer will gain heat more rapidly than salt, will become buoyant with respect to its surroundings and accelerate upwards. Hence, alternating upward (fresh) and downward (salty) motion is produced, which transfers salt downwards across the interface. The release of convective energy by the salt fingers results in the formation of well-mixed layers above and below the interface. Such fine structure has been found under the Mediterranean outflow where salty and relatively warm water from the Mediterranean Sea overlies the less saline, but colder, North Atlantic Deep Water.

A second process, which can lead to mixing between water masses, depends on the non-linear equation of state for sea water, and it is referred to as 'caballing' in the oceanographic literature. Consider two parcels of fluid having different temperatures and salinities but equal densities, as depicted in Figure 5.18. The resultant mixture will lie along the line AB, its position

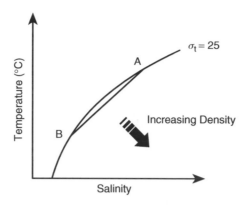

Figure 5.18   Mixing of two water masses (A, B) having similar densities ($\sigma_t = 25$). Any mixture will fall on the straight line AB and therefore will always have a density greater than $\sigma_t = 25$

being determined by the relative proportions of the two original fluids. However, because of the curvature of the lines of constant density, known as isopycnals, the mixture will be more dense than the original water parcels and therefore the mixture will sink. The curvature of the isopycnals is a maximum at low temperatures and, therefore, caballing is particularly important in Antarctic and Arctic waters. The formation of Antarctic Bottom Water has been attributed to this process. In the Weddell Sea, the surface water is generally colder and less saline than the underlying Antarctic Circumpolar Water. In winter, the formation of sea ice releases extra brine into the surface layer which, in conjunction with winter cooling, may produce a water mass of similar density to the underlying one. In this case, mixing of the upper and lower water masses may produce a mixture which has a density greater than the deep water, and which will therefore sink.

The above fine-structure processes are not the only ones by which water masses are diluted by mixing. In the case of outflows of dense water from semi-enclosed seas, the plumes may be rapidly mixed by entrainment with the surrounding water. Occasional intense gravity currents are observed over the Faeroes-Iceland ridge and in the Faeroes Channel, when dense Norwegian Sea water flows over the sills into the Atlantic Ocean. The plume accelerates down the slope because of the density differences with the ambient water and may reach speeds of $1\,\mathrm{m\,s}^{-1}$. Intense mixing of the plume occurs, which increases its volume of water and reduces its density, to form North Atlantic Deep Water. Mixing calculations have shown that the overflow water is composed of $1 \times 10^6\,\mathrm{m^3 s}^{-1}$ of Norwegian Sea water and $3 \times 10^6\,\mathrm{m^3 s}^{-1}$ of entrained Atlantic water. The dense high-salinity plume from the Red Sea also experiences mixing with Indian Ocean water as it moves off the Bab el Mandeb sill into the Gulf of Aden. Here, the plume descends to a depth of 1000 m before it reaches a similar density to the surrounding water and spreads horizontally into the Indian Ocean.

Another important horizontal mixing process is caused by energetic mesoscale eddies, typically 50–200 km in diameter. These eddies have been found in all parts of the ocean and can homogenize water masses, especially in intense frontal zones associated with the Gulf Stream and the Antarctic Circumpolar Current. It is known that the diffusion of Mediterranean water into the North Atlantic is primarily due to mixing by eddies. The Gulf Stream rings, which are about 200km in diameter, may have lifetimes of 2 to 3 years. Another example of the longevity of these eddies occurred in 1978. A water mass having a temperature and salinity characteristic of Mediterranean water was located in the western Atlantic in a small intense mesoscale circulation. It had apparently taken about three years to cross the Atlantic Ocean and in that time it had not mixed, to any significant extent, with its environment. These examples show that mesoscale eddies may mix on time scales of years, and in that time can travel large distances across ocean basins.

5.7 Tracers in the ocean

In the early 1970s, a research programme known as GEOSECS (Geochemical Ocean Sections Study) was initiated to measure the distribution of natural and anthropogenic tracers in the world ocean. Although oceanographers had used naturally occurring tracers, such as dissolved oxygen and carbon-14, to study water masses before this experiment, GEOSECS was the first comprehensive attempt to measure these tracers, and many others, on a global basis. The WOCE (World Ocean Circulation Experiment) in 1990–97 provided a complete measurement of these tracers in the global ocean as well as some new tracers.

Table 5.2 shows tracers which have been successfully used in the ocean to determine water-mass production, mixing and circulation. Dissolved oxygen has been measured in the ocean, in conjunction with salinity and temperature, since the 1930s, and much information on its global distribution is available. Oxygen is absorbed most efficiently at low temperatures and thus the largest concentrations are found in Antarctic and Arctic waters. It is a good tracer for studying the mixing of water masses, such as the North Atlantic Deep Water and Antarctic Bottom Water, but it has to be monitored continuously because it is consumed by bacteria and animals even in the deep ocean. Nitrates and phosphates are excreted by animals and bacteria, and they have been used as short-term tracers.

Silica is produced from siliceous bottom sediments such as red clays, which are plentiful in the Pacific Ocean and, to a lesser extent, in the Indian and Atlantic Oceans. Silica is consumed by biological processes in the upper

Table 5.2 Selection of artificial and natural tracers in the ocean

Tracer	Origin	Conservative	Half life	Comment
Dissolved O_2	Surface exchange	No	—	Consumed by bacteria and zooplankton
Silica SiO_2	Bottom sediments	No	—	Removed by upper ocean
Nitrate NO_3^-	Bacteria and zooplankton	No	—	—
Phosphate PO_4^{3-}	Bacteria and zooplankton	No	—	—
$[O_2] + 9[NO_3^-]$	—	Quasi-conservative	—	Not conserved in low O_2 water
$[O_2]+ 135[PO_4^{3-}]$	—	Quasi-conservative	—	—
δ^3 He	Lithosphere	Conservative		Appears close to deep ocean ridges.
^{3}H	Cosmic rays Nuclear testing and reactors	Conservative	12.4 years	Natural tritium very small compared with nuclear bomb tritium
δ^{14} C	Cosmic rays Nuclear testing	Conservative	5730 years	Appears in solution and particulates
^{90}Sr	Nuclear testing	Conservative	28.9 years	Concentrated in tropical waters. Recorded in coral growth rings
^{137}Cs	Nuclear testing and reactors	Conservative	30.0 years	Requires large samples for measurement
^{39}Ar	Cosmic rays	Conservative	269 years	Requires very large samples
^{228} Ra	Decay product of uranium	Conservative	5.75 years	Concentration highest in surface layers
^{85}Kr	Nuclear testing and reactors	Conservative	10.7 years	Sophisticated analysis
Chlorofluoro-carbons (CFC)	Anthropogenic Sources from 20th Century.	Conservative	—	Sophisticated analysis required as contamination is a hazard
δ^{18} 0	Global hydrological cycle	Conservative	—	Sophisticated analysis Important for paleo-oceanographic studies

layers of the ocean and it is, therefore, most useful for studying bottom and intermediate water circulations. In the North Atlantic Deep Water, the silica concentrations are low compared with those in the Antarctic Circumpolar Waters and Antarctic Bottom Waters, and it is therefore an ideal tracer when studying the mixing of North Atlantic Deep Water into the Antarctic Circumpolar Current. North Atlantic Deep Water is composed of a number of different water types and so the θ-S characteristics vary accordingly.

However, the silica concentration is remarkably uniform and it is therefore a more consistent tracer than the θ-S profile. The North Atlantic Deep Water (low silica concentrations) flows south against the western boundary and the Antarctic Bottom Water (high silica concentration) flows northwards along the bottom. It is noted that in bottom waters, where the θ-S properties usually show only a small variation, the use of silica as a tracer is particularly useful.

The information obtained from the some of the natural tracers (dissolved O_2; silica; nitrate and phosphate) is only qualitative because of the lack of knowledge about rates of production and consumption. These tracers are therefore not conservative tracers, though by combining them they can be made quasi-conservative (see Table 5.2).

Quantitative information on the age and circulation time scale of any given water mass can be obtained using radioactive tracers. Radioactive tracers can be classed as conservative because the decay rate is known precisely. As Table 5.2 shows, some tracers, such as ^{39}Ar, are produced naturally from cosmic rays; some are by-products of nuclear tests, such as ^{90}Sr and ^{137}Cs, and others, such as tritium, ^{3}H and ^{14}C, are byproducts of both nuclear tests and cosmic rays. The long-lived, naturally produced isotopes are particularly useful because they are generally in a steady-state balance in the ocean and therefore they can be used to estimate the age of the ocean waters and the ventilation time of the ocean. ^{14}C measurements, obtained before nuclear bomb tests significantly altered the ^{14}C distribution in the ocean, enabled the first estimate of the ventilation time, T, to be obtained thus:

$$T = V/R_v \tag{5.17}$$

where V is the volume of the ocean and R_v is the ventilation rate for the deep water.

For radiocarbon, the ventilation time is also given by:

$$T = \left[\frac{\left(\frac{^{14}C}{C}\right)_S}{\left(\frac{^{14}C}{C}\right)_D} - 1 \right] T(^{14}C) \tag{5.18}$$

where $(^{14}C/C)_S$ and $(^{14}C/C)_D$ are the carbon isotope ratios for the surface and deep waters, respectively, and $T(^{14}C)$ is the mean half-life of ^{14}C, which is 5730 years.

Equation 5.18 gives a mean ventilation time of 1400 years ±250 years for the global deep-water circulation. Recent evidence has suggested that this is an overestimate, because it is often not possible to distinguish old water masses which may have been recirculated around the ocean a number of times from newly formed water masses. Measurements on North Atlantic Deep Water have indicated an apparent age of between 80 and 225 years, which shows the deep ocean circulation of the Atlantic Ocean is more vigorous than in the Pacific and Indian Ocean basins. Argon is also a potentially useful natural radioactive tracer because it has a half-life similar to the ventilation time. However, at present it is technically difficult to determine because several cubic metres of sea water are required to obtain just one measurement. The ventilation time may also be estimated using values for the production of deep ocean water masses. The NADW and AABW together are produced at a rate of $30 \times 10^6 \, m^3 s^{-1}$ and the volume of the ocean is $1.37 \, 10^{18} \, m^3$ therefore the ventilation time for the global ocean (Equation 5.17) is 1400 years. This is reasonably consistent with the independent tracer analysis.

The radioactive tracers produced from atmospheric nuclear-weapon tests between 1945 and 1963 have inadvertently provided a number of useful transient tracers. These tracers are not in equilibrium with the ocean and they cannot provide information on the age of water masses, but they have proved valuable for determining the progress of recently formed bottom and intermediate waters. The bomb isotopes best studied in the WOCE and GEOSECS experiment were ^{14}C and ^3H. Anthropogenic carbon-14 can be distinguished from natural carbon-14 because of its very large signal in the surface layer. The distribution of tritium is not known prior to the bomb tests, but it is believed that the natural concentrations are much lower than the anthropogenic ones. ^{14}C and ^3H were released into the Northern Hemisphere's stratosphere, leaked into the troposphere about one year later and entered the ocean by precipitation. By 1973 about 45% of the total bomb ^{14}C had entered the ocean and penetrated to an average depth of 300 m. In the surface layers, the largest values of both isotopes are located in the Pacific and Atlantic Oceans at about 50°N. The concentrations decrease towards the Southern Hemisphere but the concentration gradient is not uniform. Near 15°N there is a remarkable reduction in both ^{14}C and ^3H concentrations, as shown in Figure 5.19, which is associated with a frontal boundary between the sub-tropical gyre to the north and the equatorial waters to the south. The low concentrations and limited vertical penetration between 15°N and 15°S suggest that water is being upwelled from depths of at least 500 m. The equatorial upwelling flux is comparable with the total production of deep and bottom water in the global ocean.

The most remarkable and unexpected finding, shown in Figure 5.19, is the relatively high concentrations of ^3H in the North Atlantic Deep Water only 20 years after its surface input, thus indicating a more rapid ventilation of the deeper ocean than had previously been anticipated. In the western Atlantic, the southward bottom boundary current has been identified by its high tritium concentrations.

In addition to radioactive tracers, chlorofluorocarbons (CFC) released into the atmosphere over the 20th century have found their way into the world's oceans. Individual species of CFC's can be determined reliably and are

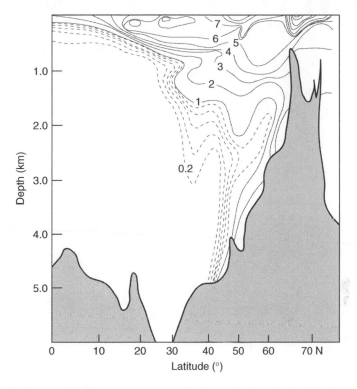

Figure 5.19   Tritium section along the western basin of the North Atlantic obtained during the GEOSECS programme. The tritium was introduced into the ocean from atmospheric testing of nuclear weapons in the 1950s and early 1960s. After the test ban treaty, no further tritium was introduced into the ocean, and therefore it provides a unique marker of water masses in contact with the atmosphere during this period. Values are in Tritium Units (TU); 1 TU = 1 × 10^{-8} atoms of tritium per atom of hydrogen. Reproduced, with permission, from Broeker, W.S. and Peng, T.H.,1982, Tracers in the Sea, Eldigo Press: page 409, figure 8.15

providing additional information on the movement of surface, intermediate and deep waters throughout the oceans.

In summary, ocean tracers can determine the time scales for the whole ocean circulation as well as time scales for the rates of mixing between the surface waters, the thermocline and the deep waters. The transient tracers show where water masses are formed and their transport pathways in the ocean. This information is complementary to the direct measurement of ocean circulation considered in Chapter 6.

6

Observations of Winds and Currents

6.1   Measurement of winds and currents

Both atmosphere and ocean are fluid environments and the basic techniques of the measurement of winds and currents have close similarities. There are two fundamentally distinct ways of specifying a flow field in the atmosphere and the ocean as illustrated in Figure 6.1. One approach, known as the Eulerian description, expresses the flow field at a fixed position in space relative to the Earth's surface. For example, a meteorologist may obtain an Eulerian description of the surface wind field from measurements of the wind direction and speed at an array of weather stations at a particular instant in time. Mathematically, an Eulerian velocity flow can be defined as a vector quantity at a position x, y, z and at a time t:

$$v = v(x, y, z, t) \tag{6.1}$$

The co-ordinates in both fluids refer to an eastward direction, x, a northward direction, y, and vertically upwards, z.

The second approach, referred to as a Lagrangian description, measures the velocity of the chosen element of fluid as a function of time. The flow is defined by the vector quantity:

$$v = v(a, t) \tag{6.2}$$

where a is the position vector of the chosen element of fluid with respect to a defined origin. An example of the Lagrangian specification is the measurement of the wind from a constant level balloon which is moving at the same velocity as the air immediately surrounding it.

The Atmosphere and Ocean: A Physical Introduction, Third Edition. Neil C. Wells.
© 2012 John Wiley & Sons, Ltd. Published 2012 by John Wiley & Sons, Ltd.

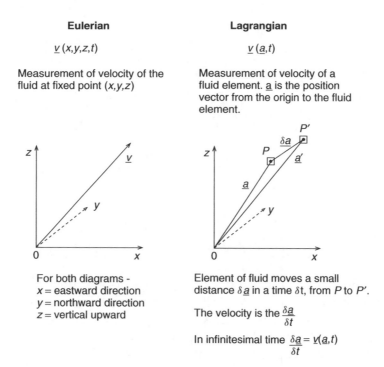

Figure 6.1   Eulerian and Lagrangian specifications of flow. N.B. Vectors in the text are denoted by bold type. Reproduced, with permission, from Wells, N.C. 'Ocean Circulation: General Processes' for Encyclopaedia of Ocean Sciences, Academic Press. 2001: page 1529, figure 1

It is important to note that the two specifications of the flow field are independent, and therefore they give different types of information on atmospheric and oceanic flows. It is not possible to obtain *exact* Eulerian velocities from the velocity of balloons or floats and, similarly, it is not possible to determine *exact* Lagrangian velocities from Eulerian measurements. However, as will be shown presently, it is possible to make approximate comparisons between the two systems of measurement if sufficient simultaneous measurements of Lagrangian and Eulerian velocities are available.

The choice of measuring system depends on the type of information required. A Lagrangian measuring system using, say, drifting ocean buoys would be appropriate for determining the movement and dispersal of a natural tracer or a pollutant. On the other hand, a weather forecaster may prefer the Eulerian system for specifying winds, since he or she is most concerned with prediction over a specific area. As noted above, the distribution of fixed synoptic stations essentially gives an Eulerian description of the circulation and therefore numerical weather prediction models have been designed on an Eulerian grid system for the numerical solution of the equations of motion (see Chapters 7 and 10).

It is important to note the different conventions for specifying the flow direction in the atmosphere and the ocean. Wind is usually defined by the direction from which it has come whilst current directions are defined by the direction in which they are going. Hence, a westerly wind is flowing from the west to the east, and this would correspond to an eastward current in the ocean. From here on, the oceanographic convention for both winds and currents is adopted wherever possible.

In both the atmosphere and the ocean, Eulerian velocities are most commonly measured by a rotary device, such as an anemometer or current meter, where the number of rotations of a set of wings or propellers in a given time is measured. The anemometer has to be secured from a tower or mast, and therefore it can only be used for wind measurements near the Earth's surface. Most cup-anemometers are sensitive to wind variations longer than 10 s and have an accuracy of $1 \, \mathrm{m \, s^{-1}}$. Smaller and lighter anemometers can be used for more accurate observations of the wind close to the ground, such as required in micrometeorological experiments discussed in Section 5.1. The main difficulty with anemometer measurements is the problem of the alteration of the flow by obstructions, in particular by the tower or mast. For example, a ship may distort the air flow and this will lead to errors in wind measurement which are dependent on the orientation of the ship to the wind direction and the exposure of the anemometer.

Current meters were the mainstay of Eulerian measurements in the ocean until the end of the 20th century, and they are still in use today. A commonly used meter is the Savonius meter, which consists of a set of cups which rotate in proportion to the flow through the meter. The direction of flow is obtained from a freely moving direction vane attached in line with the instrument. The Savonius meter can measure current speed to an accuracy of $2 \, \mathrm{cm \, s^{-1}}$. Though the original instrument was designed in the 1930s, it has been improved upon by using an electronic storage system to record velocity and direction at about eight times per rotor revolution. The instrument samples bursts of data every 15 min because of the high-frequency variability of the currents. It automatically vector averages bursts of data to give an accurate measurement. This is important close to the surface where waves produce high-frequency noise. One unfortunate problem with the Savonius rotor is its susceptibility to vertical motion induced by surface waves, and this alone is capable of inducing the rotor to move. The problem does not arise with the vector-measuring current meter that uses two horizontal axis propellers orientated in perpendicular directions. This meter automatically measures both horizontal components of the flow and it is unaffected by surface wave motion. In all current meters, the orientation of the instrument is measured by a magnetic compass.

A major difficulty in the Eulerian measurement of currents at sea is the provision of a stable platform. Current measurements from ships are subject to the ship's motion and care must be taken in their interpretation. Most current measurements are now taken from surface and bottom moorings, which are laid by an oceanographic vessel and left in position for up to two years before being recovered. The main features of the moorings are a satellite transceiver, a light and a float, which are all necessary to locate the mooring at the end of an experiment. Current meters, and other instruments, are suspended at the required intervals down to the ocean bottom. Moorings usually have between five and ten current meters attached to the wire or the nylon cable. Buoyancy is necessary because of the weight of the instruments and cable, and it is provided by glass spheres encased in plastic containers. Finally, the wire is anchored to the bottom. A feature of most moorings is the acoustic release system which detaches the cable from the bottom anchor when it is activated by a research ship at the end of the experiment. In the early days of development, the recovery rate was about 50%, but now, with the advent of more robust systems, the recovery rate for moorings is better than 95%.

Some of the problems that have had to be overcome were from fish severing the mooring lines, from corrosion of the mooring wires and from intense surface currents pulling the floats under. Surface moorings have a higher failure rate than sub-surface moorings because of the considerable stresses induced by surface wave motion. Although moorings have proved a cost-effective way of obtaining Eulerian current observations, the movement of the mooring, particularly at or close to the surface, produces errors. This type of motion can be measured by acoustic tracking of the mooring and it is therefore possible to compensate for it when analyzing the current measurements.

Sound velocity measurements can be used to determine Eulerian velocities by two different methods. The first method requires the measurement of the travel time of sound between two hydrophones, A and B, as shown in Figure 6.2a. The travel time between A and B is $L/(c+v)$, where v is the average current velocity, c is the speed of sound and L is the distance between A and B. The travel time between B and A is $L/(c-v)$. In an acoustic current meter L is known and thus v may be determined, without knowledge of the velocity of sound c, from the difference between travel times. The system as shown will only measure the current velocity in the direction of sound waves and therefore a second pair of hydrophones is required in the perpendicular direction to measure both components of the horizontal flow.

The Acoustic Doppler Current Profiler (ADCP) has become a widely used instrument for the determination of ocean currents (Figure 6.2b) The ADCP can either be mounted on the hull of a ship or on a moored buoy. The sound waves emitted by the ADCP are back-scattered by particles in the water column to the instrument. The relative movement of the particles, and

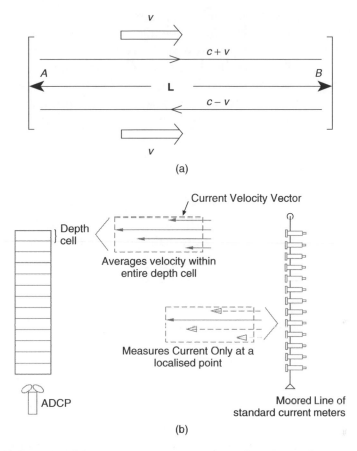

(a)

(b)

Figure 6.2   (a) Principle of the acoustic current meter: c is the velocity of sound and v the water velocity. (b) A comparison of the Acoustic Doppler Current Meter with a standard current meter

therefore the current which contains the particles, causes a Doppler shift in frequency between the emitted and back-scattered sound.

The velocity of the patch of water, v, is given by:

$$v = \frac{c\Delta F}{F_0} \tag{6.3}$$

where ΔF is the Doppler shift in frequency, F_0 is the frequency of the emitted sound and c is velocity of sound.

The ADCP can provide both horizontal and vertical components of the current. It relies on the backscattering of sound waves from a range of depths, and automatically averages the signal into 'bins', typically of 1 m to 10 m thickness and therefore it does not provide an exact velocity at a fixed position. The ADCP is most useful for the measurement of the vertical variation of the horizontal current, from estuaries to the deep ocean, and therefore has become the preferred method of measurement for many oceanographers. In practice the vertical velocity is a measure of the turbulence created, for example, by surface wind and swell waves (see Section 8.2), which is of little value for oceanographers wishing to measure longer time scale variations of the flow. In shallow water the ADCP can be referenced to the bottom and therefore absolute currents are determined reliably from a moving ship. In deep water accurate navigation, provided by the global positioning fixing system (GPS) is required to obtain absolute currents.

In the atmosphere, the cup anemometer, deployed on a mast is the standard instrument used for measuring the wind speed. An attached vane provides the wind direction.

Another method used by aircraft is the pressure anemograph or pitot tube. This instrument measures the difference between the dynamic pressure facing the wind and the static pressure. The pressure difference is

$$\frac{1}{2}\rho_A v^2$$

where ρ_A is the density of the air and v is the wind velocity. This technique is very accurate and it is sensitive to rapid fluctuations in the wind. It is able to measure fluctuations of up to 1000 Hz and is particularly useful for turbulence measurements. It can be used *in situ* on a mast or tower, or from an aircraft. Three components of the flow can be obtained by using instruments in perpendicular directions.

The propeller bi-vane anemometer can also provide wind components at high frequency measurements in the three directions.

For high frequency measurements the sonic anemometer, which can measure very high frequency wind variations in all three directions, is used. Similar to the acoustic current meter, it measures transit times in three directions, from which the components of the velocity can be determined. The sound velocity can be deduced from the simultaneous measurements of air temperature. This instrument can be used from a mast on land or at sea.

Above the Earth's surface, horizontal wind measurement depends on the use of Lagrangian techniques; the most widely used being the tracking of a balloon by radar, theodolite or satellite. During the Second World War, when balloon tracking became important for obtaining measurements of the winds in the upper troposphere for military aircraft operations, small pilot balloons of known positive buoyancy were released from weather stations and tracked by theodolite. The observer could calculate the ascent rate from the known buoyancy of the balloon at the surface and, by accurate timing, could therefore obtain the x, y, z coordinates of the balloon at any time. This type of operation could be carried out by single, skilled observers but it was limited by the height of the cloud base and the assumption of a constant ascent rate. Radar tracking systems permit the range, elevation and azimuth to be obtained automatically and thus the exact trajectory of the balloon can be traced in all weathers throughout the entire depth of the troposphere and the lower stratosphere up to a height of ~20 km. These balloon ascents are not, by definition, exact Lagrangian measurements because the balloon is sampling different parcels of air as it moves upwards through the atmosphere. Furthermore, because it is taking a profile of the atmosphere, the balloon may not be moving exactly with the horizontal velocity of the air at any particular level. However, despite these problems, the balloons provide a valuable source of information on the upper-level winds for weather forecasting purposes.

The winds are obtained by taking differences in horizontal positions at, for example, 1 min intervals. A balloon may take about an hour to attain a height of 20 km and, during that time, it may travel a horizontal distance of 50–100 km. depending on the strength of the upper winds. Meteorologists involved in weather prediction are mainly interested in mapping synoptic motions with scales greater than 100 km. By comparison, the horizontal distances traveled by the balloons are relatively small and the measured winds are, therefore, a reasonable approximation to the Eulerian winds at the weather station.

Lagrangian measurements may be obtained from constant-level balloons which are designed to float at a specified pressure level in the atmosphere by appropriate adjustment of the buoyancy. These balloons are much larger than their synoptic counterparts and individual balloons have been tracked around the globe using satellite position-fixing systems. In the 1970s, a large number of these balloons were released in the Southern Hemisphere to obtain information on winds over remote regions of the Southern Ocean, where conventional balloon data were unavailable. The balloon trajectories provided valuable information on the circulation of the upper troposphere but their impact on weather prediction has not been so marked. As mentioned earlier, a major difficulty with a Lagrangian system is that the observer has no control over the position of the balloon and so the velocity information may not be in the area required by the meteorologist. Another problem was that the formation of ice on the balloon's surface resulted in a high failure rate during the research programme.

A technique for obtaining synoptic-scale wind information over remote areas is to measure the velocity of clouds. High-resolution cloud images, in the visible and the infra-red, from satellites allow the regular monitoring of patches of cloud on an hourly basis, from which a time-averaged wind speed and direction can be obtained. Furthermore, because many cloud elements can be monitored, a pattern of the large-scale wind circulation can be obtained from a few images. The main drawback of the technique is that it is not easy to determine the exact height of the clouds. To obtain the height of the clouds, additional information on the vertical temperature profile and the cloud-top temperature is required.

From the above discussion, it is clear that the main requirement for wind observations has come from the desire to obtain global wind data in order to understand and predict the weather. The ascents of pilot and radiosonde balloons, coupled with satellite techniques, now provide daily observations of wind at all heights of the troposphere and the stratosphere.

In the ocean, it is only in the last 30 years that measuring systems have become sufficiently reliable and frequent to provide time dependent flows on ocean basin scales. In the late eighteenth and the nineteenth centuries, with the development of colonial trade routes, information on surface currents was invaluable for minimizing sailing times. Major Rennell, in 1778, was the first to determine the surface currents of the Atlantic Ocean. Providing that the position of the ship, its heading and speed are accurately known, the surface current can be calculated. Later, Maury (1853), the superintendent of the Naval Observatory in Washington, used this technique to produce the first global charts of surface currents. These charts are still the most commonly presented data on surface currents and, with the aid of improved position-fixing systems for shipping using satellite navigation aids, the technique is still valuable for determining large-scale flow patterns, especially where the currents are strong.

Another early Lagrangian technique, popular for the determination of currents in the nineteenth and early twentieth century, was the drift bottle. Its main disadvantage was that it gave no information about the trajectory between the launch and the pick-up point and the current determination would always be the minimum velocity of the flow between the two points. Today, drift cards and bottom drifters are still used to determine flow patterns over small distances, particularly in coastal and estuarine environments.

The major revolution in modern oceanography has come with the development of remotely tracked ocean buoys. The concept of an ocean buoy system is quite simple but it is only recently, with improved electronic systems, telemetry and satellite navigation using GPS, they have come into widespread use. A surface drifting buoy consists of a transmitter with batteries and an antenna for radio location by satellite, aircraft or by a shore station. It is desirable to have most of the volume of the buoy below the water to reduce the direct effect of wind stress. They usually have a drogue, or parachute, at a depth of 20–30 m to enable the drifter to move with the ocean current rather than with the direct wind-induced current. Many surface drifters also record the surface atmospheric pressure and temperature. Some have a thermistor chain attached which enables the sub-surface temperature profile to be monitored. Surface drifters are less expensive than current moorings and they also give more information on the large-scale flow patterns than an array of current meters. They can be tracked relatively easily for many months, sometimes for years, and are ultimately limited by their batteries' duration.

It was recognized by Henry Stommel and other oceanographers in the 1950s that a drifting buoy could also yield important information on deep-ocean flows, which were believed to be difficult to measure by conventional techniques. John Swallow, working at the National Institute of Oceanography in the United Kingdom, was the first to design a deep-ocean float which could be tracked by a ship using hydrophones. His success lay in the observation that a body which is less compressible than sea water will gain buoyancy as it sinks. Furthermore, if the excess weight of the body at the surface is small, then at a given depth it will become neutrally buoyant and no further sinking will occur, and thus the float would become the oceanographic equivalent of the constant-level balloon in the atmosphere. The float consists of a 6 m length of aluminum tube, half of which is used for buoyancy, and the other half for housing an acoustic transmitter and batteries. The original float weighed about 10 kg and weights were added to obtain a negative buoyancy of 38 g at the surface, thus allowing the float to become neutrally buoyant at a depth of about 1000 m. Two hydrophones, one at each extremity of the research ship, were used to determine the distance and direction of the float. One of the difficulties is that it is not possible for a ship to track more than a few floats at any one time. Furthermore, the ship is totally involved in locating the floats and this is a very expensive operation. A subsequent development, by Rossby *et al.* in the USA in the early 1970s, involved the location of neutrally buoyant floats in the sound channel, between 500 m and 1500 m, which enabled continuous tracking of large numbers of floats by remote, shore-based hydrophones up to 1000 km away. These floats, originally located in the western Atlantic (Figure 6.3), were able to collect a considerable quantity of information of the large-scale ocean circulation.

It is important to realise that even neutrally buoyant floats only give a picture of the horizontal Lagrangian flow at a specified depth, and that they can give no information on the vertical velocity field. In addition, a large density of floats is required to give a reliable, statistical picture of the flow field. For example, in the MODE 1 (Mid-Ocean Dynamics Experiment), 20 floats were required to adequately monitor a 300 km diameter region of the North Atlantic Ocean west of Bermuda.

In the 1990's the ALACE (Autonomous Lagrangian Circulation Explorer) float was developed and deployed in the South Pacific and Southern Oceans. This float is essentially a Swallow float that can drift at a predetermined depth. The float comes to the surface at intervals of say 10–25 days, where its data and position can be transmitted to a satellite.

A development from the ALACE float is the ARGO float. The main purpose of the ARGO float is to obtain a vertical profile of temperature and salinity from a depth of 2000 m to the sea surface every 10 days. It sits at a parking

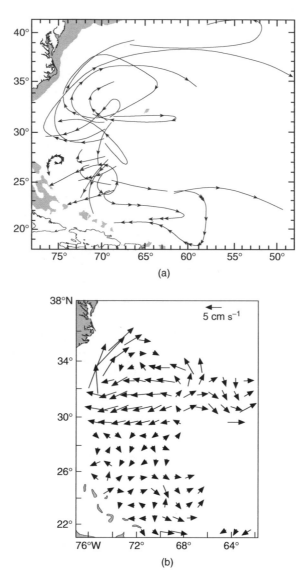

(a)

(b)

Figure 6.3 (a) A plot of 700 m SOFAR float trajectories. Arrows are 100 days apart. Note the high velocity of floats caught in the Gulf Stream. (b) The mean equivalent Eulerian circulation estimated from 700 m SOFAR float data. Reproduced, with permission, from Rossby, H.T. *et al.*, 1983, In *Eddies in Marine Science*, ed. Robinson, A.R., Springer-Verlag: pages 68 and 70, figures 1 and 3

depth, typically 1000 m, for about nine days where it has neutral buoyancy. Inside the instrument is a small bladder which is filled with an oil, and when it is evacuated by a pump, it loses buoyancy and sinks slowly to a depth of 2000 m. On reaching this depth the bladder is inflated and the instrument rises to the sea surface, measuring the temperature, salinity and pressure from 2000 m to the sea surface. The data is transmitted to a satellite whilst the instrument is at the sea surface and then the float returns to its parking depth. The international ARGO programme has been providing measurements in all the world's ocean basins from 1999 and over 3000 floats are currently deployed.

In the meteorological context, both radiosondes and dropsondes (i.e. radiosondes parachuted from aircraft) give a measurement of the vertical variation of horizontal velocity that is useful for the detection of large local velocities associated with jet streams and internal waves. When used in conjunction with other wind measurements, from other radiosonde stations, a three-dimensional picture of the horizontal flow field can be obtained. An oceanographic equivalent of the dropsonde is the velocity profiler, which is simply an acoustic float with temperature and pressure sensors. The float is allowed to sink slowly over an array of hydrophones on the bottom. As with the neutrally buoyant floats, the time delays between the float emission and the hydrophone reception allow the float's position to be determined to an accuracy of $+/-1\,\mathrm{m}$. By differencing the position plot, the horizontal velocity can be obtained with an accuracy of $+/-2\,\mathrm{cm\,s^{-1}}$.

The Acoustic Doppler Current Profiler (ADCP) can also provide an accurate and reliable method to measure velocity profiles as discussed above.

A summary of the main methods for current and wind measurement are presented in Tables 6.1 and 6.2.

All the Lagrangian techniques described above only measure the horizontal part of the three-dimensional flow pattern. Furthermore, the Global Positioning System measures the position and the time of the float accurately, therefore a good estimate of the Lagrangian velocity can be determined.

Table 6.1   Some direct methods for measuring ocean currents

Instrument	Type of measurement	Use	Platform	Sampling frequency
Rotor Current Meter	Eulerian horizontal velocity	Fixed single position	Ocean mooring	Minutes
Acoustic Doppler Current Profiler	Eulerian Horizontal and vertical velocity	Profile depth up to 1000 m	Ship mounted or ocean mooring	Seconds
SOFAR Sound source on float	Position of float	Sound channel	Hydrophones from ship or ocean mooring	Seconds
RAFOS Hydrophone On Float	Position of float	Sound channel	Sound source on mooring	Seconds
Argo	Position of float	Parking depth ~1000 m	Satellite position	10 days

Table 6.2   Some methods for measuring winds in the atmosphere

Instrument	Type of measurement	Use	Platform	Sampling frequency
Anemometer	Eulerian velocity	Fixed position	Mast, Tower	Seconds
Pilot / Radio Sonde Balloons	Lagrangian position	Track by surface radar	Released from surface	1 hour
Doppler Radar	Eulerian velocity	Profile of wind velocity	Surface radar	Seconds

It has long been recognized that chemical or radioactive tracers in a turbulent environment like the ocean will move with the individual parcel of water into which they are inserted, provided that they are in a sufficiently low concentrations to leave the physical properties of the water mass unaltered and hence the tracers can be used to determine the circulation. In Section 5.7 it was shown how naturally occurring chemical and radioactive tracers can be used to infer circulation patterns but the monitoring and diagnosis of a Lagrangian tracer experiment are quite different to the techniques previously described.

Consider the evolution with time of a patch of a tracer on the ocean surface. The tracer spreads in all directions as the result of a large number of small-scale fluctuations in the surface velocity induced, perhaps, by the surface wind or by small eddies in the flow. This is an example of diffusion. Furthermore, the centre of gravity of the patch of tracer will show a translation. Thus, the tracer gives two types of information:

(i)   The area spread of the tracer gives information on the mixing by time-dependent currents.
(ii) The movement of the centre of gravity of the tracer patch gives information on the mean current.

To monitor a tracer experiment, it is necessary to sample both spatially and with time. The tracer can be sampled from a ship and can give useful information over the area of interest.

Chemical and radioactive tracers require more time-consuming onboard laboratory techniques to measure concentrations. Care has to be taken to prevent the contamination of samples during collection and transfer to the laboratory, but such tracers do have the advantage of being able to yield high-precision results. Tritium, produced by atmospheric hydrogen-bomb tests in the late 1950s and early 1960s, has been used to trace both horizontal and vertical ocean currents over large areas of the North Atlantic. In the waters around the British Isles, circulation has been determined by the regular monitoring of caesium-137, which is a low-level radioactive waste product from the nuclear reprocessing plant on the Cumbrian coast.

6.2 Scales of motion in the atmosphere and ocean

Figure 6.4 shows four individual time series of wind speed measured by an anemometer. The first series shows the fluctuations in velocity covering a period of one minute, the second shows one minute average values for a period of one hour, the third shows one hour average values for a period of four days and the fourth shows one day average values for a period of one year. The main conclusion that can be drawn from these four time series is that fluctuations in wind velocity occur on a whole range of time scales, from a few seconds to years. Time series, not dissimilar to those of the atmosphere,

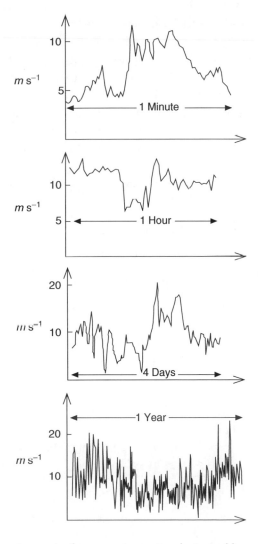

Figure 6.4 Four time series from an anemometer; 1 minute, 1 hour, 4 days and 1 year

can also be obtained for the ocean. A meteorologist, or an oceanographer, has to be aware of these fluctuations and their magnitudes.

For example, a synoptic meteorologist who wanted to obtain wind information to discern large-scale weather systems would be interested in obtaining 10 min or hourly averaged winds, in order to remove the high-frequency variations which have little direct relevance to the large-scale flow patterns. On the other hand, a micro-meteorologist who wanted to know how the smoke from a chimney was going to behave in the immediate vicinity of the discharge would be interested in the high-frequency fluctuations in the wind, because it would be the gusts of wind which would be responsible for the mixing of the smoke and the spread of the plume. A climatologist would probably average winds over a month, a season or even a few years in order to ascertain long period fluctuations and trends.

There are a number of methods with which to visualize a time series of Eulerian velocity measurements at a single point in the atmosphere or ocean. Velocity is a vector quantity, as shown in Figure 6.1, and therefore it can be shown as a time series of speed and direction as illustrated in Figure 6.5a. In this example the current is varying both in speed and direction over a period about 45 days. This same information may be plotted as a progressive vector diagram (see Figure 6.5b). The first observation is plotted as a vector from the origin with North towards to the top of the page. In this example the current is heading southwestwards. The next vector is plotted from the end point of the first vector, and so on. In Figure 6.5b it can be seen that the current is heading generally in the SW direction over the 45 days. This information could also have been obtained by converting the directions and speeds into the u (eastward) and v component (northward) of flow, and then time averaging the two components. Figure 6.5c shows that vectors can be plotted individually as a time series, with north towards the top of the page. This shows the current is changing in all directions of the compass on a time scale of days to months.

A very useful way to represent these fluctuations is to convert the time series shown in Figure 6.4 into time frequency spectra, as illustrated in Figure 6.6. The simplest approach to this kind of analysis is to consider a time series as the summation of a series of sine and cosine functions having a range of discrete frequencies. The highest frequency is $\pi/\Delta t$, where Δt is the sampling period of data, and the lowest frequency is $2\pi/T$, where T is the period of the observations. Thus, a time series of the current or wind (in the eastward direction) can be represented by:

$$u(t) = u_0 + \sum_{n=1}^{N} a_n \cos nt + b_n \sin nt \qquad (6.4)$$

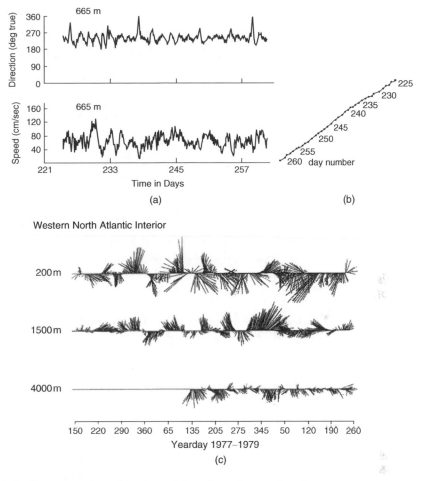

Figure 6.5 Representation of horizontal flow from time series of current measurements in the Denmark Strait (a) Direction and speed. (b) Progressive vector diagram. (c) Current vectors (North directed to top of page) shown at 200 m, 1500 m and 4000 m depth, in the Western North Atlantic Ocean. This is known as a stick diagram

where u_0 is the mean value of the time series. The coefficients a_n and b_n can be determined from the following relationships:

$$a_n = \frac{2}{T} \int_0^T u(t) \cos\left(\frac{2\pi n t}{T}\right) dt$$

$$b_n = \frac{2}{T} \int_0^T u(t) \sin\left(\frac{2\pi n t}{T}\right) dt \tag{6.5}$$

where T is the length of record.

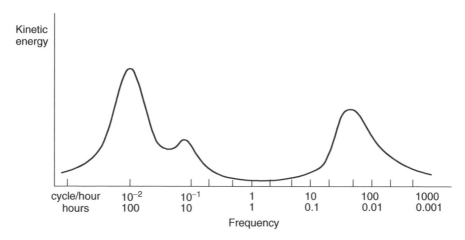

Figure 6.6 Frequency spectrum of the horizontal wind speed. Reproduced, with permission, from van der Hoven, J., 1957, Journal of Meteorology, 14: page 161, figure 1

This Fourier decomposition reveals two categories of information. First, the modulus of the coefficients, given by $\sqrt{(a_n^2 + b_n^2)}$, yields a measure of the amplitude of the fluctuations at frequency n, whilst the ratio of the coefficients, $\tan^{-1}(a_n/b_n)$, determines the phase in radians of the fluctuations at frequency N. Though the Fourier method and also other spectral methods produce a discrete set of frequencies, there is a tendency in the literature to show the spectra as continuous, as in Figure 6.6. As can be seen from this figure, there is no single frequency which dominates the spectrum, although there are certain frequencies where the fluctuations are large. In this atmospheric spectrum there is a peak of fluctuations with a period of 1 min associated with gusts produced by turbulent processes in the surface boundary layer. There is also a tendency for high-energy fluctuations at a period of a day, which is caused by diurnal variations in the wind, whilst between the period of a few minutes and one day the fluctuation energy is much smaller. For periods longer than a day there is a dramatic increase in energy and a maximum occurs at a period of two to six days. This is due to the large-scale synoptic variability of the atmosphere. Transforming the time series into a frequency spectrum thus makes it possible to perceive different time scales of motion in the atmosphere.

Figure 6.7 shows a frequency spectrum obtained from the analysis of readings from moored current meters in over 2000 m of sea water. It is evidently quite different from the atmospheric spectrum, having two identifiable peaks. One peak arises from the semi-diurnal tide (12.4 hours) and the other peak is an inertial oscillation which has a period of 19 hours. The inertial oscillation will be discussed in Chapter 7, but it is essentially a function of latitude and

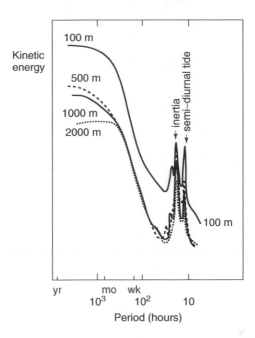

Figure 6.7 Frequency spectrum of kinetic energy of horizontal currents from four depths at Site D (39°N, 70°W) north of the Gulf Stream. Note the two high-frequency peaks, coinciding with the inertial period (19 hours) and the semi-diurnal tidal period (12.4 hours). The kinetic energy associated with the tidal and inertial frequencies is smaller than the low-frequency variability between seven and 20 days. Reproduced, with permission, from Rhines, P.B., 1971, Deep Sea Research, 18: pages 21–26, figure 1

the rotation of the Earth. Its period, T, is given by:

$$T = \frac{\pi}{\Omega \sin \phi} \tag{6.6}$$

where Ω is the rotation rate of the Earth and φ is the latitude.

At longer periods the spectrum shows no obvious peak, but does display a large amount of variability. This variability is mainly associated with mesoscale eddies which have time scales of weeks to months, and is caused, in this case, by instabilities in the Gulf Stream.

The time variability is only one of several possible dimensions in which to analyse motions. The spatial equivalent is the wavenumber spectrum, the wavenumber being the inverse of wavelength. In this case, unlike the infinite time series, there are quite definite limits to the spectrum which are defined by the finite size of the ocean basins and the Earth's atmosphere. Figure 6.8 demonstrates the wavenumber frequency spectrum of the wind obtained from the analysis of the large-scale atmospheric flow. It can be seen that a large part of the variability is on very long scales, with peak energy in waves which have five to eight cycles around the a latitude circle. The lowest wavenumbers

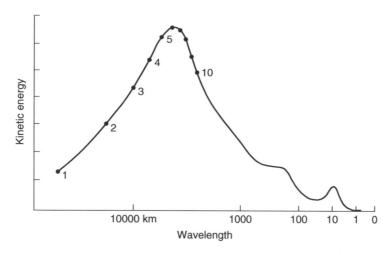

Figure 6.8 Wavelength spectrum of wind fluctuations in temperate latitudes. Numbers indicate circumpolar wavenumbers. Area under segment of curve is proportional to eddy kinetic energy in corresponding band of the spectrum. Reproduced, with permission, from Lorenz, E.N., 1969, In The Global Circulation of the Atmosphere, ed. Corby, GA., Royal Meteorological Society: page 11, figure 4

(1–4) are long Rossby waves and they are associated with the distribution of the continents and oceans, and with topography (Figure 6.9). Higher wavenumbers, particularly between 8 and 10, are associated with synoptic weather systems, which have space scales of a few thousand kilometres. At higher wavenumbers (i.e. shorter wavelengths), the energy of the spectrum drops very quickly, indicating that the large-scale systems are, energetically, the dominant flow in the atmosphere. As discussed above, in the ocean basins it is found that the energetically dominant space scales are associated with mesoscale circulations having wavelengths of approximately 50–200 km (Figure 6.10) The dynamics of these mesoscale eddies are similar to those of synoptic scale eddies in the atmosphere.

Velocity time series can be displayed and computed in many ways to reveal the most significant time and space scales present in the data. Furthermore, from current or wind measurements at different geographical positions the spatial variation of the flow *at a particular time* may be described.

6.3   Time averaged circulation

6.3.1   *The time-averaged circulation of the atmosphere*

The flow in the atmosphere and ocean varies with time and position over the globe. The simplest way to map the flow is to calculate time means of

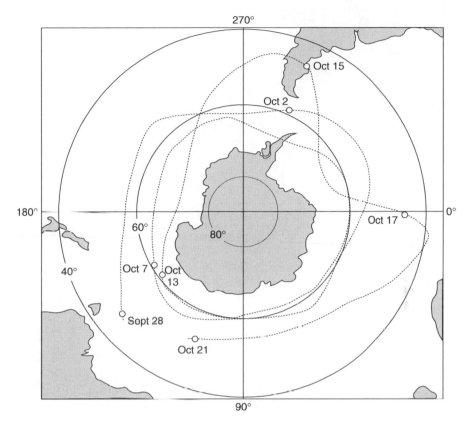

Figure 6.9 A portion of the trajectories of constant-level balloons at height of 150 hPa ($\approx$ 13 km) in the Southern Hemisphere. Successive positions on the dates indicated are also shown. The last circuit indicates a Rossby wave with northward excursions over Australia (October 13th), S. America (October 15th), S. Africa (October 17th). Reproduced, with permission, from The TWERLE Team, 1977, Bulletin of the American Meteorological Society, 58: page 942, figure 6a

the velocity measurements over a sufficiently long period of time. The solar radiation cycle has a major influence on the circulation of the atmosphere and therefore it is normal to consider monthly or seasonal averages of the wind circulation. For example, to obtain the mean winter circulation in the Northern Hemisphere we could average (December, January, February) observations from a number of years. For the purposes of clarity, the atmospheric mean circulation will be defined as the monthly average taken over a period of 5–10 years. For surface observations, a 30-year average is usually taken as the climatic average, but there is no *a priori* reason to suppose that this will be more representative of the circulation than a 10-year average. In the troposphere good quality observations have been available over a large part of the globe since the 1950s and therefore the main mean circulation is well established.

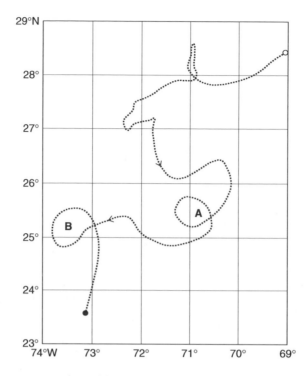

Figure 6.10   The trajectory of one SOFAR float from 11 July 1973 to 30 June 1974. It is caught in each eddy for about one month. Note the general westward drift of the float of about 1 cm s^{-1}. Reproduced, with permission, from Rossby, H.T. *et al.*, 1975, Journal of Marine Research, 33: pages 366 and 369, figures 6 and 10

Figure 6.11 shows the global winds as a function of height and latitude for the winter and summer, obtained from daily observations of the atmosphere from weather stations; ships; aircraft and balloons. The observations have been averaged around latitude circles.

The predominant feature in both hemispheres is a region of high wind speed, reaching 30 m s^{-1}, in the upper troposphere between 30° and 40° latitude. These climatological, eastward-flowing sub-tropical jet streams are the most energetic part of the troposphere circulation. It is noted that the Northern Hemisphere sub-tropical jet stream moves poleward in the summer and weakens by about 50%. A similar poleward movement occurs in the Southern Hemisphere summer but the seasonal variation in intensity is smaller. The magnitude of the jet stream is related to the horizontal meridional temperature gradient and this is discussed in Chapter 7. The mean flow is dominated by eastward winds (westerly) throughout most of the upper troposphere, but westward (easterly) winds do occur in the tropical troposphere and near the surface in Polar Regions. At the surface it is necessary to have both eastward and westward winds to avoid a net torque on the Earth, which would result

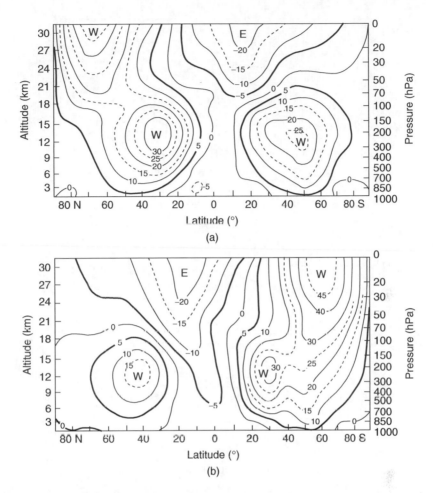

Figure 6.11   Mean zonal wind for (a) December-February, (b) June-August. Units: $m s^{-1}$ Positive (negative) values denote westerly (easterly) flow. Reproduced, with permission, from Newell, R.E. *et al.*, 1969, In The Global Circulation of the Atmosphere, ed. Corby, G.A., Royal Meteorological Society: page 69, figures 14a and b

in a change in the Earth's rotation (see 9.4). In the stratosphere there is a very significant seasonal variation in the wind. In winter a strong jet develops at high latitudes (60° latitude) with wind speeds up to $70\,m s^{-1}$. This is known as the polar night jet. It is caused by the large meridional temperature gradient between the dark polar and the sunlit stratosphere (Figure 6.12). As discussed in Chapter 3, the temperature of the stratosphere is controlled by the absorption of solar radiation by ozone. During the polar night in winter, in the absence of solar heating, there is rapid decline in the temperature of the polar stratosphere. In the summer hemisphere the eastward jet, weakens immediately above the troposphere, and is replaced by westward jets at

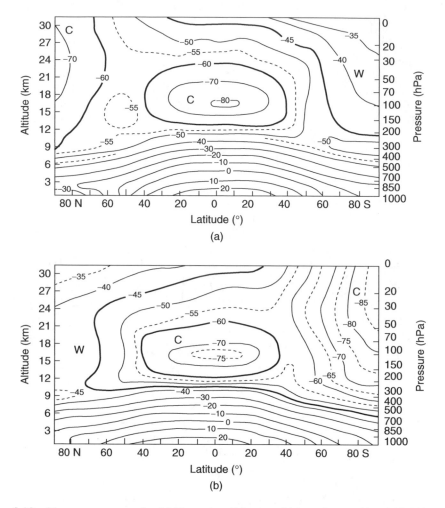

Figure 6.12 Mean temperature for (a) December-February, (b) June-August. Units: °C. W = warm; C = cold. Reproduced, with permission, from Newell, R.E. *et al.*, 1969, In The Global Circulation of the Atmosphere, ed. Corby, GA., Royal Meteorological Society: page 65, figures 9a and b

about 20° latitude. During the summer the meridional temperature gradient is reversed, and therefore the winds will also change direction. This is shown in Figure 6.11 and 6.12.

A different picture of the mean circulation is obtained from Figure 6.13, which shows the vertical and meridional components of the flow. A large cell, known as the Hadley cell, is seen to cover the tropical region and two much weaker cells, known as the Ferrel cell and polar cell respectively, are also present.

In the Hadley cell, air is lifted up over the equatorial zone, mostly in the ITCZ, where large amounts of latent heat are released by heavy

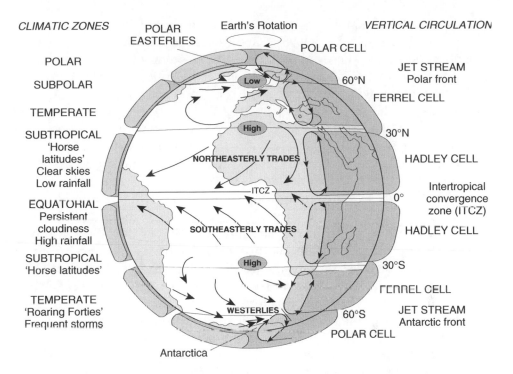

CLIMATIC ZONES

POLAR

SUBPOLAR

TEMPERATE

SUBTROPICAL
'Horse
latitudes'
Clear skies
Low rainfall

EQUATORIAL
Persistent
cloudiness
High rainfall

SUBTROPICAL
'Horse latitudes'

TEMPERATE
'Roaring Forties'
Frequent storms

POLAR
EASTERLIES

Earth's Rotation

POLAR CELL

Low

High

NORTHEASTERLY TRADES

ITCZ

SOUTHEASTERLY TRADES

High

WESTERLIES

Antarctica

VERTICAL CIRCULATION

JET STREAM
Polar front

FERREL CELL

60°N

30°N

HADLEY CELL

Intertropical
convergence
zone (ITCZ)

HADLEY CELL

30°S

FERREL CELL

JET STREAM
Antarctic front

POLAR CELL

0°

60°S

Figure 6.13 Schematic representation of features of the general circulation of the atmosphere. Reproduced, with permission, from Charnock, H. In: Oceanography, Edited by S.A. Thorpe and C. Summerhayes. Manson Publishers. 1995: page 34, figure 2.9. See plate section for a colour version of this image

precipitation. This air rises to the tropopause, moves polewards to the sub-tropics (30° latitude) close to where the subtropical jet stream is located. The air then descends, warms by adiabatic compression, and its relative humidity decreases. The circuit is completed by a return flow in the lowest 2 km of the atmosphere towards the equator. This return flow is very steady and it is associated with the north-east and south-east trade-wind systems. Some of the descending air in the subtropics travels polewards and ascends in the middle latitudes. This ascent occurs in the region of the polar front along which sub-polar low-pressure systems develop. A weak return flow to low latitudes completes the circuit. One feature of the mean meridional circulation is that it is much weaker than the corresponding zonal circulation, having mean velocities of less than $3 \, \mathrm{m \, s^{-1}}$, and the maximum meridional velocity occurs in the low-level trade winds. The vertical velocities are generally less than $1 \, \mathrm{cm \, s^{-1}}$. The mean meridional circulation is, therefore, a much weaker feature than the zonal circulation, which is related to the dominance of the equatorial - polar temperature gradient in controlling the flow. The relationship between the winds and the atmospheric temperature gradient will be explained in Chapter 7.

The horizontal climatological surface flow is shown in Figure 6.13. The subtropical high pressures systems are the largest in area of the surface atmospheric systems. These are regions where descending air from the upper branch of the Hadley cell enters the lower troposphere. These regions are largely cloud free, with only low level broken cloud present. The surface trade winds can be seen flowing towards the equatorial zone where they meet, from both hemispheres, at the Inter Tropical Convergence Zone.

This convergence zone is generally located in the Northern Hemisphere in the Pacific and Atlantic Oceans along the thermal equator (i.e. the region of maximum sea surface temperature). In the Indian Ocean and the western Pacific Ocean, seasonal monsoonal variations in the wind circulation dominate the flow. The large mass flow from the southern Indian Ocean high pressure centre to the lower pressure centre over the Asian continent is evident in summer, from about June to September. The curving of the winds from a south-east flow in the Southern Hemisphere to a south-west flow in the Northern Hemisphere is related to the rotation of the Earth, as discussed in Chapter 7. In the Northern hemisphere winter, a low-pressure centre is located over the Indonesian maritime continent and convergence of air from both the Indian and Pacific Oceans occurs. The regions of low level convergence are areas of upward motion and heavy precipitation, as seen in the ITCZ (Figure 6.13). Weaker monsoonal circulations occur over Africa and South America.

6.3.2   Time mean circulation of the ocean

Apart from the flow in the surface layer of the ocean (which has been reasonably well determined over the main shipping routes since the 19th century), the flow in the deeper ocean has only been measured directly in the last few decades and the methods used have been described in Section 6.1.

These measurements show a very complex behaviour, previously discussed in Section 6.2, but the measurements can be averaged over time to obtain a more coherent view of the flow for example as shown in Figure 6.3b. This is the time averaged flow at one depth for a region of the Western North Atlantic Ocean measured in the MODE experiment in the 1970s. Subsequently oceanographers planned an international experiment to measure the whole global ocean circulation using the highest quality data in the 1990s. It was known as the World Ocean Circulation Experiment (WOCE). WOCE (1990–97) had the goal of measuring this circulation over all the major ocean basins during its seven year period. Surface drifters, to which drogues were attached, were deployed to measure the flow at $10\,m-15\,m$ below the ocean surface. The influences of surface waves are therefore reduced significantly.

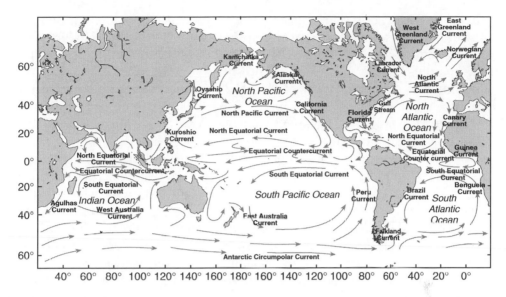

Figure 6.14 A schematic diagram of the horizontal surface circulation in the global ocean: February. Reproduced, with permission, from American Meteorological Society. 2005

Figure 6.14 shows surface time averaged flow in the ocean. This flow is very much influenced by the surface wind stress and therefore the direction of these flows has similar patterns to those of the surface wind.

The flow in the upper ocean and thermocline (0–1000 m) is influenced by the wind stress at the ocean surface, the Coriolis force and horizontal pressure gradients and these will be discussed in Chapter 7.

The main features of this flow are large scale horizontal circulations or gyres which occupy the whole width of the ocean basins (Figure 6.14). The subtropical gyres (15–40°) are the deepest of these circulations and circulate in a clockwise direction in the Northern Hemisphere and anti-clockwise in the Southern Hemisphere as shown in Figure 6.14. Between 40–60°N are the subpolar gyres which circulate in an anti-clockwise direction. In the Southern Hemisphere, the gyre circulations interact with the Antarctic Circumpolar Current (ACC) which flows eastward around the globe between 40° and 60°S.

The flow around these gyres is not a simple circular flow. The flows in the western basins are much stronger than those in the eastern basins and they are also concentrated into a narrower region. For example at 30°N in the Atlantic and Pacific Ocean the northward flows of the Gulf Stream and Kuroshio are found close to the western boundaries of the basins, having widths of roughly 100 km. These western boundary currents are found in all the ocean basins and are the most energetic currents in the world. They have velocities of up to $1–2\,\mathrm{m\,s}^{-1}$, compared with ocean currents over most of the gyres which typically have velocities of about $0.1\,\mathrm{m\,s}^{-1}$.

The Indian Monsoon has a major influence on the ocean circulation in the equatorial region and in the Northern Indian Ocean (about 10°S to 20°N) and this is shown in Figure 6.15a, b. The reversal of the wind circulation leads to major changes both in the direction and speed of the surface currents.

The flows in equatorial regions tend to be aligned along lines of latitude, with only significant N-S (meridional) flows close the western boundaries. The equatorial flows are relatively strong flows with speeds of ~0.2 to $0.3 \, \text{m s}^{-1}$ in the surface and thermocline waters.

In discussing ocean circulation another measure called the volume transport is useful. The volume transport is the transport of water across a defined cross sectional area, and has units of $\text{m}^3 \, \text{s}^{-1}$. For example a river would have a cross sectional area defined between the river banks and depth of the river. Though the velocity would vary across the river and with depth, the volume transport is one number which provides an integrated measure of the flow. The volume transport removes some of the complexity of the smaller scale flow and allows oceanographers to measure the larger scale exchanges between and within ocean basins. For ocean circulation a unit of 1 Sverdrup is used, which is $10^6 \, \text{m}^3 \, \text{s}^{-1}$.

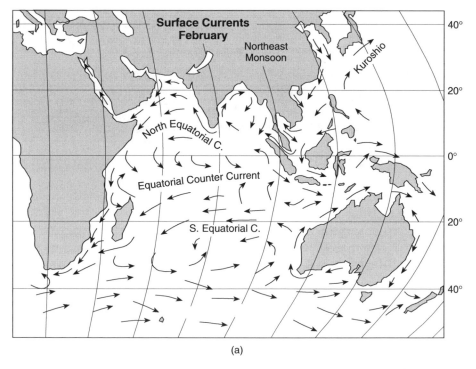

(a)

Figure 6.15 (a) Monsoonal surface currents: February. (b) Monsoonal surface currents: August. Reproduced, with permission, from Grant-Gross, M., 1987, Oceanography: A View of the Earth, Prentice-Hall: page 180, figure 7.5

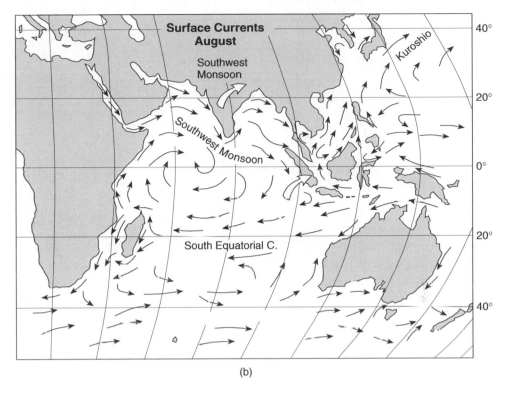

Figure 6.15   (*continued*)

The deep flows below the thermocline, particularly in the vicinity of the western boundary, can be quite energetic. They are often rather complicated because of their interaction with topography at the ocean margins.

On the largest scales we have a reasonable picture of the time-mean ocean circulation based on the temperature, salinity, chemical and radioactive tracers as discussed in Chapter 5. Transient tracers from bomb tritium and various chlorofluorocarbons have a known time at which they entered the ocean. For example, most tritium entered the ocean between 1950 and 1963. Deep ocean measurements record the spread of these tracers and provide a description of the deep ocean circulation on a global scale. This deep circulation forms part of the global thermo-haline circulation (Figure 6.16) but it is difficult to make sufficient direct measurements. In WOCE temperature, and salinity measurements together with velocity measurements across all the ocean basins were measured, and the transports at different levels in the ocean were determined. These will be discussed in Chapter 7.

Hydrographic data on temperature, salinity and pressure can be used to obtain an indirect estimate of the mean current. The details of this method will also be discussed in Chapter 7 but, in its simplest form, it involves the

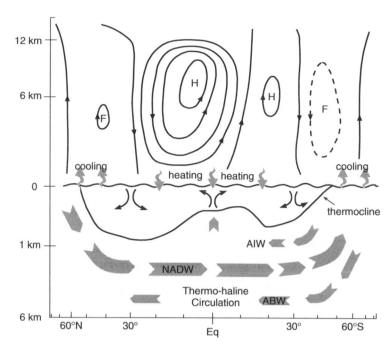

Figure 6.16 A meridional section through the atmosphere and ocean, in the Atlantic sector, for December-February. The atmosphere circulation shows the Hadley cell (H) and Ferrel Cells (F), and the strength of the cells are represented by the contours. The small arrows represent the wind driven Ekman flow in the upper ocean. Downwelling occurs ~30° latitude and upwelling at the equator. The thermocline is the boundary between the relative warm upper ocean and the colder deep ocean. The larger arrows represent the direction of the Antarctic Intermediate Water (AIW), N. Atlantic Deep Water (NADW), and the Antarctic Bottom Water (ABW). Reproduced, with permission, from Wells, N.C. 'Ocean Circulation: General Processes' for Encyclopaedia of Ocean Sciences, Academic Press. 2001: page 1536, figure 9

calculation of the horizontal pressure field from the density of sea water and the use of the hydrostatic equation.

The horizontal flow tends to follow lines of constant pressure, and hence the circulation can be calculated. This flow is known as a geostrophic current. However, only relative horizontal motion can be obtained because it is not possible to calculate the absolute horizontal pressure field. Generally, a reasonable estimate of the upper ocean flow can be obtained by referencing it to a level of, say, 1 km below the surface, where the flow and the horizontal pressure gradients are generally weaker. This method can utilize the substantial hydrographic database for the global ocean to obtain an estimate of the geographical distribution of ocean currents.

A schematic of the Atlantic Ocean circulation is shown in Figure 6.16. The meridional overturning circulation is a relatively weak circulation and the volume transport across the ocean basin is about 20 Sverdrups. This is equivalent to a northward flow across the North Atlantic Ocean of about 1 mm/s. By contrast the northward flow in the Gulf Stream across the Florida Straits has a transport of 30 Sverdrups.

The surface and thermocline waters move northwards in the Atlantic as part of the upper limb of the thermo-haline circulation. They are cooled in the Labrador, Greenland and Norwegian seas, where they sink to depths of 1–3 km. These new water masses are known as North Atlantic Deep Water and Labrador Sea water. These water masses move southward as the lower limb of the thermo-haline circulation. They enter the Southern Ocean where the Antarctic Circumpolar Current transports the water eastwards into the Indian and Pacific Ocean. The Antarctic Circumpolar current is the largest ocean current in the world and transports about 135 Sverdrups. This water gently upwells and mixes into the surface and thermocline waters in the Indian and Pacific oceans. These upper waters are transported from the Pacific through the Indonesian archipelago into the Indian Ocean, and then transported around South Africa into the South Atlantic. This forms the complete loop of the thermo-haline circulation. The time for a parcel of water to circulate is about 2000 years if it has a mean velocity of 1 mm/s. This time is similar to that deduced by measurements for the decay of the isotope C14 in the deep ocean.

In summary we have described both the global mean circulation of the atmosphere and ocean.

6.4   Time-dependent motion

6.4.1   Time-dependent atmospheric motion

It has been shown that time-dependent motions occur on a variety of space and time scales. The larger scales of motion constitute the most energetic and longest time-scale phenomena in the atmosphere. Figure 6.9 shows the trajectory of a constant-level balloon at 150 hPa (a height of approximately 14 km) observed over a period of 23 days. If the flow was purely zonal and constant in time, concentric circles would be anticipated with the balloon repeating a similar trajectory on each circuit. Instead, considerable variations in time and space are observed. On the last circuit of the hemisphere there are two large excursions over South America and South Africa, and a smaller excursion close to Australia, which were not apparent on the previous circuits. These deviations from circular motion are caused by Rossby waves. The number of these Rossby loops in the circumpolar flow can vary from one or two to as many as six or seven waves. If they were sinusoidal waves, a spectrum such as the one shown in Figure 6.8 would indicate their wavenumber.

For example, a wavenumber 1 flow would be a single wave around the hemisphere, whilst a wavenumber 4 flow would indicate four wave loops around the hemisphere. The central axis of the waves is often associated with jet streams and, unlike the climatological sub-tropical jet stream with mean winds of $30 \, \text{m s}^{-1}$, wind speeds of up to $100 \, \text{m s}^{-1}$ can occur. The observation and forecasting of these jet streams are important priorities for aircraft operators. The low wavenumber patterns tend to move only slowly with respect to the Earth's surface, with the circumpolar flow moving eastward through the pattern. Occasionally, the Rossby wave pattern may become stationary for a few days, or even weeks, and then it is known as a 'blocked flow'. The waves are generated by a dynamic instability of the strong circumpolar flow, which grows on a time scale of a few days. However, the waves may be enhanced by large mountain barriers, such as the Andes or the Rocky Mountains, and by the thermal contrast between ocean and continent.

Embedded within these large Rossby wave loops are the familiar, synoptic-scale cyclones and anticyclones which have length scales of the order of 1000–3000 km and time scales of one to five days. There are typically 20–25 cyclones and anticyclones in the lower part of the atmosphere (i.e. below 5 km) in the Northern Hemisphere at any one time. There tend to be larger numbers of cyclones than anticyclones, but the latter tend to have larger horizontal scales. The extra-tropical, or middle-latitude, cyclone is also associated with stronger wind circulation than the anticyclone. The positions of the cyclones and anticyclones do not tend to be random with respect to the upper air flow. The cyclones form on the eastern side of the trough of the Rossby wave, whilst anticyclones are generally located within, or slightly downstream of, the ridge. Within a stationary or 'blocked' flow, the anticyclones and cyclones will grow and decay repeatedly over the same regions, thus resulting in persistent regions of 'good' and 'bad' weather. The surface pressures of anticyclones are generally in the range of 1020–1050 hPa, whilst extra-tropical cyclones have a range of 930–1010 hPa. Both highest and lowest pressures tend to occur in the winter hemisphere. Vertical motions within synoptic systems tend to be between 1 and $10 \, \text{cm s}^{-1}$.

Within the extra-tropical cyclones are frontal systems which range from 50 to 200 km in width. These frontal regions tend to spiral out from the cyclone's centre and they may extend for many thousands of kilometres, and even into another cyclonic system. Frontal regions are boundaries separating air masses and are, therefore, associated with large temperature gradients, especially in winter. Temperature changes are usually between 5 and 10°C, but occasionally, over continental regions, temperature changes of up to 20°C are observed. These frontal regions are often dynamically active and they are associated with upward motion, often with continuous precipitation. The strongest upward motion in an extra-tropical cyclone is usually associated with the frontal systems and it may reach values of $0.5 \, \text{m s}^{-1}$ in intense fronts.

The next important scale in atmospheric motions is the mesoscale, which includes a variety of phenomena ranging from shower clouds, thunderstorms and sea-breeze circulations to larger systems such as squall lines and rain bands. The former group of phenomena has scales of 1–10 km and time scales at most of a few hours, whilst the latter may reach 100 km in size and persist for up to one or two days. The squall line is often associated with a line of thunderstorms and strong, gusty winds, and it is common in tropical latitudes. Rain bands are regions of heavy precipitation which exist in, and propagate through, larger mid-latitude frontal systems. The tropical cyclone, typhoon or hurricane is a large mesoscale system, of the order of 100 km in diameter, associated with intense winds of up to $50 \, m \, s^{-1}$ and heavy precipitation. Vertical velocities in mesoscale systems may vary from $1 \, m \, s^{-1}$ in a tropical cyclone to more than $10 \, m \, s^{-1}$ in intense thunderstorms.

Smaller-scale systems, less than 1 km in size, tend to last no longer than a few minutes. The most ferocious small-scale phenomenon is the tornado which, although less than 1 km in diameter, may attain circulation velocities of between 50 and $100 \, m \, s^{-1}$. Vertical velocities tend to be of a similar order of magnitude. Waterspouts are small-scale funnel clouds which form over the sea. They are usually less intense but more numerous than tornadoes. Both tornadoes and waterspouts tend to have a cyclonic rotation which is derived from the circulation of the parent thunderstorm cloud. Small-scale whirlwinds, including the dust devil, may attain wind speeds of $10 \, m \, s^{-1}$ in a column a few tens of metres wide and about a hundred metres high, persisting for a couple of minutes. They are formed in the surface layer when the prevailing wind is light and the surface is being intensely heated. The temperature gradient close to the surface becomes very large, or superadiabatic, and the surface layer becomes unstable. The strong upward acceleration results in the spectacular growth of a vortex. Other small-scale phenomena include the circulation within a cumulus cloud, and the local eddying motion around buildings and obstructions. Beyond the small-scale phenomena are the microscale motions associated with the final dissipation of kinetic energy into thermal energy by viscosity.

6.4.2   Time-dependent oceanic motion

Ever since surface current maps were first produced in the nineteenth century, the variability of ocean currents has also been recognized. Indeed, on surface charts information is often included on the steadiness of the currents, in addition to their mean speed and direction. As detailed in Section 6.2, both current meter and float measurements have indicated the presence of variability not only in the surface layers of the ocean but also in the deepest abyssal layers. A small part of the variability is accounted for by predictable frequency variations such as tides and by inertial motion. Reference to Figure 6.7 will show that the majority of variations occur on time scales greater than a few days, with a peak energy between 50 and 200 days. Beyond this time scale there is an indication of a reduction in energy but suitable time series (i.e. time series sufficiently long to distinguish these low-frequency currents) are rare. However, there are seasonal fluctuations, for example the Somali current, which reverses with the monsoon, and thus variations at the annual period may be anticipated. Low-frequency wind forcing, on interannual and decadal time scales, is associated with very long-period fluctuations in the ocean.

The first recognition of the variability of the deep ocean came with the Aries experiment west of Bermuda in 1959 when 72 neutrally buoyant floats were individually tracked for 4–10 days over an experimental period of 14 months. It had been expected, from indirect measurements of hydrographic properties, that the floats would travel northwards with a steady velocity of approximately $1 \, \text{cm s}^{-1}$. However, the floats drifted away with velocities of $10 \, \text{cm s}^{-1}$ in all directions, much to everyone's surprise. From the apparently random pattern of the buoy tracks, it was possible to deduce that energetic mesoscale eddies with length scales of 50–100 km were present.

Subsequent to this early experiment, many neutrally buoyant and SOFAR floats have been tracked, and a more general picture of the mean deep-ocean flow has emerged. Figure 6.10 shows the trajectory of a single SOFAR float over a period of one year in the North Atlantic during the MODE experiment. It can be seen that the float is caught in Eddy A for 27 days and Eddy B for 28 days. The diameter of Eddy A is about 60 km, and therefore the mean speed around this eddy is about $10 \, \text{cm s}^{-1}$. This float also shows a tendency to drift towards the south-west with a gentler speed of about $1 \, \text{cm s}^{-1}$ over the year.

When many more floats were tracked and their trajectories plotted, although the picture was complicated, the westward drift of the floats was very discernible. Current meter data from moorings in the same region show current speeds of between $20 \, \text{cm s}^{-1}$ close to the surface to $10 \, \text{cm s}^{-1}$ near the bottom. The variability of the current has periods ranging from three to about six weeks, which is typical of mesoscale eddies. These measurements, taken in the 1970s in a relatively small region of the Atlantic Ocean, have shown that not only is there variability in the ocean, but also that the current variability is about one order of magnitude larger than the mean velocities of the flow (see Figure 6.5b) This is due to mesoscale eddies and thus the kinetic energy per unit volume of the eddy is probably about 100 times larger than that of the mean circulation. Similar studies in other regions of the

North Atlantic Ocean and in other ocean basins have revealed the presence of energetic mesoscale eddies.

Long-term float measurements, which have been averaged over 100 days to remove the eddy field, have shown details of the large-scale circulation. Figure 6.3a clearly shows the trajectories of floats in the anticyclonic flow of the large sub-tropical gyre (i.e. clockwise, in the Northern Hemisphere). A few floats have been caught up in the intense jet-like flow of the Gulf Stream. The velocity of one such float has a mean velocity of $25 \, \text{cm s}^{-1}$ in a 100-day interval.

Other important scales of variability exist in the ocean. On the larger scale, seasonal and long-period wind variations can give rise to time variations in currents, such as those observed in the Indian Ocean. Only recently have long-period current measurements been obtained, but they have already shown that remarkable oscillations in supposedly steady currents do exist. In 1974, current moorings on the equator showed significant 15-day oscillations in the flow of the equatorial undercurrent. The meanders are apparently the result of a hydrodynamic instability of the undercurrent and it has also been found that it varies on an interannual time scale, in response to changes in wind circulation. In 1982–83, the equatorial undercurrent in the eastern Pacific Ocean virtually disappeared for a period of two months as a result of a reversal in the surface wind from westwards to eastwards. Meanders in surface equatorial currents, about 1000 km long, have been observed by infrared satellites measuring sea surface temperatures. Furthermore, large eddies have recently been discovered in the Somali current during the south-west monsoon. These eddies are rather larger than the mesoscale eddies described previously. They are about 500 km in diameter and tend to be slower moving.

On smaller scales, less than 100 km, surface frontal systems are observed which are energetic for their scale. Fronts form around coastal upwelling features and form the boundary between the cold upwelled water and the warmer surrounding water. They are also found along the edge of the continental shelf break at the boundary between oceanic water and shelf water. Convergent and divergent surface flows will also tend to result in frontal discontinuities and may be associated with surface wind patterns as, for example, the Antarctic divergence and the sub-tropical convergence zones in the Southern Ocean. Mesoscale fronts may also be associated with a developing eddy such as a Gulf Stream ring, where a frontal discontinuity develops about the edge of the ring separating the warm Sargasso Sea water from colder, fresher shelf water.

Small regions of deep convection have been discovered in the Mediterranean and Labrador Seas and the Antarctic Ocean, and result from intense surface cooling. The convection occurs on horizontal scales of 10 km and may extend down to the ocean bottom, thus providing a source of deep, cold water masses. Measurements with a vertical current meter have shown vertical velocities of $10^{-2}\,\mathrm{m\,s^{-1}}$ in these convection regions. Such velocities are of the order of 100 000 times larger than the mean vertical velocity of the general ocean circulation, as inferred from tracers. However, though these deep convective events are relatively infrequent and very localised, they are very important for their contribution to the global thermal-haline circulation.

7

The Influence of the Earth's Rotation on Fluid Motion

7.1 An introduction to the Earth's rotation

The Earth is a rapidly rotating planet. A point on the Earth's surface at the equator has a velocity of $463 \, \mathrm{m \, s^{-1}}$ relative to an absolute frame of reference such as the position of the stars. By comparison, our nearest planetary neighbours Venus and Mars have absolute velocities on their equators of 1.3 and $239 \, \mathrm{m \, s^{-1}}$ respectively. The velocities of the atmosphere and ocean relative to the rotating Earth are small, and therefore a parcel of air or water will only travel a small distance on the globe compared with the movement of the Earth. Under these circumstances, it is to be expected that the rotation of the Earth will have a profound influence on the circulation of both atmosphere and ocean.

First, consider the effect of rotation on a stationary particle on the Earth's surface. The particle will be subjected to the acceleration due to gravity, g, and a centrifugal acceleration in the direction perpendicular to the axis of rotation, as illustrated in Figure 7.1. At the point P the outward centrifugal acceleration is $\Omega^2 r_1$, where $r_1 = r_e \cos \phi$. At the equator this acceleration is $0.034 \, \mathrm{m \, s^{-2}}$, compared with an average acceleration due to gravity of $9.81 \, \mathrm{m \, s^{-2}}$. The resultant vector, known as the 'effective gravity', is directed at an angle of $0.1°$ to the true direction of gravity. In both meteorology and oceanography, the direction of gravity is that of the resultant gravity vector and therefore the centrifugal acceleration is implicitly taken into account. The local vertical is also assumed to be aligned with the resultant gravity vector.

In Chapter 1 it was shown that a geopotential surface is a surface of constant gravitational energy and, by definition, the component of gravity is perpendicular to this surface. Due to the variation of the local component of g with position, it is necessary to define a level surface in terms of a geopotential surface rather than a geometric surface. The geopotential, Φ, is the work per

The Atmosphere and Ocean: A Physical Introduction, Third Edition. Neil C. Wells.
© 2012 John Wiley & Sons, Ltd. Published 2012 by John Wiley & Sons, Ltd.

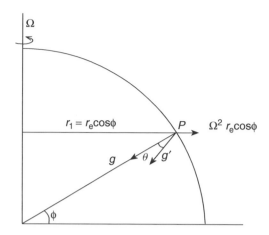

Figure 7.1   Centrifugal acceleration on a particle (P) stationary with respect to the Earth. g' (apparent gravity vector) is the resultant of the centrifugal acceleration and gravity, g. The radius of the Earth is r_e and the latitude is Φ

unit mass in moving a parcel of fluid a vertical distance, z, against gravity and is given by:

$$\Phi = \int_0^z g\,dz' \qquad (7.1)$$

where z is the geometric height relative to mean sea level and z' is a dummy variable because it does not appear in the final evaluation of the integral. It is noted that the geopotential has units of energy. The acceleration due to gravity varies horizontally from the pole to the equator by about 0.5% and in the vertical by 0.23% per 10 km from sea level.

In both oceanography and meteorology, a common unit for the geopotential is the dynamic meter. In the atmosphere it is defined as:

$$Z = \Phi/10 \qquad (7.2)$$

and in the ocean as:

$$D = -\Phi/10 \qquad (7.3)$$

The negative sign appears because it is measured downwards from the surface of the ocean.

The geopotential, or dynamic height, is related to the pressure, p, by the hydrostatic equation (see 2.4):

$$\alpha dp' = -g\,dz' = -d\Phi \qquad (7.4)$$

Note α is the specific volume which is defined as the inverse of density $(\alpha = \frac{1}{\rho})$.

For the ocean, the dynamic height at pressure p with reference to sea level in a stationary ocean is:

$$\int_{p}^{0} \alpha dp' = -\int_{h}^{0} g dz' = -\Phi \tag{7.5}$$

Therefore:

$$D = \frac{1}{10} \int_{p}^{0} \alpha dp' \tag{7.6}$$

It is assumed in the ocean that the pressure at sea level is zero. The geopotential depth, D, can therefore be calculated from numerical integration of the profile of the specific volume, α, obtained from *in situ* measurements of temperature, salinity and pressure.

In Section 7.3 the dynamic height is used, in the dynamic method, to calculate ocean currents.

The influence of rotation on a moving particle on the Earth's surface will now be considered. A moving particle will be subjected to two forces: the centrifugal force and the Coriolis force. The centrifugal force arises for both fixed and moving particles relative to the Earth's rotation as discussed previously. The Coriolis force arises only when the particle is moving relative to the rotating Earth.

The Coriolis acceleration is a vector that acts perpendicular to both the direction of motion of a particle and to the direction of the axis of rotation of the Earth. The vector acceleration, a, is given by the vector product:

$$a = 2\Omega \times v \tag{7.7}$$

On a rotating planet the axis of rotation is only aligned in the direction of the vertical axis at the poles, whilst elsewhere it is at an angle of $90 - \Phi$, where Φ is the latitude, as shown in Figure 7.1. Therefore, to obtain the Coriolis acceleration on a locally horizontally moving particle, it is necessary to take the component of rotation in the direction of the local vertical axis. From Figure 7.1, this component is $\Omega \sin \theta$ and thus, for horizontal motion, the Coriolis acceleration is dependent on the latitude.

At the equator the local vertical is perpendicular to the axis of rotation and therefore, for horizontal motion, the Coriolis acceleration is absent. It will be noted that, in general, vertical motion will be subjected to a Coriolis acceleration as a result of the component in the local horizontal direction. However, because of the small vertical accelerations observed in both atmosphere and ocean, and the dominance of gravitational and vertical pressure accelerations, this component of the Coriolis force is usually neglected.

For a particle, moving relative to a rotating Earth, an additional centrifugal component will arise. However, because of the relatively low velocity of the atmosphere and ocean, this centrifugal component is small and it is often neglected. For a horizontal motion of $10\,\mathrm{m\,s^{-1}}$ at 45° latitude, the Coriolis acceleration is $1.03 \times 10^{-3}\,\mathrm{m\,s^{-2}}$, which compares with a centrifugal acceleration of $2.21 \times 10^{-5}\,\mathrm{m\,s^{-2}}$. It can therefore be seen that the centrifugal force is only about 2% of the Coriolis acceleration. However, both accelerations are small compared with the acceleration due to gravity.

The Coriolis force on a horizontally moving particle always acts at right angles to the direction of motion. In the Northern Hemisphere it acts to the right of the motion whilst in the Southern Hemisphere it acts to the left of the motion.

7.2   Inertial motion

A practical demonstration of the existence of the Earth's rotation can be made by observing the rotation of a Foucault pendulum.

Figure 7.2 shows a pendulum oscillating along a line AA' at the North Pole. Assume that there is an observer a small distance from the pole on the Greenwich or Universal Meridian. The pendulum will be unaffected by the Earth's position but the observer will be rotated with the Earth. To the

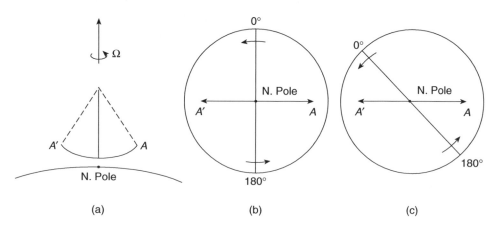

Figure 7.2   Foucault pendulum: (a) oscillating at the North Pole; (b) and (c) apparent rotation of the pendulum relative to a rotating Earth

observer, the plane of the pendulum's oscillation will apparently rotate in the opposite direction to the rotation of the Earth. After six hours the observer will have rotated by 90° and will be aligned with the pendulum, and after 12 hours the pendulum will be observed to oscillate in the original direction. This period of the rotation of the pendulum is known as the pendulum day or the inertial period.

At the poles the period is 12 hours but, because it is determined by the component of rotation along the local vertical axis, it will increase as the latitude decreases. At a latitude Φ, the period of rotation will be $12/\sin\Phi$ hours. At the equator, an east-west oscillating pendulum will be aligned in the same direction as the rotation of the Earth and thus there will be no apparent rotation relative to the Earth. However, if a pendulum oscillated in a north-south plane across the equator, it will tend to rotate in a clockwise direction in the Northern Hemisphere and in an anti-clockwise direction in the Southern Hemisphere. It will thus produce a figure of eight pattern.

It will now be shown that parcels of fluid on a rotating Earth will exhibit very similar behaviour to that of the pendulum. In a non-rotating system, from Newton's second law, a body will continue in a state of uniform motion unless acted upon by a force. For inertial motion in an absolute frame of reference relative to the stars, the rate of change of velocity of the fluid parcel is given by:

$$\frac{D_a}{Dt}(v) = 0 \tag{7.8}$$

where the subscript a refers to an absolute frame of reference. In Section 7.1 it was shown that there is an additional Coriolis acceleration in a rotating frame of reference, where the equivalent relation to equation 7.8 may be written as:

$$\frac{D_r}{Dt}(v) + 2\underline{\Omega} \wedge \underline{v} = 0 \tag{7.9}$$

where $(D_r/Dt)(v)$ is the total acceleration of a fluid element as measured on the rotating Earth. The second term is the Coriolis acceleration, which always acts at 90° to the fluid motion.

Equation 7.9 may be written for the two horizontal components of motion:

$$\frac{Du}{Dt} - fv = 0 \tag{7.10}$$

$$\frac{Dv}{Dt} + fu = 0 \tag{7.11}$$

where u is the velocity in the eastward direction, v is the velocity in the northward direction and $f = 2\,\Omega \sin\theta$. The subscript r, denoting a rotating frame of reference, has been dropped and, from now on, it will be assumed that all motion is relative to the rotating Earth. The main point to be noted here is that a fluid element, or any other body, cannot continue in a state of uniform motion on a rotating Earth, even if all frictional forces and pressure gradient forces could be eliminated.

Equations 7.10 and 7.11 will now be used to deduce the trajectory of a fluid element initially moving northwards with a velocity v in the Northern Hemisphere where $f > 0$. From equation 7.10 it will be seen that the northward motion will induce an acceleration towards the east, because the Coriolis acceleration in the Northern Hemisphere always acts to the right of the direction of motion. The parcel's trajectory will curve towards the east, because of the eastward acceleration, and it will obtain a positive u component of velocity. From equation 7.11 the positive u component of velocity will cause a Coriolis acceleration southwards, and this negative acceleration will decrease the original northward component. At some stage the parcel will have zero northward velocity and it will then have only an eastward component. After this time it will continue to accelerate southwards and it will acquire a negative v component of velocity which, from equation 7.10, will induce a westward acceleration in its turn. Thus, the parcel will follow a circular trajectory and, providing that the change of latitude of the parcel is very small, it will return to its original position. This circular trajectory is known as an inertia circle.

Eliminating v from equations 7.10 and 7.11 yields:

$$\frac{D^2u}{Dt^2} + f^2u = 0 \tag{7.12}$$

This is the equation for simple harmonic motion with the frequency f, the Coriolis parameter. The period of oscillation is:

$$T_i = \frac{2\pi}{f} \quad \text{or} \quad \frac{\pi}{\Omega \sin \phi} \tag{7.13}$$

It can be shown that the period of the motion is $12/\sin \Phi$ hours, as for the Foucault pendulum.

Inertial oscillations are frequently observed in the ocean and they can always be identified by spectral analysis of a current meter time series, as shown in Chapter 6. Figure 7.3 shows an example of an inertial oscillation recorded by a current meter in the Baltic Sea, identified by plotting a progressive vector diagram. The theoretical period of 14 hours 8 min was confirmed by these observations. The radius of the inertia circle is $R = \frac{V}{f}$, where V is the current speed and f is Coriolis parameter. For a current of $V = 0.2\,\mathrm{m\,s^{-1}}$ and using the theoretical period (equation 7.13), the radius of the circle is 1.6 km, which is a similar scale to that shown in Figure 7.3.

In the atmosphere, inertial observations are only usually obtained at times when pressure gradients are weak and there is a stable vertical stratification

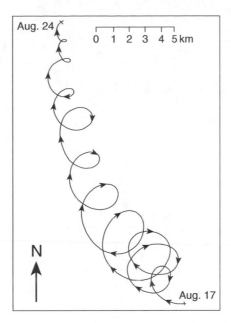

Figure 7.3   Progressive vector diagram illustrating an inertia current superimposed on a northwards motion, observed in the Baltic Sea between 17 and 24 August 1933. Reproduced, with permission, from Neumann, G. and Pierson, W., 1966, Principles of Physical Oceanography, Prentice-Hall: page 159, figure 7.19

which reduces the effects of surface friction. These conditions tend to occur in anticyclonic flow around a high pressure system at night. The oscillations will increase the local wind speed when moving in the direction of the large-scale flow but they will decrease it when moving in the opposite direction. Such oscillations can produce local nocturnal jets in the atmospheric boundary layer which have velocities of up to twice the large-scale geostrophic flow (see Section 7.3).

Figure 7.4 shows theoretical examples of inertial flow in the atmosphere where the motion is assumed to extend over a large enough range of latitude to produce a significant change in the Coriolis parameter, f. It is noted that a variety of patterns across the equator occur, depending on the initial speed and direction of the fluid element. None of the trajectories form closed loops because of the latitudinal variation of f.

7.3   Pressure gradients and geostrophic motion

The principal force acting in both atmosphere and ocean is the pressure-gradient force. It has already been shown (Section 2.4) that, in the vertical

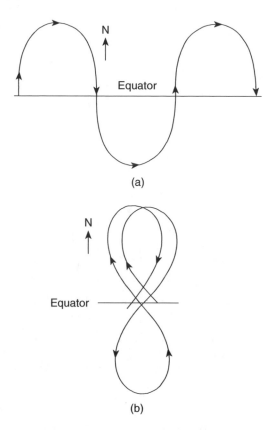

Figure 7.4   Theoretical examples of inertial oscillations about the equator. (a) Initial northward impulse. (b) Initial north-west impulse

direction, the dominant balance of forces, known as the hydrostatic balance, is between the upward pressure-gradient force and the downward gravitational force. In the horizontal plane for large scale motion the principal balance of forces is between the Coriolis force and the horizontal pressure-gradient force. This balance is known as geostrophic equilibrium.

The horizontal equations of motion, assuming that there is no friction, are now given by:

$$\frac{Du}{Dt} - fv = -\frac{1}{\rho_0}\frac{\partial p}{\partial x} \tag{7.14}$$

$$\frac{Dv}{Dt} + fu = -\frac{1}{\rho_0}\frac{\partial p}{\partial y} \tag{7.15}$$

The right-hand side terms are the horizontal components of the pressure-gradient acceleration and they can be derived in a similar way to that shown in Section 2.4.

For non-accelerating flow on a rotating Earth, Du/Dt and Dv/Dt are both zero, and hence:

$$-fv_g = -\frac{1}{\rho_0}\frac{\partial p}{\partial x} \qquad (7.16)$$

$$+fu_g = -\frac{1}{\rho_0}\frac{\partial p}{\partial y} \qquad (7.17)$$

where u_g and v_g are the eastward and northward components of the geostrophic velocity respectively. Note the density, ρ_0, is assumed constant. This is because density variations are much smaller than velocity variations in the horizontal plane.

Consider a situation in the Northern Hemisphere where the atmospheric pressure at sea level decreases southwards, as shown in Figure 7.5a. The pressure-gradient acceleration will also be directed southwards since $(1/\rho_0)\partial p/\partial y$ is positive and to produce a balance of forces, an equal and opposite directed Coriolis force is required acting northwards. A westward geostrophic flow will result, because the Coriolis force is perpendicular to the direction of flow (i.e. to the right in the Northern Hemisphere). In the Southern Hemisphere the Coriolis force acts to the left of the motion and therefore a southward pressure gradient will result in an eastward geostrophic flow as shown in Figure 7.5b. In general, clockwise flow occurs around Northern Hemisphere high-pressure and Southern Hemisphere low-pressure systems, whilst anticlockwise flow occurs around Northern Hemisphere low-pressure and Southern Hemisphere high-pressure systems. Curvature of the flow causes an additional centrifugal force but for large scale synoptic systems, such as flow around an anticyclone, the influence of the centrifugal force is small. Normally the term anticyclonic flow refers to

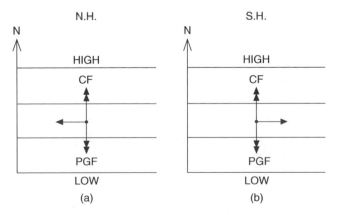

Figure 7.5 Balance of forces in geostrophic flow: (a) in the Northern Hemisphere, (b) in the Southern Hemisphere. CF is the Coriolis force and PGF is the horizontal pressure-gradient force

the flow around a high-pressure centre and cyclonic flow to low-pressure centres. A simple rule for geostrophic motion is that the Coriolis force is always directed, following the motion, 90° to the right in the Northern Hemisphere and 90° to the left in the Southern Hemisphere.

It must be remembered that geostrophic flow in the atmosphere and in the ocean is only an approximation to the actual horizontal flow. For example, rapid changes in the pressure gradient may occur which would induce inertial accelerations and non-geostrophic flow. In addition, frictional drag of the flow over the Earth's surface will induce deviations from geostrophic flow. On the equator there is no Coriolis force, and therefore geostrophic flow is impossible. Above the atmospheric boundary layer (i.e. approximately 1 km above the Earth's surface) away from the equator, in large- and synoptic-scale systems, geostrophic flow is a good approximation, being accurate to about 10% of the observed flow. In smaller-scale systems, such as thunderstorms and sea breezes, there are large departures of the winds from geostrophic motion. In the upper troposphere in intense jet streams, where velocities can reach $100 \, \text{m s}^{-1}$, the deviations from geostrophic flow may be large. An extreme example where the flow is not geostrophic is the horizontal flow in a tornado. Here the horizontal pressure gradient tends to be balanced by the local centrifugal force rather than the Coriolis force.

In the ocean, in contrast to the atmosphere, the geostrophic approximation can be applied to smaller-scale systems such as mesoscale eddies with a length scale greater than 50 km as well as to large-scale flows. However, significant departures from geostrophic motion do occur in the surface mixed layer of the ocean due to local wind-driven currents and also within 100 m of the bottom, particularly where there are large variations in bottom topography. Intense western boundary currents (e.g. the Gulf Stream) may also show departures from geostrophic balance.

Some examples of geostrophic motion in the atmosphere and in the ocean will now be considered. Equations 7.16 and 7.17 can be used directly in conjunction with observations of pressure on a horizontal surface to obtain the geostrophic wind in the atmosphere. Accurate measurements, to within about 0.1 hPa, can be made of the surface pressure using accurate barometers. Horizontal pressure gradients in middle latitudes are of the order of 10–50 hPa over a distance of 1000 km. Geostrophic winds can therefore be calculated to an accuracy of at least 1%. In practice, pressure observations have to be corrected to mean sea-level pressure (1013 hPa) using the hydrostatic equation. This correction may introduce further sources of error because a temperature has to be assumed for the fictitious air mass between the ground height and sea level. Above the Earth's surface, pressure measurements are obtained from aneroid barometers inserted into a radiosonde package. In these expendable packages, the accuracy is about 1 hPa. Of course, the height of the balloon at a particular pressure has also to be determined before the

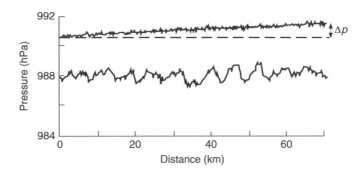

Figure 7.6   Horizontal pressure gradients measured by an aircraft over the ocean before (lower curve) and after (upper curve) correction for altitude variations. The resulting pressure difference, Δp. can be used in geostrophic wind calculations. Reproduced, with permission, from Nicholls, S. *et al.*, 1983, in Results of the Royal Society Joint Air-Sea Interaction Project (JASIN), The Royal Society: page 299, figure 3

geostrophic wind equation can be used. From the simultaneous temperature measurements made by the radiosonde package, the hydrostatic equation 7.5 can be numerically integrated from the surface at a known pressure p_0 to the observed pressure p. The geopotential height of each pressure measurement is thus obtained.

Another method of deducing the horizontal pressure gradient above the ocean surface, illustrated in Figure 7.6, involves making simultaneous measurements of pressure and height from an aircraft. Radar altimeters allow geometric height measurement to an accuracy of 0.02 m, but this technique is ultimately limited by deviations of the sea surface from the mean sea level as a result of horizontal variations in the geoid and of ocean currents. The sea level would be a geopotential surface if there were no ocean currents.

In the ocean geostrophic currents could be measured by similar methods to those in the atmosphere, i.e. by measuring the pressure gradient on a horizontal surface, which is the geoid. The geoid has been measured by satellites but not at the fine resolution of about 10 km and accuracy required for ocean currents. The horizontal change in sea level across the Gulf Stream is about 1 m in a distance of 100 km, and this variation can be measured relative to the unknown geoid using a satellite altimeter to an accuracy of 2 cm (Figure 7.7). The relative variations in sea level from month to month can be measured, but in regions where we have permanent currents the problem of the measurement of the geoid remains.

Another method of measuring absolute geostrophic velocity, is to measure the horizontal pressure gradient at the bottom of the ocean. In recent decades, bottom-pressure recorders have been deployed in the Atlantic and Indian Oceans. Their principal objective has been to obtain short-period records of pressure variations associated with changes in sea level caused by ocean

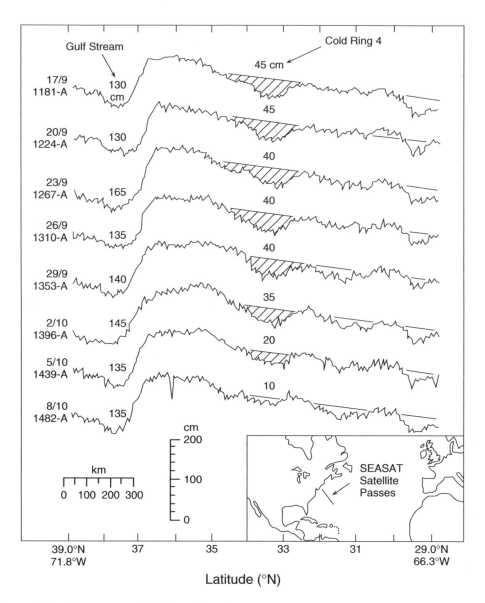

Figure 7.7 Altimeter residuals for eight collinear passes obtained over a three-week period by SEASAT in 1978 in the Gulf Stream region. Reproduced, with permission, from Cheney, R.E. and Marsh, J.G., 1981, Journal of Geophysical Research, CI, 86: page 480, figure 9

tides. The bottom pressure gauges can be deployed for up to a few years and potentially provide horizontal pressure gradients. The difficulty here is the position of the gauges. The gauges would have to be positioned to a high accuracy, of about 1 cm (1 hPa), and a very large number of gauges would have to be deployed.

Oceanographers have developed an indirect method of obtaining geostrophic currents known as the dynamic method. Unfortunately, this method has one important deficiency in that it can only be used to determine relative geostrophic currents. The method is based on the fact that, although it is not practical to obtain an absolute measurement of the pressure gradient at one level in the ocean, it is possible to determine the pressure gradient at one level *relative* to a second level by use of the hydrostatic equation. The vertical pressure difference between the two levels is dependent on the density, or specific volume, of the sea water which can be determined from temperature, salinity and pressure measurements. It has been shown in Section 7.1 that the geopotential difference between two pressure surfaces can be obtained from the integration of the specific volume, α, between the two surfaces. In order to use this information to calculate geostrophic currents, the pressure-gradient acceleration along a constant geopotential surface has to be transformed into the geopotential slope of a constant pressure, or isobaric, surface.

The horizontal pressure-gradient acceleration can be written as:

$$\frac{1}{\rho}\frac{\delta p}{\delta x} \tag{7.18}$$

From the hydrostatic equation (equation 2.4) it can be shown:

$$\delta p = -g\,\rho\,\delta z \tag{7.19}$$

From the definition of the geopotential $g\delta z = \delta\Phi$, and therefore:

$$\delta p = -\rho\delta\Phi \tag{7.20}$$

Substitution of equation 7.20 into 7.18:

$$\frac{1}{\rho}\frac{\delta p}{\delta x} = \frac{\delta\Phi}{\delta x} \tag{7.21}$$

In the limit as $\delta x \to 0$:

$$\frac{1}{\rho}\left(\frac{\partial p}{\partial x}\right)_z = \left(\frac{\partial\Phi}{\partial x}\right)_p \tag{7.22}$$

where $\left(\frac{\partial\Phi}{\partial x}\right)_p$ is the slope of the constant pressure surface, p.

The geostrophic current along a pressure surface p_1 is therefore given by:

$$fv_1 = \left(\frac{\partial\Phi_1}{\partial x}\right)_{P_1} \tag{7.23}$$

However, it is not possible to determine $\left(\frac{\partial\Phi_1}{\partial x}\right)_{P_1}$ because there is no reference geopotential surface. Although the sea level of a stationary ocean would be a geopotential surface, currents and tidal variations produce significant departures from the geoid. The geostrophic current at a second pressure level $(p_2 > p_1)$, is

$$fv_2 = \left(\frac{\partial \Phi_2}{\partial x}\right)_{p_2} \tag{7.24}$$

Subtracting equation 7.24 from 7.23:

$$v_1 - v_2 = \frac{1}{f}\frac{\partial}{\partial x}(\Phi_1 - \Phi_2) \tag{7.25}$$

From equation 7.5 the difference in geopotential, $\Phi_2 - \Phi_1$, between the two pressure surfaces is:

$$\Phi_1 - \Phi_2 = \int_{p_1}^{p_2} \alpha dp \tag{7.26}$$

or alternatively:

$$\Phi_1 - \Phi_2 = 10\Delta D \tag{7.27}$$

where ΔD is the dynamic height between the two pressure surfaces measured in dynamic meters which has units $m^2 \ s^{-2}$. A dynamic meter is numerically equivalent to 100 hPa or approximately 1.02 m depth. Thus:

$$v_1 - v_2 = \frac{10}{f}\frac{\partial}{\partial x}(\Delta D) \tag{7.28}$$

A similar equation can be derived for the u component of flow.

The dynamic height ΔD is determined from the integrated specific volume between two pressure surfaces. The specific volume can be determined from measurements of temperature, salinity and pressure from a CTD profile.

Many measurements of the dynamic height over the ocean basins have been made and its spatial variation is known reasonably well. To obtain absolute ocean currents requires independent measurement of v_1 or v_2. This can be done by:

(i)   Measurement of the current v_2 at a given depth.
(ii) Determination of a depth of no motion, i.e. $v_2 = 0$, for example, from a tracer experiment.
(iii) Measurement of the total transport through a confined passage using, say, an underwater cable (see Section 6.1). This requires a vertical profile of the relative geostrophic velocity from the surface to the bottom, and the profile is obtained by repeated applications of equation 7.28. The

geostrophic velocity at each depth can then be adjusted by a constant velocity to give the observed transport over the total depth.

Method (i) can only be used where long-period measurements are available. Method (ii) has been used widely, but it can give contradictory results. In some regions, such as the Antarctic Circumpolar Current, the current is eastward to the bottom and therefore there is an absence of a level of no motion. Method (iii) has been used to determine the geostrophic flow in the Gulf Stream across the Florida Straits but it is not of general use in the open ocean.

Another method, used in the MODE experiment, is to use a combination of float information, current meter readings and dynamic heights to obtain an objective analysis of the absolute geostrophic flow. Figure 7.8 shows a

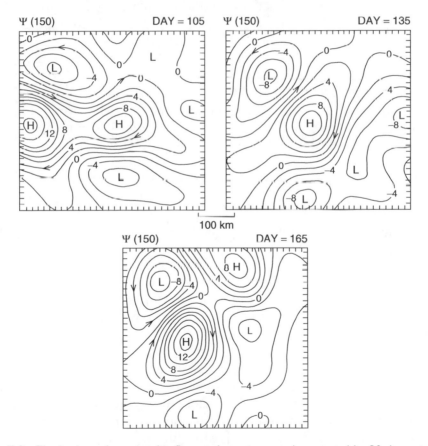

Figure 7.8   The horizontal geostrophic flow at three times, each separated by 30 days, at a depth of 150m. Tick marks on the side indicate the interpolated grid interval (15 km) and the map centre is at 69°40'W, 28°N. Reproduced, with permission, from McWilliams, J.C., 1976, Journal of Physical Oceanography, 6: page 813, figure 1

sequence of mesoscale eddies over a period of 60 days in a 500×500 km box in the western North Atlantic Ocean. The flow direction can be determined in exactly the same way as for atmospheric geostrophic flow, i.e. anticyclonic flow around the high-pressure areas and cyclonic flow around the low-pressure areas.

Finally, mathematical models (Chapter 10) can assimilate hydrographic data and produce reasonable estimates of ocean currents or inverse models can be used.

7.3.1 Thermal wind

In Section 6.4 the relationship between the wind circulation and the horizontal wind circulation was mentioned without any explanation. Having discussed the geostrophic wind it is now appropriate to present the 'thermal wind'. The thermal wind is the vertical variation of the geostrophic wind (or the geostrophic wind shear) and it is directly related to the horizontal temperature gradient in the atmosphere. In the atmosphere, as in the ocean, the distance between two pressure surfaces is the geopotential distance or height. This is shown in equation 7.25 which will be shown to be applicable to the atmosphere. The specific volume is the inverse of density and can be related, for the atmosphere, to the temperature and pressure by the ideal gas equation 2.4. We can therefore replace the specific volume by its temperature and pressure and integrate between the two pressure surfaces.

$$\Phi_2 - \Phi_1 = - \int_{p_1}^{p_2} RT \frac{dp}{p} = R\overline{T} \ln(\frac{p_2}{p_1}) \qquad (7.29)$$

where $\overline{T}$ is the average temperature between the two pressure surfaces.

In similarity to the ocean, the difference in the geostrophic wind between two pressure surfaces is related to the gradient in the dynamic height. In this case the dynamic height is related directly to the mean temperature between the two pressure surfaces. Therefore the horizontal gradient in mean temperature is proportional to the vertical difference in the geostrophic wind between two pressure surfaces. This is known as the geostrophic wind shear or thermal wind. The large wind shear between the surface and 200 hPa shown in Figure 6.11 is directly related to the large meridional temperature gradient in the troposphere at the middle latitudes.

7.4   Vorticity and circulation

A useful property of fluid flows in both ocean and atmosphere is the spin, or vorticity, of a fluid element. Vorticity is defined as the spin of an infinitesimal element of fluid about its own axis, as illustrated in Figure 7.9. It is a vector quantity and directed along the axis of rotation perpendicular to the surface of the fluid element. Positive vorticity is defined as anticlockwise rotation. In both the ocean and the atmosphere, the large-scale flow is predominantly horizontal because vertical velocities are, in general, much smaller than horizontal velocities. It is, therefore, the vertical component of vorticity that arises from the horizontal motion which is most useful for aiding an understanding of the larger-scale motions.

The vertical component of vorticity, ζ, is defined as

$$\zeta = \frac{\partial v}{\partial x} - \frac{\partial u}{\partial y}$$

Where u and v are the horizontal components of flow in the x (eastward) and y (northward) directions respectively.

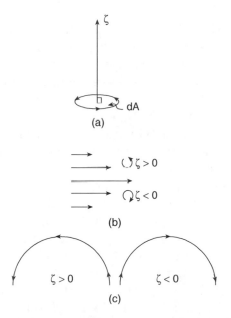

Figure 7.9   (a) Vertical component of vorticity, ζ, in a horizontal flow. (b) Vorticity in a sheared flow. (c) Vorticity in a curved flow

Figure 7.9 shows that vorticity can arise from two distinct flows:

(i)   Circular flow.
(ii)   Shear flow.

In an absolute frame of reference, in the absence of vorticity sources and sinks, a tube of vorticity is conserved as it moves through the fluid:

$$\frac{D_a \eta}{Dt} = 0 \tag{7.30}$$

In the rotational frame of reference, the vertical component of absolute vorticity, η, is the sum of the relative vorticity (i.e. the vorticity seen by an observer on the Earth's surface) and the planetary vorticity due to the spin of the Earth. For horizontal fluid motion the planetary vorticity is the component of vorticity in the direction of the local vertical, which is the Coriolis parameter, f, where $f = 2\Omega \sin \Phi$.

In the rotational frame of reference, equation 7.30 becomes

$$\frac{D_r}{Dt} (\zeta + f) = 0 \tag{7.31}$$

where ζ is the relative vorticity.

Consider a non-rotational tube of fluid on the equator. The tube has neither relative nor planetary vorticity. However, if the tube is transferred polewards, it must gain planetary vorticity appropriate to its new latitude. According to equation 7.30, its absolute vorticity, as perceived by a non-rotating observer, must remain zero and so the tube of fluid must have gained a component of relative vorticity of the same magnitude as its planetary vorticity but opposite in sign. In the Northern Hemisphere the Coriolis parameter, f, is positive, and therefore the acquired relative vorticity is negative and the rotation of the tube of fluid is clockwise. Hence, changes in relative vorticity will be induced by meridional flow due to the conservation of absolute vorticity.

The following two examples will illustrate the application of this conservation principle to flows in the ocean and in the atmosphere.

First, consider a simple poleward movement of fluid which initially has zero relative vorticity at latitude Φ_0. As shown in Figure 7.10, the fluid will gain negative relative vorticity as it progresses polewards in the Northern Hemisphere. The negative vorticity will induce a clockwise curvature in the parcel's trajectory. This will cause the parcel to turn first eastwards, then southwards. As the parcel moves equatorwards it will tend to lose its negative relative vorticity and, on reaching its original latitude, it will again have zero relative vorticity and its direction will be equatorwards with no curvature. At lower latitudes it will gain positive relative vorticity, in order to maintain the absolute vorticity at its original latitude, and this will, in turn, induce

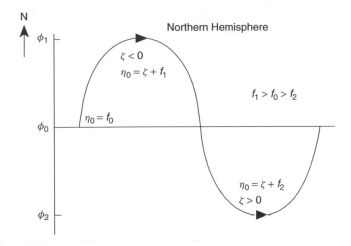

Figure 7.10   Conservation of absolute vorticity in a Rossby wave (see text)

an anti-clockwise curvature to the trajectory which will eventually rotate the parcel's motion back to a poleward direction. It can therefore be seen that the interchange of relative and planetary vorticity, required to conserve the original absolute vorticity, will produce a horizontal oscillation about the original latitude. These oscillations are known as Rossby waves, or planetary waves, and they can be readily identified in the trajectories of constant-level balloons in the atmosphere (Figure 6.9).

The Lagrangian derivative in equation 7.31 can be expanded into the Eulerian derivative form thus:

$$\frac{\partial \zeta}{\partial t} + u\frac{\partial \zeta}{\partial x} + v\frac{\partial \zeta}{\partial y} + \beta v = 0 \tag{7.32}$$

where $\frac{\partial \zeta}{\partial t}$ is the local change in relative vorticity, and the second and third terms represent the transport of relative vorticity by the horizontal flow. $\beta = \frac{\partial f}{\partial y}$ and thus βv is the change in relative vorticity caused by meridional velocity, v. The β term is the change of f (Coriolis parameter) with latitude, which is assumed constant. This is true provided the north-south displacements are small. Furthermore β has a positive sign in both northern and southern hemispheres.

Consider a distribution of meridional velocity given by:

$$v = v_0 \sin(\sigma t - kx) \tag{7.33}$$

where σ is the frequency and k is the wavenumber in the x direction. The period of the wave, (T) is $2\pi/\sigma$. and its wavelength (L) is $2\pi/k$. Equation 7.33 represents a sinusoidal variation of the meridional velocity at a particular latitude (See Section 8.1 for more discussion on waves and their properties). The relative vorticity, ζ, is given by:

$$\zeta = \frac{\partial v}{\partial x} = -kv_0 \cos(\sigma t - kx) \tag{7.34}$$

Substituting into equation 7.32:

$$+k\sigma v_0 \sin(\sigma t - kx) + 0 + 0 + \beta v_0 \sin(\sigma t - kx) = 0$$

Thus:

$$\sigma k + \beta = 0$$

or

$$\sigma = -\frac{\beta}{k} \tag{7.35}$$

The phase velocity is:

$$c = \frac{\sigma}{k} = -\frac{\beta}{k^2} = -\frac{\beta L^2}{4\pi^2} \tag{7.36}$$

The phase velocity of the Rossby wave is always westwards (as shown by the negative sign in equation 7.36) and it increases in magnitude as the wavelength (L) of the wave decreases. Therefore waves having a long wavelength will travel faster than those having a short wavelength.

Table 7.1 shows the phase velocity for different wavelengths at 45° latitude. In the middle-latitude atmosphere, the mean tropospheric flow is eastwards and therefore the Rossby waves will move either eastwards or westwards relative to the Earth, depending on the relative magnitudes of the tropospheric flow velocity and the Rossby wave velocity. If the eastward tropospheric flow is greater than the wave speed, the Rossby wave will move eastwards relative to the Earth and if it is less, then the Rossby wave will travel westwards. Very long planetary waves will, therefore, tend to travel westwards and short

Table 7.1 Phase speeds and periods of Rossby waves for typical wavelengths, derived from equation 7.36

Wavelength (km)	Period (days)	Phase speed (m s⁻¹)
100	284.0	−0.004
1000	28.0	−0.41
5000	5.6	−10.25
10 000	2.8	−41.00

waves will tend to move eastwards. Should the eastward tropospheric flow match the westward wave speed, the wave will become stationary. The most likely stationary wavelengths lie between 5000 and 10 000 km because their wave speeds correspond to the typical mean wind speeds in the troposphere, as shown in Figure 6.11. In the ocean, the large-scale Rossby waves (i.e. those with wavelengths greater than 1000 km) will generally move westwards because any opposing flow in the ocean will be slower than that in the atmosphere.

Rossby waves have been observed in the ocean, in tropical and mid-latitudes, from satellites measuring variations of sea level. They are fastest at low latitudes where their velocities are about $0.2\,\text{m s}^{-1}$. These waves are associated with vertical variations of the main thermocline, similar to the internal waves discussed in Section 8.4. At higher latitudes shorter Rossby waves ($\sim$100 km) wavelength interact with eastward flows, which can lead to 'blocking of the flow' as already discussed for the atmosphere. The eastward flow of the Antarctic Circumpolar Current is an example where there are strong interactions between the Rossby waves and the mean flow.

A second example of the application of the vorticity equation (equation 7.31) yields an explanation of the intense boundary currents which are located at the extreme western edges of the ocean basins and which include the Gulf Stream, the Kuroshio and the Agulhas Currents (see Figure 6.14). First consider the ocean circulation induced by the surface wind distribution. In subtropical latitudes, over all of the major ocean basins, there are semi-permanent, anti-cyclonic atmospheric flows. This anticyclonic wind stress will also drive an anticyclonic circulation in the upper 500 m of the ocean, by virtue of the vorticity input at the sea surface. The expected response of the ocean would be a mirror image of the overlying wind system and the maximum dynamic height would be in the centre of the ocean basin, with a broad poleward flow in the western half of the basin and an equatorward flow in the eastern half of the basin. Clearly, these expectations do not correspond to reality. However, if the relative vorticity changes caused by the initial 'mirror image' flow are considered, it will be seen that the poleward flow in the Northern Hemisphere always gains clockwise relative vorticity, and the equatorward flow will always lose clockwise relative vorticity. The equatorward flow in the eastern half of the basin will therefore tend to negate the clockwise vorticity input by the wind system. In the western half of the basin the poleward flow will always enhance the clockwise vorticity input of the wind. In time, the clockwise vorticity will become more concentrated in the western part of the basin and the gyre circulation will move from its central position towards the west. This increase in clockwise vorticity will enhance the poleward flow in the western portion of the basin and the cancelling of the clockwise vorticity in the eastern part of the basin will weaken the equatorward flow, as illustrated in Figure 7.11. Finally, an intense poleward

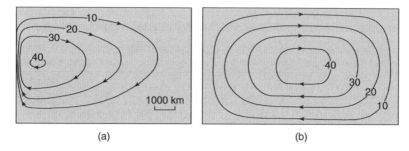

Figure 7.11   The circulation in a rectangular ocean of constant depth, induced by an anticyclonic surface wind stress. Streamlines show volume transport in 10^6 m^3 s^{-1} (a) The Coriolis parameter, f, changes with latitude and hence westward intensification of wind-driven currents. (b) The Coriolis parameter, f, remains constant and the gyre is symmetrical. Reproduced, with permission, from Stommel, H., 1966, The Gulf Stream, Cambridge University Press/University of California Press: page 92, figure 58

boundary current will form, with very large clockwise vorticity, against the western boundary of the basin. The question may be asked – what stops the poleward current from becoming more and more intense, as clockwise vorticity continues to be pumped into the ocean? In fact, as more vorticity is added, the intense western boundary current becomes dynamically unstable and energetic mesoscale eddies form. These eddies eventually move out of the intense system and dissipate the energy and vorticity of the western boundary current.

It can also be shown that an equatorward-flowing western boundary current will form for a cyclonic or anti-clockwise wind circulation in the northern hemisphere. In this case, the poleward flow on the eastern part of the wind-driven gyre will gain clockwise vorticity and it will therefore tend to oppose the anti-clockwise vorticity input by the wind. At the same time, equatorward flow on the western flank of the gyre will gain vorticity of the same sign as that of the wind. The narrow, southward Labrador current, on the western boundary of the North Atlantic sub-polar gyre, is an important example of an equatorward-flowing western boundary current. This intense, cold current is responsible for the rapid movement of icebergs from the Greenland glaciers into one of the major shipping lanes between North America and northern Europe.

The vorticity equation (equation 7.31) is only applicable to homogenous, horizontal flows of constant depth. However, in both the atmosphere and the ocean, vertical motion can have a significant influence on the pattern of vorticity and therefore on the circulation. First, consider a simple example in which a cylinder of fluid is rotating with an angular velocity Ω. If the cylinder is now stretched along the vertical axis, the diameter of the cylinder will decrease. This will reduce the moment of inertia of the cylinder and, to

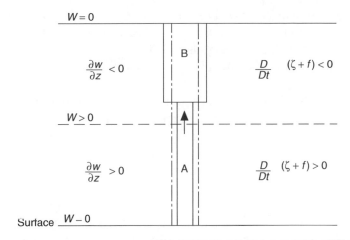

Figure 7.12 Absolute vorticity changes caused by upward motion in the troposphere. It is assumed that the tropopause acts as a lid to vertical motion

conserve angular momentum, the cylinder must increase its angular velocity. An analogous situation occurs when a vertical tube of vorticity, having an absolute vorticity η, is stretched along its vertical axis by the vertical velocity gradient ($\partial w/\partial z$) as shown in Figure 7.12. This can be written as:

$$\frac{D\eta}{Dt} = \eta\frac{\partial w}{\partial z} \tag{7.37}$$

It is noted that stretching, $\partial w/\partial z > 0$ will always increase the original vorticity, whilst compression, represented by $\partial w/\partial z < 0$, will reduce it. In the absence of vorticity, i.e. $\eta = 0$, vertical motion can neither generate nor destroy vorticity.

On a rotating Earth it has been shown that absolute vorticity $\eta = \zeta + f$ and hence:

$$\frac{D}{Dt}(\zeta + f) = (\zeta + f)\frac{\partial w}{\partial z} \tag{7.38}$$

An interesting aspect of this vorticity equation is that relative vorticity can be produced by vertical stretching or compression even in the absence of an original relative vorticity because of the vorticity of the rotating Earth.

Consider first an example from the atmosphere. Let us assume there is an upward motion field in the middle troposphere (Figure 7.12). If the upper troposphere acts as a 'lid' to upward motion (a simplistic assumption), then it can be seen that stretching of the lower troposphere and compression of the upper troposphere would occur. Let us assume that there is no original relative vorticity (i.e. $\zeta = 0$), then there will be an increase of positive absolute vorticity in the lower layer and a decrease in positive absolute vorticity in the upper layer of the troposphere. For small synoptic-scale motions, the change in f with latitude can be ignored. Since f is constant, only the relative vorticity (ζ) changes, and therefore an increase in anti-clockwise vorticity in the lower layer is expected. Similarly, an increase in clockwise vorticity in the upper troposphere may also be anticipated.

This type of behaviour is observed over continental regions in summer when surface heating induces a surface cyclonic circulation. Furthermore, the upward motion implies that air near the surface will converge into the circulation and, if this air is very moist, latent heat will be released on ascent. This will increase the buoyancy of the air and intensify the vertical and horizontal motions. Such behaviour also occurs in tropical cyclones over warm seas where very moist air converges into the cyclone and aids its development. Although other factors are responsible for initializing the tropical cyclone, this effect becomes dominant in its subsequent growth. It is noted that a sinking motion in the troposphere will have the opposite effect, producing an anti-cyclonic flow in the lower layer and a cyclonic flow in the upper layer. It can therefore be seen that vertical motion can generate both cyclonic and anticyclonic motions in the atmosphere.

On the larger, planetary scale, the change of the Coriolis parameter with latitude becomes an important term in the vorticity balance. In a steady state, where there is no growth with time of the circulation and the relative vorticity is smaller than the Coriolis parameter, then equation 7.38 can be reduced to:

$$\beta v = f \frac{\partial w}{\partial z} \qquad (7.39)$$

This equation says that the relative vorticity produced by vertical stretching is balanced by the relative vorticity lost by meridional motion, v. Hence, for steady poleward motion, vertical stretching of the lower troposphere will occur and, for equatorward motion, vertical compression will take place. Therefore, large-scale poleward movement of warm air will be associated with rising motion and the equatorward motion of cold air will be associated with descending motion in the middle troposphere. This simple picture goes some way towards explaining the patterns of vertical motion seen in large-scale Rossby waves, usually called baroclinic Rossby waves. Regions of upward motion occur on the upwind side of a pressure ridge and they are associated with the development of surface cyclonic circulations. Surface anticyclones tend to develop on the downwind side of the ridge where downward motion occurs.

Equation 7.39 has been applied in the ocean by Stommel and Arons (1960) to deduce the deep-water flow. It has been remarked earlier that regions of deep sinking motion in the ocean are very limited in horizontal extent and therefore, to balance this local sinking, upward motion is required over the remainder of the ocean. The observation of a permanent thermocline is evidence of the vertical motion of cold water opposing the downward mixing of warmer water. Hence, in the abyssal layers of the ocean, $\frac{dw}{dz}$ is positive and, from equation 7.39, poleward motion is expected. This result ran contrary to earlier ideas of deep-water circulation which had suggested that cold, dense water, formed by sinking in high latitudes, would spread uniformly equatorwards over the width of the ocean basin. Stommel and Arons reasoned that, if cold water moved polewards in the interior of the ocean as predicted by equation 7.39, then it must move equatorwards in a deep western boundary current.

Stommel (1958) produced a theoretical model of the abyssal circulation of the world's ocean, driven by sources of deep water in the North Atlantic Ocean and the Weddell Sea. This is shown in Figure 7.13. Subsequent evidence from all of the ocean basins has shown that these deep western boundary currents do exist. Their velocities are, however, modest when compared with surface western boundary currents. The flows are generally less than $10\,\mathrm{cm\,s^{-1}}$ and the widths of the currents are generally between about 100 and 200 km.

There are regions where the circulation shown in Figure 7.13 is misleading. In the South Atlantic Ocean, the North Atlantic Deep Water moves southwards along the western boundary, but this southward flow overrides the colder and more dense Antarctic Bottom Water which flows northwards along the same western boundary from the Weddell Sea. The bottom topography

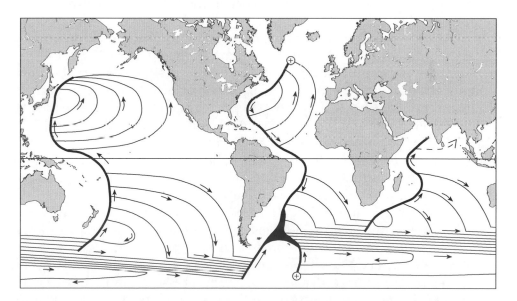

Figure 7.13 A schematic diagram of the abyssal circulation of the world oceans, driven by cold deep-water sources in the North and South Atlantic. If the two sources, $\oplus$, in the North and South Atlantic are taken as $20 \times 10^6\,\mathrm{m^3\,s^{-1}}$ each, then transport between streamlines is $5 \times 10^6\,\mathrm{m^3\,s^{-1}}$. Reproduced, with permission, from Stommel, H., 1958, Deep Sea Research, 5: page 81, figure 1

also influences the direction of the flow to a marked extent and the Antarctic Bottom Water, near the equator, moves from the western boundary eastwards to the mid-Atlantic ridge. There is also evidence of very cold water in the eastern basin of the Atlantic Ocean, near the equator, which may be derived from Antarctic Bottom Water flowing through a gap in the mid-Atlantic ridge. Recent studies, using floats and deep-current moorings below the Gulf Stream, have demonstrated that a deep southward boundary current exists in the region, but its flow is complicated by a series of permanent eddy circulations. These eddies are driven by the highly unstable flow in the Gulf Stream.

7.5   The atmosphere and ocean boundary layers

In the previous sections of this chapter the flows of both the ocean and the atmosphere have been discussed in terms of the geostrophic equation. However, close to the interface between the atmosphere and the Earth's surface, the flow becomes turbulent (see Box 5.1). The predominantly horizontal flow becomes broken up into vertical eddies which carry momentum down from the free atmosphere (above the influence of surface friction) towards the surface. These eddies produce a tangential stress on the land (see Box 5.1) or the ocean (see equation 5.4) when close to the surface. The surface stress on the ocean produces a surface wave field and surface current which, in turn, produce vertical motions which mix the surface momentum down into the ocean. These turbulent regions of atmosphere and ocean are known as the boundary layers and they extend to a height of about 1000 m above the Earth's surface and to a depth of approximately 20 m to 100 m in the ocean.

In this section the effects of the boundary layers on flow will be considered. The two boundary layers are different because, in the atmosphere, the boundary layer is removing momentum from the flow and it is therefore acting to retard the flow of the free atmosphere. However, in the ocean the boundary layer is the main source of momentum to drive not only the surface circulation but also the deeper geostrophic circulation.

First, consider the influence of a frictional force on the geostrophic wind. Figure 7.14 shows the balance of forces between the pressure-gradient force, the Coriolis force and a frictional force. Basically, the frictional force opposes the motion and reduces the wind to below its geostrophic value. The Coriolis force is reduced and, in order to balance the pressure gradient, the wind must rotate towards low pressure to allow a component of the frictional force to oppose the pressure gradient. The angle of rotation is typically 30° over a land surface and 10–15° over the ocean. A component of the flow is directed down the pressure gradient and therefore mass is transferred

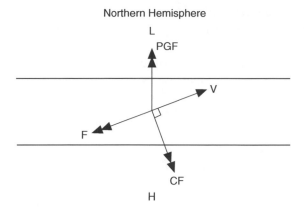

Figure 7.14 Balance of forces in the atmospheric boundary layer for a steady flow, *V*. PGF represents the pressure-gradient force, CF is the Coriolis force and F is the frictional force

between high- and low-pressure systems in these frictional boundary layers. By this process, friction acts to reduce the horizontal pressure gradients and to spin down circulation systems in the atmosphere. An extra-tropical cyclone will take about six days to dissipate its kinetic energy through the surface boundary layer. It is also noted that the frictional layer will, in turn, produce surface convergence and upward motion in a low-pressure system, and surface divergence and downward motion in a high-pressure system. These circulations are known as secondary circulations. However, it must be pointed out that vertical motion in large-scale and synoptic systems is predominantly induced by the flow in the upper troposphere rather than by frictional boundary layers.

A second example of the effect of friction on motion is that in the surface ocean boundary layer in the open ocean. Consider a region where the horizontal pressure gradient is negligible. The surface wind stress will initially impart momentum into the surface layer and drive a current downwind. The current will, in turn, transfer momentum into the deeper layer, which will oppose the surface wind stress. Without any other forces acting, the surface wind stress will be balanced by the frictional stress at the bottom of the surface layer and a steady surface current will be produced. Now consider the effect of the Coriolis force which acts perpendicular to the direction of the motion. This will induce a clockwise rotation of the surface flow in the Northern Hemisphere (and an anticlockwise rotation in the Southern Hemisphere). Eventually a balance of forces between the surface wind stress, the frictional force below the surface layer and the Coriolis force is achieved. For steady flow, frictional drag by the water below the surface layer is necessary to produce a three-way balance of forces. This clockwise rotation of the surface current with respect to the wind stress was first noted by Nansen, when he observed that pack ice drifted at an angle between 20 and 40° to the right of the wind during the Arctic Ocean expedition (1893–96) on board the research vessel *Fram*.

Consideration of the balance of forces below the surface shows that the surface-layer flow will induce a stress on the layer below which, in turn, will produce a further clockwise rotation of the flow. This rotation of the flow with depth was first theorized by Ekman (1905), and it is now known as the Ekman spiral. Below about 50 m these Ekman currents are generally weak. Despite a large number of attempts to observe the Ekman spiral in the ocean, only a handful of measurements have shown its existence. There are probably two reasons for this difficulty in observing Ekman spirals. First, a fluctuating wind would induce strong inertial oscillations in the surface flow. Second, the surface layer is often very well mixed as the result of surface waves, direct wind mixing and convection.

The effect of the ocean surface boundary on the deeper ocean is of crucial importance in the explanation of the climatological current systems above the main thermocline. The Ekman balance between the tangential surface wind stress, τ, and the Coriolis force is given by:

$$-fv = \frac{1}{\rho} \frac{\partial \tau_x}{\partial z} \tag{7.40a}$$

and

$$+fu = \frac{1}{\rho} \frac{\partial \tau_y}{\partial z} \tag{7.40b}$$

The frictional force is represented by the vertical gradient of the tangential stress. Integrating equation 7.40a between the surface and the bottom of the frictional layer, h, yields:

$$-\int_{-h}^{0} \rho v dz = \frac{\tau_x^o}{f} - \frac{\tau_x^{-h}}{f}$$

At the bottom of the frictional layer, h, the tangential stress is zero ($\tau_x^{-h} = 0$), and therefore:

$$-\int_{-h}^{0} \rho v dz = \frac{\tau_x}{f} \tag{7.41a}$$

where τ_x and τ_y are the surface components of the wind stress.

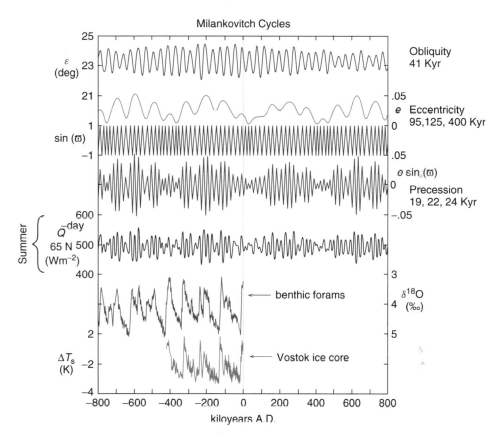

Plate 1.4 Past and future Milankovitch cycles ε is obliquity (axial tilt), e is eccentricity, ϖ is longitude of perihelion. $e \sin(\varpi)$ is the precession index, which together with obliquity, controls the seasonal cycle of insolation. $\overline{Q}^{day}$ is the calculated daily-averaged insolation at the top of the atmosphere, on the day of the summer solstice at 65 N latitude. Benthic forams and Vostok ice core show two distinct proxies for past global sea level and temperature, from ocean sediment and Antarctic ice respectively. Vertical gray line is current condition, at 2 ky A.D

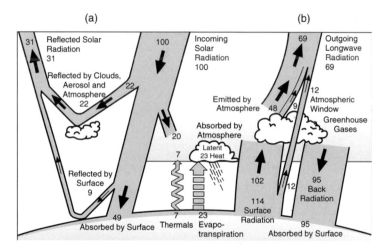

Plate 3.10 The globally and annually averaged components of the solar (a) and planetary (b) radiation streams. The proportion of the radiation intercepted by various components of the atmosphere and the Earth's surface is normalized to the incident radiation at the top of the atmosphere (100 units). 100 units equivalent to $342\,\mathrm{W\,m^{-2}}$. Reproduced, with permission, from Trenberth, K.E., 1997, Bulletin of the American Meteorological Society, **78**: page 206, figure 7

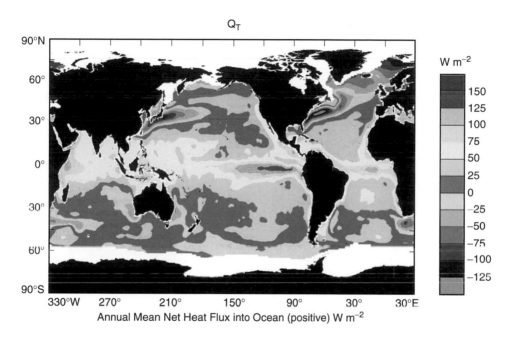

Plate 5.3 Annual mean net heat flux over the global ocean W m^{-2}. Reproduced, with permission, from Ocean circulation and climate Seidler, G., Church, J. and Gould, J. Ocean Heat Transport by H.L. Bryden and S. Imawaki Plate 6.1.4 715pp

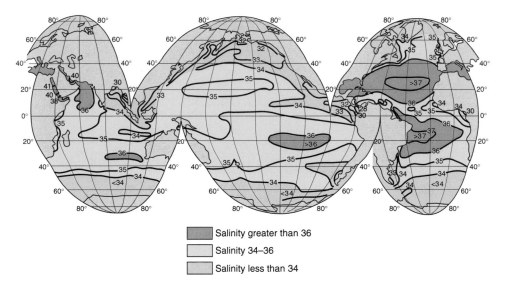

Salinity greater than 36

Salinity 34–36

Salinity less than 34

Plate 5.12   Surface salinity in the Northern Hemisphere summer. Reproduced, with permission, from Charnock H.1996, In Oceanography – An Illustrated Guide, ed. Summerhayes, C.P. and Thorpe, S.A.T. Manson Publishing: page 30, figure 2.5

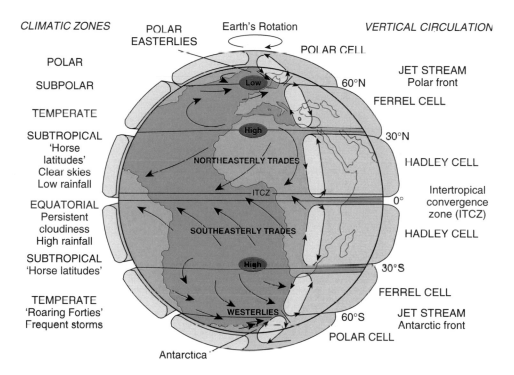

CLIMATIC ZONES

POLAR
EASTERLIES

Earth's Rotation

VERTICAL CIRCULATION

POLAR CELL

POLAR

SUBPOLAR

Low

60°N

JET STREAM
Polar front

FERREL CELL

TEMPERATE

SUBTROPICAL
'Horse
latitudes'
Clear skies
Low rainfall

High

30°N

NORTHEASTERLY TRADES

HADLEY CELL

ITCZ

EQUATORIAL
Persistent
cloudiness
High rainfall

0°

Intertropical
convergence
zone (ITCZ)

SOUTHEASTERLY TRADES

HADLEY CELL

SUBTROPICAL
'Horse latitudes'

High

30°S

FERREL CELL

TEMPERATE
'Roaring Forties'
Frequent storms

WESTERLIES

60°S

JET STREAM
Antarctic front

POLAR CELL

Antarctica

Plate 6.13 Schematic representation of features of the general circulation of the atmosphere. Reproduced, with permission, from Charnock, H. In: Oceanography, Edited by S.A. Thorpe and C. Summerhayes. Manson Publishers. 1995: page 34, figure 2.9

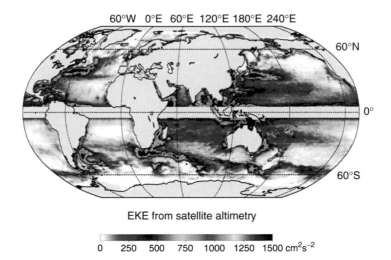

EKE from satellite altimetry

0   250   500   750   1000 1250 1500 cm^{2}s^{-2}

Plate 9.4 EKE in global ocean. Reproduced, with permission, from Barnier, B, Medec, G. *et al.* Ocean Science, 59, page 9, Figure 5b

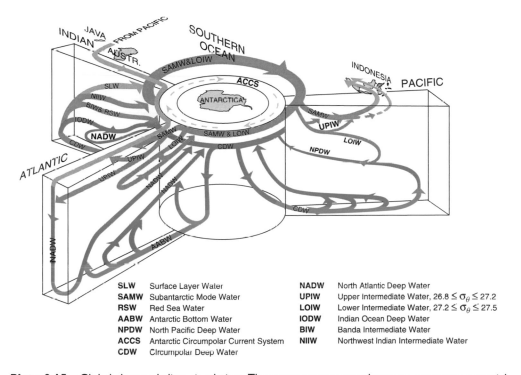

SLW	Surface Layer Water	**NADW**	North Atlantic Deep Water
SAMW	Subantarctic Mode Water	**UPIW**	Upper Intermediate Water, $26.8 \lesssim \sigma_\theta \lesssim 27.2$
RSW	Red Sea Water	**LOIW**	Lower Intermediate Water, $27.2 \lesssim \sigma_\theta \lesssim 27.5$
AABW	Antarctic Bottom Water	**IODW**	Indian Ocean Deep Water
NPDW	North Pacific Deep Water	**BIW**	Banda Intermediate Water
ACCS	Antarctic Circumpolar Current System	**NIIW**	Northwest Indian Intermediate Water
CDW	Circumpolar Deep Water		

Plate 9.15 Global thermo-haline circulation. The water masses are shown on constant potential density surfaces denoted by σ_θ. This is sea water density -1000, and is evaluated using potential temperature rather than in-situ temperature. Reproduced from Siedler, figure 1.2.7, Schmitz, and Gordon A.L., 1996

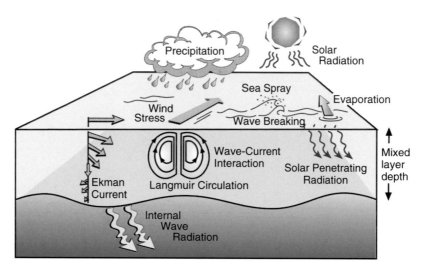

Plate 10.1 A schematic of the physical processes influencing the surface ocean mixed layer in the Northern Hemisphere

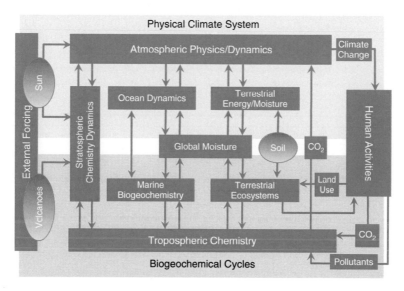

Plate 10.9 Components of an IPCC climate model. Reproduced, with permission, from Glantz, M.H. and Krenz, J.H. 1992 Human components of the system in Climate System Modelling. Ed: Trenberth K.E. Cambridge University Press, Plate 2

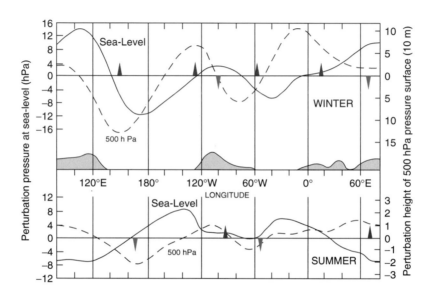

Plate 11.4 Zonal profiles of normal perturbation pressure at sea level (solid line) and normal perturbation height of the 500 hPa surface (dashed line). Continental elevations are indicated by a shaded area. Positions of observed relative heat sources and sinks are indicated by spires pointed upwards and downwards, respectively. Upper part: winter profiles averaged over 20° latitude centred at 45°N. Lower part: summer profiles averaged over 20° latitude centred at 50°N. Reproduced, with permission, from Smagorinsky, J., 1953, Quarterly Journal of the Royal Meteorological Society, 79(341): page 362, figure 7

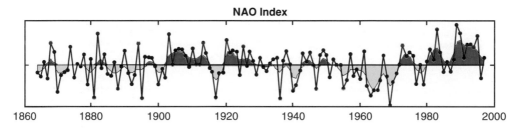

NAO Index

1860　1880　1900　1920　1940　1960　1980　2000

Plate 11.10 NAO is a mainly winter phenomena (Dec-Mar) and is an index of the strength of westerly winds between the Azores (40N) and Iceland (60N). The index is measured by the mean winter surface pressure difference between the Azores and Iceland. The westerly winds are always prevailing at the middle latitudes and are at their strongest each winter. NAO positive is when the winter westerlies are particularly strong, whilst NAO negative is when the westerlies are much weaker. NAO positive has storm systems developing in the Atlantic and tracking ENE towards NW Europe. We have above average rainfall in the UK especially in Northern Britain, and warmer than average winter. NAO negative the storms track northwards towards Greenland, and eastwards into the Mediterranean Sea. The UK is colder than average, with weaker westerly winds, and blocked flow in the northern N. Atlantic and Scandinavia. Reproduced from Lamont Doherty Earth Observatory, Columbia University

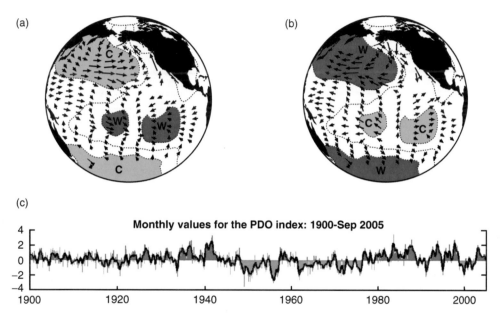

Plate 11.11   (a) Positive Pacific Decadal Oscillation is warm phase, (b) negative is cold phase. Colours are winter SST, Sea level pressure are contours, arrows are surface wind and (c) time series of the PDO. Reproduced from JIASO, University of Washington

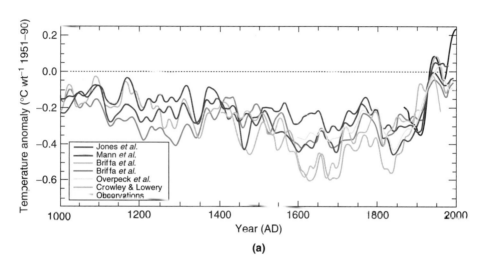

(a)

Plate 12.1   The variation of the Earth's surface temperature on three time scales: (a) estimates of global temperature for last 1000 years. Observations are based on temperature measurements and other estimates from indirect proxies. The dashed line is the global average 1951–1990. From Wikipedia Temperature record of past 1000 years

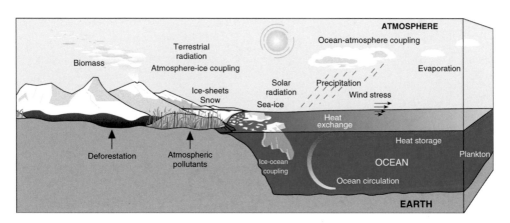

Plate 12.3 A schematic view of the Earth's climate system showing the roles of land, atmosphere, oceans, sea ice, glaciers, and ice sheets. Reproduced, with permission, from Wells, N.C., Gould, W.J. and Kemp, A.E.S., 1996, In Oceanography – An Illustrated Guide, ed. Summerhayes, C.P. and Thorpe, S.A.T. , Manson Publishing: page 42, figure 3.1

Global and Continental Temperature Change

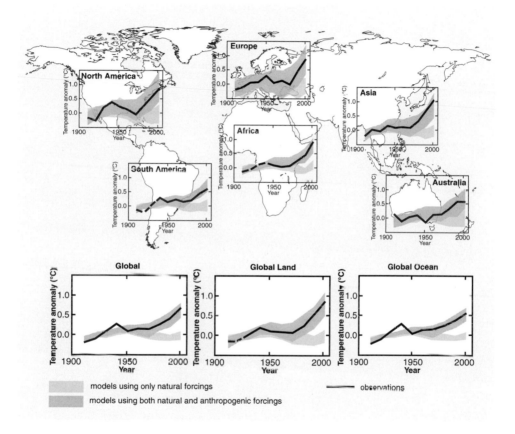

Plate 12.5 Global, ocean and continental temperature change in 20th Century from observations and climate models. Lower curves indicates models which include only natural radiative forcing due to solar and volcanic activity, whilst upper curves indicates models which include both anthropogenic (e.g. greenhouse gases) forcing and natural forcing. Reproduced, with permission, from Intergovernmental Panel on Climate Change (IPCC) 2007 WG1-AR4

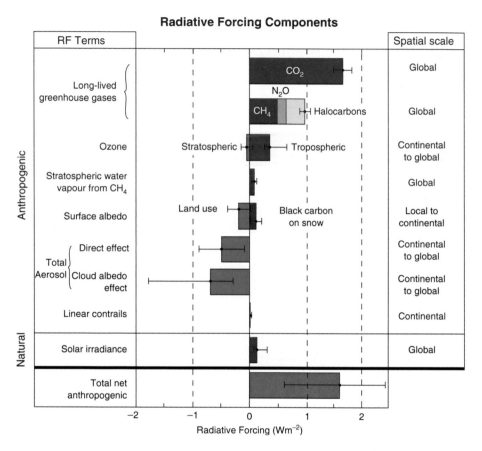

Plate 12.6 Anthropogenic and anomalous Solar Radiative Forcing ($W\,m^{-2}$) of climate system in 2005. Volcanic aerosols are not included. The net radiative forcing in 2005 is $1.6\,W\,m^{-2}$. This can be compared with a long term mean Solar Radiative forcing of $342\,W\,m^{-2}$. Reproduced, with permission, from Intergovernmental Panel on Climate Change (IPCC) 2007 WG1-AR4

Similarly from equation 7.40b:

$$\int_{-h}^{0} \rho u dz = \frac{\tau_y}{f} \tag{7.41b}$$

The left-hand sides of equations 7.41a and b are the mass transports in the frictional Ekman layer. These Ekman mass transports are not dependent on the detailed structure of the Ekman spiral and are **only** dependent on the magnitude and direction of the surface wind stress.

The Ekman mass transports are perpendicular to the direction of the stress and, in the Northern Hemisphere, they act to the right of the wind stress. Figure 7.15 shows the effect of an anticyclonic and a cyclonic atmospheric wind circulation on the mass transport in the surface layer. In the anticyclonic case, the mass transport is directed towards the centre of the circulation and, because there can be no net accumulation of mass over a long period of time, mass can only be removed from the surface layer by downward vertical motion. The downwelling, in turn, tends to bow down the pycnocline and thermocline in the central region of the anticyclonic circulation and

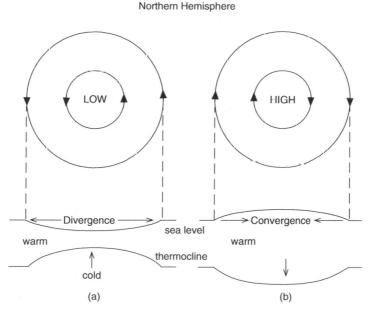

Figure 7.15 Ekman transport and vertical motion in the surface layer induced by (a) cyclonic wind stress and (b) anticyclonic wind stress in the Northern Hemisphere

the accumulation of warm, less dense water will produce a region of high dynamic topography. By the geostrophic equation (equations 7.16 and 7.17), this will cause a surface anticyclonic circulation relative to the deep water. In the cyclonic case, surface divergence will cause upwelling of the pycnocline and thermocline, and produce a region of relatively dense, cold water below the surface boundary layer, which will induce a surface cyclonic flow relative to deep water.

On a smaller scale, tropical cyclones can have a dramatic influence on sea surface temperatures as the result of strong upwelling induced by the cyclonic circulation in the rear of the storm path.

In summary steady winds over many days will produce Ekman mass transports directed perpendicular to the direction of the surface wind stress. They are important for ocean circulation because they cause vertical motion at the base of the frictional layer which extends into the main thermocline. This causes a vertical displacement of the layers of constant density, with the largest displacements in the centre of the ocean basins. This will, in turn, induce horizontal pressure gradients and geostrophic flow in the ocean interior. The surface wind stress is therefore an important quantity for the generation and maintenance of the major ocean currents.

7.6   Equatorial winds and currents

For both the atmosphere and ocean, the equatorial region is a special case. In the atmosphere the winds are steadier and weaker than the winds at higher latitudes. Synoptic variability of the equatorial atmosphere is less than in middle latitudes, although the seasonal variations associated with the monsoonal circulations can be strong. In the Pacific Ocean, interannual variations of the equatorial tropospheric and stratospheric winds are an important feature of its variability. In contrast, ocean currents tend to be stronger in equatorial regions than at higher latitudes and they are generally orientated in an east-west direction. They also respond more quickly to wind variations, as seen in the Indian Ocean where current reversals occur in response to the monsoonal surface-wind circulation (Figure 6.15).

These features of the equatorial region can be related to the reduction in the Coriolis force close to the equator and to its absence on the equator. At $4°$ latitude, the Coriolis parameter ($2\Omega \sin \Phi$) is an order of magnitude less than at $45°$ latitude. The geostrophic and the Ekman balances, therefore, will no longer be good approximations for winds and currents close to the equator. The question then arises as to how close to the equator can one go before the geostrophic wind equation is of no further use? Generally, it is not acceptable to apply the approximation to large-scale atmospheric flows within about $15°$ of the equator and to ocean flows within about $4°$ of the equator, for reasons which are too complex to be discussed in an introductory text. However, the above limitations do not imply that the Coriolis acceleration can be neglected. Rather it is that other terms in the equations of motion become increasingly important.

Figure 7.16 shows a theoretical calculation of the surface mass transport caused by a steady westward wind stress in a region spanning the equator. On the equator there is no Coriolis force and therefore a current transport in the direction of the wind stress is induced. This current will continue to accelerate until a frictional balance between the wind stress and the bottom-layer stress is achieved (see Section 7.5). Within $1°$ of the equator the Coriolis force begins to have an effect, and it turns the current to the right in the Northern Hemisphere and to the left in the Southern Hemisphere. Therefore, there is a divergence of surface flow from the equator which, by the conservation of mass, must be balanced by an upward motion of water into the surface layer. This upwelling is of the order of $1\,\mathrm{m\ day}^{-1}$ and it causes a lifting of the thermocline. The equatorial thermocline usually lies between 100 and $200\,\mathrm{m}$ depth but, in local areas where upwelling is strong, it may be less than $50\,\mathrm{m}$ below the surface. At about $10°$ latitude the surface transport is nearly perpendicular to the wind stress, in agreement with the Ekman balance

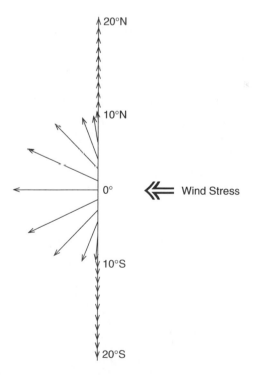

Figure 7.16 Surface transport in the equatorial ocean, caused by a westward wind stress. Reproduced from Rowe, MA., 1985, University of Southampton: PhD thesis

(equation 7.41). In regions where the surface-wind stress is directed eastward, such as in the northern Indian Ocean during the south-west monsoon, then the surface transport will converge onto the equator. In this case, downward motion out of the surface layer will occur and this will depress the equatorial thermocline.

The meridional tilt of the thermocline, caused by equatorial upwelling, will induce a meridional pressure gradient with high dynamic height polewards of the equator. A few degrees of latitude from the equator, the flow will be close to geostrophic and therefore a westward flow within the thermocline will occur. It is this meridional pressure gradient which is responsible for the north and south equatorial currents which are found in all ocean basins, except for the northern Indian Ocean during the south-west monsoon. In this latter case, because of downwelling on the equator, the meridional pressure gradient is reversed and eastward equatorial currents occur. Downwelling regions in the eastern equatorial Pacific and Atlantic Oceans, near 10°N, result in a similar reversal of the pressure gradient and induce a narrow geostrophic countercurrent towards the east. These examples show the intimate relationship between the surface wind, upwelling and pressure gradients, and the production of strong, zonal equatorial currents.

Along the equator the primary balance of forces in the upper ocean is between the pressure-gradient force and the surface-wind stress. Figure 7.17a shows the zonal pressure gradient on the equator in the Atlantic Ocean for February and September. In February, the surface winds are generally weak westward winds and there is only a small eastward pressure gradient. However, in September, very strong westward winds extend across the equator and a large eastward pressure gradient is required for balance. The highest dynamic heights occur in the west and here the thermocline is depressed, relative to the eastern part of the region. Figure 7.17b shows the depth of the 23°C isotherm, which is indicative of the depth of the thermocline and which changes its slope in response to the surface-wind stress. The upward slope of the thermocline towards the east accounts for the frequent appearance of cold, upwelling water in the eastern equatorial Atlantic and Pacific Oceans. In the Indian Ocean during the south-west monsoon, the zonal tilt of the thermocline is reversed because of the change in direction of the surface wind stress.

If the horizontal pressure gradient exactly balanced the surface-wind stress, there would be no motion along the equator. However, the balance of forces is not exact and the currents respond quickly to the imbalance. Below the surface layer, unimpeded by the wind stress, the eastward pressure gradient in the thermocline drives an eastward undercurrent along the equator. In the Pacific Ocean, the equatorial undercurrent is 14 000 km long, but only about 400 km wide. It is confined to within 2° of the equator and has a maximum velocity of $2\,\mathrm{m\,s^{-1}}$ close to its core in the thermocline. In both the Pacific and

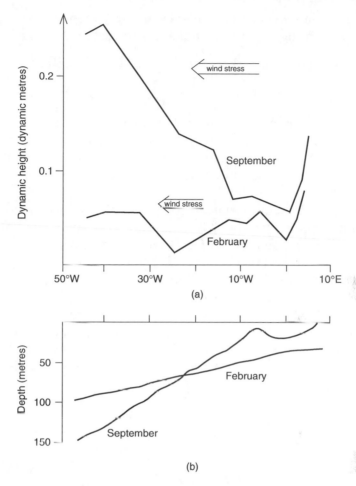

Figure 7.17 (a) Zonal pressure gradient in the equatorial Atlantic in February and September. (b) Depth of 23°C isotherm (equivalent to depth of the thermocline) in February and September. The zonal wind stress is much stronger in September than in February

Atlantic Oceans, the undercurrent slopes upwards towards the east and it can, occasionally, surface in the eastern basins.

The following calculation will give an indication of the maximum eastward acceleration which can occur as a result of the imbalance of the observed pressure gradient with the wind stress. From Figure 7.17a, it can be seen that the maximum difference in dynamic height between 40°W and 0°E in the Atlantic Ocean is 0.2 dynamic metres, and therefore using equation 7.3, the geopotential difference is $2\,\mathrm{m}^2\,\mathrm{s}^{-2}$.

The horizontal distance is 40° longitude, or 4.4×10^6 m and therefore the eastward pressure gradient is $(0.2 \times 10\,\mathrm{m}^2\ \mathrm{s}^{-2})/(4.4 \times 10^6\ \mathrm{m})$, or $4.5 \times 10^{-7}\ \mathrm{m\,s}^{-2}$.

The velocity, u, after constant acceleration, a, through a distance, s, is given by $u^2 = 2as$. Assuming uniform acceleration across the Atlantic Ocean:

$$u = \sqrt{(2 \times 4.5 \times 10^{-7} \times 4.4 \times 10^6)}\,\mathrm{m\,s^{-1}}$$
$$= 2.0\,\mathrm{m\,s^{-1}}$$

This calculation therefore gives a reasonable estimate for the maximum observed velocity of the equatorial undercurrent. In practice, frictional turbulence with the deeper ocean and horizontal mixing associated with meanders in the undercurrent will tend to reduce the observed eastward acceleration below that obtained in the above calculation.

Some features of the equatorial atmosphere will now be considered. One of the dominant seasonal variations is the Asian monsoonal circulation which, during the summer, causes a surface atmospheric flow from the southern Indian Ocean to the Asian continent. This flow can be intense, with winds of up to $20\,\mathrm{m\,s^{-1}}$ occurring off the Somali coast. Figure 7.18 is a simplified map of surface pressure for the south-west monsoons and shows that the main feature is the curvature of the winds as they cross the equator. There is a pressure gradient directed from the sub-tropical high pressure in the southern Indian Ocean towards the Indian sub-continent. In the Northern Hemisphere, away from the equator, the geostrophic wind is eastwards,

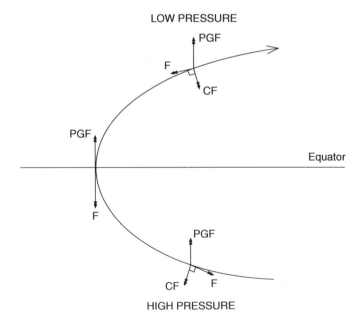

Figure 7.18 Balance of forces in a monsoonal flow (summer monsoon) across the equator. The horizontal pressure-gradient force is PGF, the Coriolis force is CF and the frictional force is F

whilst in the Southern Hemisphere it is westwards. However, because of surface friction, the westward flow will rotate clockwise as shown.

Consider a parcel of air in the Southern Hemisphere moving towards the equator. As the parcel approaches the equator, the deflecting Coriolis force is weak and the parcel will accelerate northwards down the pressure gradient. Only friction would be able to reduce the acceleration at this stage. Once the parcel crosses the equator into the Northern Hemisphere, it will begin to feel the effect of the Coriolis force now acting to the right and it will gradually be deflected towards the east.

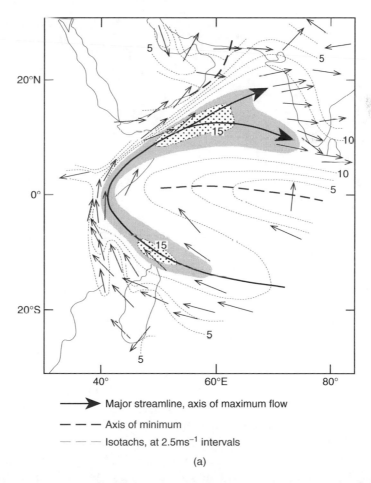

Major streamline, axis of maximum flow

— — — Axis of minimum

— — — Isotachs, at 2.5ms^{-1} intervals

(a)

Figure 7.19 (a) East African jet in the atmosphere: monthly mean air flow at 1 m in July. Wind speeds (isotachs) at 2.5 m s^{-1} intervals. Reproduced, with permission, from Find later, J., 1971, Geophysical Memoirs, 115: page 2, figure 1. (b) Surface currents in the Indian Ocean during the summer monsoon. The intense flow adjacent to the African coast is the Somali jet. The dynamic height (dotted lines) at 0.25 dynamic meters intervals. Reproduced, with permission, from Swallow, J.C. and Bruce, J.G., 1966, Deep Sea Research, 13: page 866, figure 2

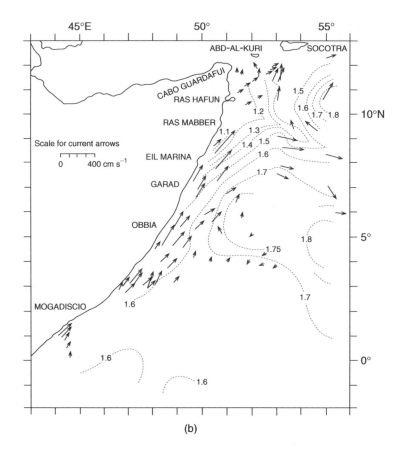

Figure 7.19   (*continued*)

A second feature of the monsoonal circulation is the appearance of an intense, low-level jet stream, as shown in Figure 7.19a. It appears off the coast of East Africa during the south-west monsoon. This jet stream, unlike those in the middle latitudes, is directed from the south to the north across the equator and velocities of up to $25\,\mathrm{m\,s^{-1}}$ are reached between a height of 1 and 2 km. It was discovered by Findlater, a British meteorologist, who mapped the complete jet stream after receiving reports of very strong, anomalous winds from civil aircraft operating from East Africa. The formation of this jet stream appears to be analogous to that of the western boundary currents in the ocean (Figure 7.11). In the atmospheric case, the East Africa tablelands provide a convenient western boundary to the low-level monsoon flow.

The relationship between the low-level jet stream, the Somali current and the intense upwelling off the Somali coast has been a major area of research for both physical oceanographers and meteorologists. In 1979, during the First Global GARP Experiment (FGGE), a detailed oceanographic and atmospheric study of the summer Indian monsoon was undertaken. Observations showed that the atmospheric jet was exceptionally strong and surface stresses of over $0.4\,\mathrm{N\,m^{-2}}$ were recorded near the Somali coast at 10°N in June and July. Sea surface temperatures near the coast dropped from 27°C at the beginning of June (i.e. the time of the monsoon onset) to 18 °C by the end of the month as the result of intense upwelling. Measurements of surface currents by ships' drift in August showed north-eastward currents of between 3 and $4\,\mathrm{m\,s^{-1}}$ (Figure 7.19b).

In summary, the equatorial ocean circulation has a fast response to seasonal variations in wind. In the Atlantic and Indian Oceans, the response time is less than one month, whilst for the Pacific Ocean, because of its large zonal scale, the response time is between two and three months. In the Indian Ocean, as discussed above, the Somali current responds to the onset of the south-west monsoon within two to three weeks. In contrast, the large sub-tropical gyres in the middle latitudes respond on time scales of years to decades and they therefore show less seasonal variation. The atmosphere and ocean circulations in low latitudes respond on similar time scales (i.e. seasonal and interannual) and therefore it is in these regions that strongly coupled interactions between atmosphere and ocean can take place. This will be discussed in more detail in Chapter 11.

8

Waves and Tides

8.1 The spectrum of surface waves

Before discussing the spectrum of surface waves, it is necessary to introduce the following definitions:

(i)   The wave height, H, is the vertical distance between the wave trough and the wave crest, and it is twice the amplitude, a, of the wave.

(ii) The wave period, T, is the time interval between the passage of two consecutive wave crests, or troughs, at a fixed point.

(iii) The wavelength, L, is the distance between two wave crests measured in the direction of propagation.

(iv) The individual wave speed, c, is given by $c = L/T$.

(v)   The frequency, $\sigma = 2\pi/T$.

(vi) The wavenumber, $k = 2\pi/L$.

(vii) The phase speed, is given by $c = \sigma/k$. The speed of energy propagation of a wave is related to the group velocity of the waves, c_g where $c_g = \frac{d\sigma}{dk}$.

(viii) If the wave speed, c, is not dependent on the wavenumber of the wave, then the wave is said to be non-dispersive and $c = c_g$.

Figure 8.1 shows the energy spectrum of surface waves in the ocean for periods ranging from 0.01 to 10^6 s. The highest-frequency components of the spectrum are the capillary waves which have wavelengths of a few centimetres and periods of less than 0.1 s. They are generated, almost instantaneously, by gusts of wind which produce the familiar 'cats' paws' ripple pattern on an otherwise smooth water surface. They are also dissipated in less than 30 s by

---

The Atmosphere and Ocean: A Physical Introduction, Third Edition. Neil C. Wells.
© 2012 John Wiley & Sons, Ltd. Published 2012 by John Wiley & Sons, Ltd.

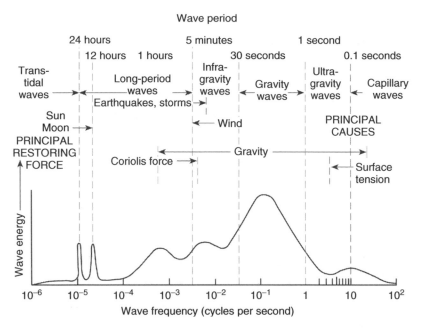

Figure 8.1 Schematic representation of the relative amounts of energy in waves of different frequencies. Reproduced, with permission, from Kinsman, B., 1984, Wind Waves: Their Generation and Propagation on the Ocean Surface, Prentice-Hall: page 23, figure 1.2.1

viscosity. The restoring forces for these capillary waves are, in general, both gravity and surface tension, and the wave speed, c, is given by:

$$c = \sqrt{\left(\frac{gL}{2\pi} + \frac{2\pi S}{\rho L}\right)} \tag{8.1}$$

where S is the surface tension of water, which is typically $74 \times 10^{-3}\,\mathrm{N\,m^{-2}}$. It can be shown that the minimum wave velocity, c_m, occurs when $gL_m/2\pi = 2\pi S/\rho L_m$, i.e. when $L_m = 2\pi\sqrt{S/\rho g}$), where L_m is the wavelength corresponding to the minimum wave velocity.

For sea water, $L_m = 1.7\,\mathrm{cm}$ and $c_m = 23\,\mathrm{cm\,s^{-1}}$. At wavelengths shorter than L_m, the surface tension in equation 8.1 becomes increasingly dominant and c increases as L decreases. For wavelengths longer than L_m, gravity becomes more important than surface tension. In this case, c increases as L increases. An interesting peculiarity of these waves is that the group velocity is greater than the phase velocity and so a packet of capillary waves will move more quickly than the individual waves. The profile of a capillary wave is quite unlike that of a gravity wave in that it has a flat crest and a sharp trough. Capillary waves, because of their dependence on surface tension, are strongly affected by surface films of oil or other organic matter and, because the absence of capillary waves on the surface changes the reflection, surface slicks are easily seen. More recently, it has been shown that, because gravity – capillary waves are a sensitive indicator of wind speed, the magnitude of the reflection of microwave radiation ($\lambda = 2.2$ cm) can give a quantitative measurement of wind speed and direction. The instrument, known as a scatterometer, has been shown to give estimates of wind speed accurate to $0.6 \, \mathrm{m \, s^{-1}}$.

The most significant energy band in the spectrum is that of wind waves and swell, which have periods between 1 and 30 s. In this frequency band, gravity is the only restoring force and therefore these waves are true gravity waves generated by a turbulent surface-wind stress. The particles in these waves execute slightly elliptical orbits. The particles tend therefore to return to a position close to the earlier position over the period of the wave. Because of the slight elliptical nature of the orbit they do not return exactly to the earlier position but are displaced in the direction of the surface waves. This displacement is known as Stoke's drift. This drift speed is much less than the particle velocity associated with an individual surface wave and it will be assumed that they are circular orbits.

The radius of the orbits of the particles decreases with depth according to the relationship $r = a \, exp \, [-(2\pi /L)z]$, where a is the amplitude of the wave. At a depth of half the wavelength of the surface wave, the particles' orbit radius is $a/23$. For example, a surface wave having a 1 m amplitude and a 10 m wavelength would have a 4 cm amplitude of particle motion at a depth of 5 m. Therefore, wave motions can be seen to be quickly damped with depth. It is noted that the profile of the wave is not exactly sinusoidal. The wave crests are sharper and the wave troughs are flatter than in a true sinusoidal wave. For the larger-amplitude waves this effect becomes more marked, with wave crests reaching a limiting angle of 120°, corresponding to a wave slope of about 1 in 7. At this stage, wave breaking occurs.

For a wave whose amplitude is small compared with its wavelength, the wave speed, c, is given by:

$$c^2 = \frac{gL}{2\pi} \tanh \left(\frac{2\pi h}{L} \right) \tag{8.2}$$

where h is the depth of the water. Such waves are known as linear gravity waves and for waves whose wavelength is smaller than twice the depth of the water (i.e. $h/L > 0.5$), then $\tanh (2\pi h/L) \approx 1$ and $c^2 = gL/2\pi$. Since $c = L/T$ it can be easily shown that:

$$c = \frac{g}{2\pi}T \quad \text{and} \quad L = \frac{g}{2\pi}T^2 \tag{8.3}$$

Table 8.1 shows numerical values of c, L and T for typical ocean waves. It is noted that long waves have higher wave velocities than short waves and that the wavelength is proportional to the square of the period. These

Table 8.1   Characteristics of wind waves and swell derived from equation 8.3

	Wind waves	Swell	Long swell
Period $T(s)$	1–5	10	20
Wavelength $L(m)$	1.6–39	156	624
Phase velocity c $(m\,s^{-1})$	1.6–7.8	15.6	31.2
Group velocity c_g $(m\,s^{-1})$	0.8–3.9	7.8	15.6

gravity waves are dispersive since their speed is dependent on wavelength and it would therefore be expected that the group velocity of a packet of wind waves would be different from the phase speed. Since $c = \sigma/k$ and $T = 2\pi/\sigma$, it follows that $\sigma^2 = gk$ or $\sigma = \sqrt{(gk)}$. Differentiating:

$$c_g = \frac{\partial \sigma}{\partial k} = \frac{1}{2}\sqrt{\frac{g}{k}} = \frac{1}{2}c \qquad (8.4)$$

Therefore, the group velocity is equal to half of the wave speed, c. This deduction can be easily verified by studying the propagation of a group of waves produced by a ship or a boat. The individual wave crests will be seen to move towards the observer through the group of higher waves and then they will be seen to die out.

As mentioned earlier, apart from tides, the largest energy density in the surface-wave spectrum is associated with wind waves and swell. A simple example illustrates the typical energy in ocean swells. The energy density of a sinusoidal wave is $\rho g a^2/2$ or $\rho g H^2/8$, and therefore a wave with a height of 2 m would have an energy density per unit area of $5 \times 10^3\,J\,m^{-2}$. If it is assumed that the ocean swell has a 10 s period, then its group velocity is $1/2 \times 15.6\,m\,s^{-1}$ or $7.8\,m\,s^{-1}$. The energy flux per unit time, impinging on a straight coastline, is the product of the energy density and the group velocity. In the present example, this flux is $7.8\,m\,s^{-1} \times 5 \times 10^3\,J\,m^{-2}$, or approximately 40 kW for each metre of coastline. Most of this energy is dissipated into heat in the wave-breaking zone.

These wind and swell waves are known as 'deep water' waves because we have assumed in deriving equation 8.3 that depth $h > 0.5\,L$. For example a 20 s swell would need a depth of more than 312 m for equation 8.3 to be valid. In shallower depths the general equation 8.2 would be needed.

In water depth which is small compared with the wavelength (i.e. $h/L < 0.5$), $\tanh(2\pi\,h/L) \approx (2\pi h/L)$ and, from equation 8.2, $c = (gh)^{1/2}$. Unlike the ordinary gravity waves for which $h/L > 0.5$, the wave speed is independent of wavelength and the group velocity is equal to the phase speed. Such waves are known as long waves and particle motions in long waves are horizontal and do not decay with depth. These waves can be observed as they approach a coastal beach. Such waves, as they approach shallow water, will slow and the wave crest will refract and have its wave crest parallel to the beach before breaking. Long waves appear in the spectrum between swell and tides, and they have periods from seconds (waves at a beach) to the periods of tides ($\sim$24 hours).

An extreme example of a long wave is the tsunami, or seismic wave, which can produce locally high-energy densities. These seismic waves are caused by submarine volcanoes, earthquakes or landslides, and they propagate quickly at speeds of about $200 \, \mathrm{m \, s^{-1}}$ in deep ocean water (i.e. 4 km depth) and at about $30 \, \mathrm{m \, s^{-1}}$ on continental shelves where the ocean is less than 100 m deep. Their wavelengths are very long and, for a tsunami having a wave period of 10^3 s, the wavelength in deep ocean water would be approximately 200 km. Therefore the wave would be difficult to detect visually. However, in shallow water, where it propagates more slowly and where the energy is concentrated into a smaller depth, the amplitude of the wave would grow catastrophically and cause much damage to coastal communities. Though these waves are rare, early warning systems have been in place in the Pacific Ocean since the 1960s and more recently installed in the Indian Ocean.

Other examples of long waves include 'surf beat', which has a period of a few minutes and can be observed as a long-period oscillation in water level on a beach. It is caused by a long wave travelling longitudinally between the surf zone and the shore. A second example is the shelf waves which travel along the 'shelf break' at the edge of the continental shelf. They have periods between 30 min and 2 hours, and can travel for hundreds or even thousands of kilometres along the shelf break.

The extreme low-frequency long waves are the tides which are forced by gravitational attraction between the Earth, the Moon and the Sun. The tides will be discussed in Sections 8.4 and 8.5. Another long wave which may cause extreme sea levels in coastal regions is the 'storm surge'. This is a complex wave produced by storm winds and surface pressure changes and it will be discussed in Section 8.6.

8.2   Wind waves and swell

In the previous section, a number of formulae have been briefly discussed which have been derived from elegant theories of surface waves developed by mathematicians over the last two centuries. It has also been seen that they make available practical tools with which to calculate relationships between wave speed, period and wavelength. However, these formulae and theories give very little insight into how to describe the complicated surface wave patterns observed in nature, or into the understanding of how waves are generated and how they decay. It was the need to answer such questions that gave rise to the first quantitative study of ocean waves in the early 1940s. At this time, military planners required predictions of wave heights and swell for seaborne operations, in particular for beach landings.

During this early period, a number of different methods were devised to measure the height of the sea surface covering a frequency range of 1 to 50 s. One of the most successful instruments was the underwater pressure recorder which could be located either on the bottom in shallow water or below the water line of a ship. The pressure fluctuations, a, at depth, h, can be used in conjunction with wave theory to give the amplitude of the variation of the sea surface, using the following relationship:

$$a_0 = a_h \cosh(kh) \tag{8.5}$$

where a_0 is the amplitude of the sea surface, k is the wavenumber and a_h is the amplitude at depth h.

The signals can be recorded in the ship or at the shore base. This type of instrument tends to filter out high-frequency fluctuations. A second method involves the measurement of the vertical displacement of a ship or buoy using an accelerometer. Many of the measurements of ocean waves have been obtained from ships equipped with pressure recorders on the hull and accelerometers, the latter being used to correct for the ship's motion. Wave buoys are now used extensively to provide wave measurements from the coast to the deep ocean. Other methods include the use of inverted echo-sounders, either moored or on the bottom, to obtain surface displacement, and the use of radars and lasers, either on fixed platforms or on aircraft and satellites. All of these methods have their own advantages and disadvantages and in the end it depends very much on the question being asked as to what system of measurement is adopted.

A typical trace from a bottom-pressure wave recorder is shown in Figure 8.2. It confirms our casual observation that the sea surface is, in general, irregular and does not undergo simple sinusoidal variations. It is clearly not possible to define a single wave height or a single wave period for this record. In order to analyse such a record, it is necessary to decompose it into its component frequencies and amplitudes, using Fourier analysis. This is done by dividing the record into blocks of equal periods, say two minutes, and then calculating the Fourier coefficients for each block of the sea-level record, Z. If the sampling rate is 1 s, then 120 data points are obtained for each block which, in turn, will give Fourier coefficients, H_n, for 59 discrete frequencies, thus:

$$H_n = \frac{1}{60} \sum_{j=1}^{120} Z_j \exp\left(i\frac{2\pi jn}{120}\right) \quad \text{for} \quad n = 1,59 \tag{8.6}$$

The energy, E_n, for each frequency, n, is then proportional to the square of the modulus of the Fourier coefficients, $|H_n|^2$. In order to obtain a statistically

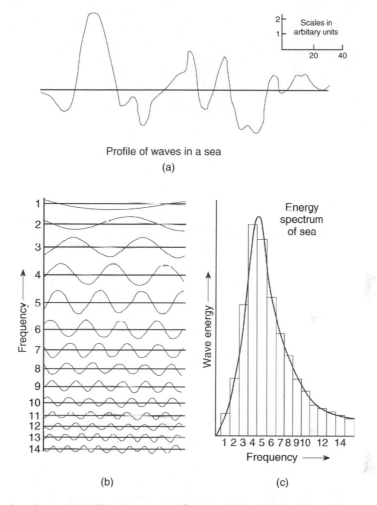

Profile of waves in a sea

(a)

(b) (c)

Figure 8.2 An observed profile of waves in the sea (a) such a complicated wave pattern can be described as consisting of many different sets of sine waves, (b) all of them superimposed. The lower-frequency waves contain more energy than the higher frequency ones, (c) the energy in a wave is proportional to the square of its height. Reproduced, with permission, from Grant-Gross, M., 1987, Oceanography: A View of the Earth, Prentice-Hall: page 208, figure 8.4

valid sample, the procedure is repeated for subsequent blocks of data and the average energy, E_n, is calculated for each frequency. The average energy is then plotted against frequency to produce an energy or power spectrum such as the one shown in Figure 8.2. First, this type of analysis shows that, in general, there is a continuous frequency distribution of waves with the maximum energy at the frequency of the dominant wave, F_0. Second, the distribution is not symmetrical in that it has a more rapid reduction towards lower frequencies. Finally, the total energy of the wave field is proportional to the area under the curve.

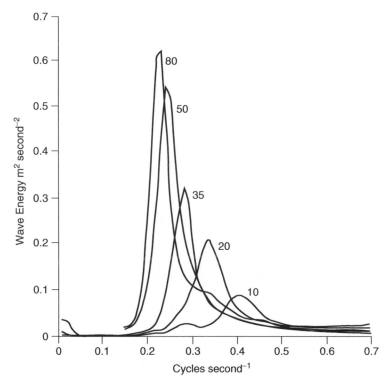

Figure 8.3 Evolution of wave spectrum with fetch for offshore winds from 11.00–12.00 hours, 15 September 1968. The fetch, in kilometres, is shown adjacent to each spectrum. Reproduced, with permission, from Charnock, H., 1981, In Evolution of Physical Oceanography, ed. Warren, BA. and Wunsch, C., MIT Press: chapter 17, page 493, figure 17.6

Two spectral examples will now be considered. Figure 8.3 shows a series of wave spectra observed off the West German coast when the wind was blowing steadily from the land. From common experience, one expects the height of the waves, and therefore their energy, to increase with the distance offshore, or the fetch. In the example, the areas under the curves increase markedly with the fetch. The spectra also show that the dominant wave frequency decreases with fetch, from 0.4 to 0.2 cycles s^{-1}. This corresponds to an increase in wave period from 2.5 to 5 s. Second, as the fetch increases, the spectra become narrower and this implies that more of the wave energy becomes concentrated in fewer frequencies. At each position the wave spectra are in a steady state. This means that, although energy is being added to the waves by the wind, energy is being dissipated at about the same rate. Such spectra are known as 'saturated' wave spectra, and their shape, dominant wave frequency and total energy can be related to two parameters, namely, the fetch and the wind stress.

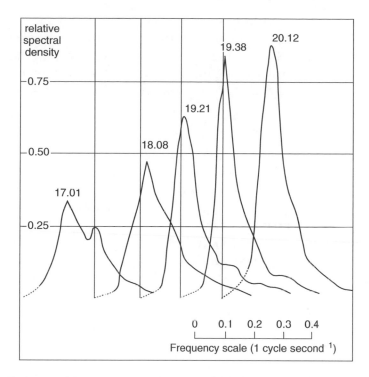

Figure 8.4 Evolution of wave spectra with time on 28 October 1976, between 17.00 and 20.00 hours. The wind increased from $3\,m\,s^{-1}$ to $10\,m\,s^{-1}$ from 15.00 to 16.00 hours and then was constant for the remaining period. Reproduced, with permission, from Revault-d'Allones, M. and Caulliez, G., 1980, In Marine Turbulence, ed. Nihoul, J.C.J., Elsevier: page 110, figure 5

The second example is illustrated by Figure 8.4, which shows the change of shape of spectra with time under steady wind conditions, i.e. prior to the attainment of a steady state. The high-frequency waves become saturated within a relatively short period of time but the lower-frequency waves continue to grow during the three-hour period. Unlike the previous example, the dominant wave frequency remains relatively constant during the period, but the spectra exhibit a similar narrowing with time. The time taken to reach a saturated spectrum depends upon the wind speed. For relatively low wind speeds of $5\,m\,s^{-1}$ saturation will be attained in about two to three hours, whilst for higher wind speeds of, say, $15\,m\,s^{-1}$, it may take 24 hours. As the fetch increases beyond a critical value, the spectra become independent of fetch and they can be directly related to the wind velocity. At low wind speeds of $5\,m\,s^{-1}$, this will occur at a distance of approximately 10 km offshore but, for a higher wind speed of $15\,m\,s^{-1}$, a fetch of more than 500 km is required.

Provided that the wave spectra are in equilibrium with the wind and they are not limited by fetch from the coast, then they can be used to relate the statistics of the wave field to the wind speed. The total energy of the wave field, E_T, at a given wind speed is equal to the area under the spectral curve, and the quantity $\sqrt{E_T}$ is a measure of the root-mean-square height of the waves. If the significant wave height, H_s, is defined as the mean wave height of the one-third highest (peak to trough) waves, then it has been shown that:

$$H_s = 4.0\sqrt{E_T}$$

Spectral decomposition is, therefore, a powerful tool for the presentation of large quantities of wave data in a succinct manner and for obtaining empirical formulae which can be used for wave prediction. Wave spectra also have considerable scientific value in the understanding of how waves are generated and dissipated.

The two previous examples show how wave fields grow with time and change with offshore distance, and thus give important clues to the actual mechanisms by which waves are generated. Two mechanisms have been proposed for the development of waves by wind. In the first mechanism, it is assumed that a wind blowing across a water surface will produce small surface pressure fluctuations as a result of turbulence in the atmospheric boundary layer. This will, in turn, cause small waves to form. If some of the waves move at a similar speed to these pressure fluctuations, then the pressure field will reinforce the wave field and the waves will grow. Theoretical analyses have shown that the wave energy will grow linearly with time. The important aspect of this theory is the assumption that waves do not affect the pressure fluctuations in the atmosphere and it is, therefore, most applicable to the initial generation of waves when amplitudes are small. The second mechanism assumes that the surface waves will affect the pressure fluctuations in the atmosphere in such a way that surface pressure fluctuations and waves grow with time. If the flow over the waves is smooth, then the pressure distribution cannot transmit energy into the waves but, if the flow is turbulent, the pressure distribution can come into phase with the vertical velocity of the sea surface and allow a rapid energy exchange. This causes the waves to grow exponentially with time.

These two mechanisms fail to explain two features of the spectrum; namely the shift of the spectral peak to low frequencies with time and the development of a saturated spectrum. The mechanisms have been studied in the JONSWAP experiment in the North Sea. Figure 8.5 shows the wave energy transfer measured for a developing wave spectrum during this experiment. It may be seen that most of the wind energy goes into the spectrum over a broad band of frequencies, whilst the dissipation is concentrated at the high-frequency end of the spectrum. In order to achieve a balance between energy sources and sinks, energy has to be transferred to high-frequency waves. This has to be done by interaction between waves of different frequencies. A qualitative

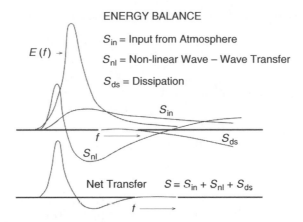

Figure 8.5 Schematic energy balance of the wave spectrum. Reproduced, with permission, from Charnock, H., 1981, In Evolution of Physical Oceanography, ed. Warren, BA. and Wunsch, C., MIT Press: chapter 17, page 493, figure 17.8

example of the process is to consider two sinusoidal waves with frequencies σ_1 and σ_2, and amplitudes a_1 and a_2. Frequency σ_1 is greater than frequency σ_2. The interaction of these two sine waves is given by $a_1 a_2 \sin \sigma_1 t \sin \sigma_2 t$. Using sum and difference formulae, the following terms are obtained:

$$\cos (\sigma_1 - \sigma_2)\, t \quad \text{and} \quad \cos (\sigma_1 + \sigma_2)\, t$$

Therefore, the interaction between the two original waves has produced one wave with a higher frequency, $\sigma_1 + \sigma_2$, than either of the original waves and a second wave with a lower frequency. Hence, by this non-linear interaction, energy is transferred to both high and low frequencies. In the wave spectrum, the energy that is transferred to the higher frequency is dissipated by wave breaking observed as white caps, and, ultimately, by viscosity. The energy that is transferred to the low frequencies is responsible for the spectral shift to low frequencies during a developing sea, as shown in Figure 8.4.

The low-frequency limit of the wave spectrum is the swell which propagates away from the region of wave generation and is dissipated only slowly. The energy of the low-frequency swell will propagate with a group velocity greater than that of wind-generated waves (see Table 8.1) and therefore, at a distance from the storm, the lowest-frequency swell waves will be the first to arrive. These will be followed by progressively slower, higher-frequency waves.

Consider a wave recording station at a distance L from a storm. From equations 8.3 and 8.4:

$$c_g = g/2\sigma$$

where σ is the wave frequency. The time, t, taken for the swell to travel a distance L is:

$$t = (2\sigma/g)\, L \tag{8.7}$$

Therefore, the frequency of the waves arriving at the recording station will increase with time at a rate proportional to L. From a graph of wave frequency against time, the gradient will give the distance traveled and the intercept will give the time of origin of the waves. If the direction of the swell is also measured, or if wave information is available from an additional station, then the position of the storm at that time of swell generation can be ascertained.

Low-frequency swell can travel for very long distances because of the small attenuation and the swell will propagate along great circle routes, as shown in Figure 8.6. This map shows the possible great circle routes to the Island of St Helena at 16°S. This island has been the subject of much interest in wave research because of the arrival of exceptionally large rollers, or swell, between December and March. From work by Cartwright *et al.* (1977), it has been shown that this swell may originate from winter storms in the Newfoundland region of the Atlantic Ocean. From wave records taken during a large-swell event on St Helena, Cartwright showed that the origin of the swell on the great circle route was coincident with an intense storm in the Newfoundland region eight days prior to the arrival of the swell. Similar studies have shown swell propagation along great circle paths from the South Pacific to Alaska. It is noted that the group velocity of the swell can be greater than the velocity of the generating storm, and so the arrival of swell can give forewarning of a tropical cyclone or a middle-latitude depression. In February 1979 some exceptional swell waves, with a period of 18 s and significant height of 7 m, propagated along the English Channel causing considerable damage to Portland. These waves originated from an intense depression in the central North Atlantic.

8.3   Long waves

The main feature that distinguishes long waves from wind waves is that their wavelength is large compared with the water depth and therefore the waves influence the whole depth of the ocean. In a long wave, water particles oscillate horizontally and are in phase over the depth of the ocean. Only close to the bottom boundary will they show any attenuation. Provided that the elevations of the waves are small compared with the mean water depth they will propagate with a wave speed $\sqrt{(gh)}$.

Long waves in the ocean may be caused by large falls in surface pressure associated with tropical cyclones or by seismic activity. These waves, like

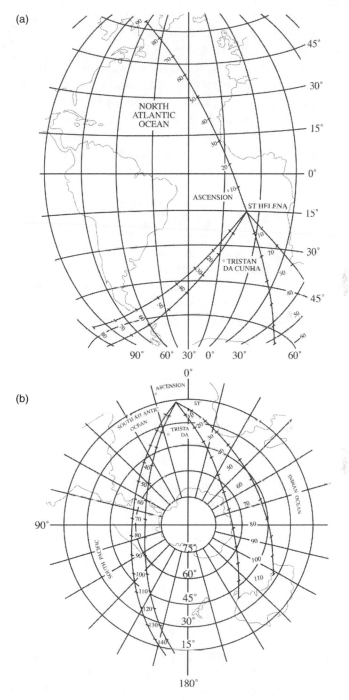

Figure 8.6 Great circle routes for propagation of swell to the island of St Helena from (a) the North Atlantic Ocean and (b) the Southern Ocean. Reproduced, with permission, from Cartwright, D.E. *et al.*, 1977, Quarterly Journal of the Royal Meteorological Society, 103(438): page 664, figures 4a and b

swell waves, tend to follow great circle routes, and the location and time of generation of the waves can be obtained by noting the time of their arrival and period at the coast. By this method the origins of major submarine earthquakes have been located in the Pacific Ocean. In deep water such as the Pacific Ocean, the variations in bottom depth are relatively small compared with the total depth of the ocean and therefore the wave speed is little affected by the bottom. However, in shallow water the bottom variations reduce the speed of the long waves considerably. As a wave approaches a shore its speed will diminish but its frequency will remain unchanged, and so the wavelength becomes shorter and the waves appear to bunch together. Mathematically:

$$L = 2\pi \frac{\sqrt{(gh)}}{\sigma} \tag{8.8}$$

The flux of energy per unit time, F, across a coast is the product of the wave-energy density and the group velocity. This flux of energy, F, will remain constant and therefore, because of the reduction in group velocity, the wave-energy density must increase. The wave energy is proportional to the square of the wave height and hence the wave height, H, will increase as the shore is approached. Thus:

$$F \propto \sqrt{gh}\, H^2$$

and

$$\therefore H \propto \frac{F^{\frac{1}{2}}}{(gh)^{\frac{1}{4}}} \tag{8.9}$$

If a wave has height H_1 in water of depth h_1, then equation 8.9 implies that its height H_2 in water depth h_2 is given by:

$$H_2 = H_1 \left(\frac{h_1}{h_2}\right)^{\frac{1}{4}} \tag{8.10}$$

For a tsunami which has an elevation of 1 m in water 4 km deep propagating into a coastal region having water 10 m deep:

$$H_2 = 4.47\,\text{m}$$

Tsunami waves between 3 and 5 m high have been measured in coastal locations in the Pacific Ocean. It is also noted that, for a tsunami wavelength of 100 km in water 4 km deep, equation 8.8 predicts that the wavelength will decrease to 5 km in the shallow water.

For wind waves and swell a similar phenomenon occurs but, because of the increase in slope of the waves, they become unstable and break to form the surf zone. Refraction effects cause the wave front approaching at an angle to the shore to bend towards the shore.

For tsunamis, the wave frequencies are approximately 10^{-3} s^{-1} or about 15 min. This is considerably less than the Coriolis frequency of 10^{-4} s^{-1} and therefore these waves are not affected by the Coriolis force. However, for longer wave periods the Coriolis force has to be taken into account. Long waves forced by tides or synoptic scale storms are all considerably influenced by the Coriolis force.

Lord Kelvin first described the behaviour of a long wave in a channel constrained by the Coriolis force. Figure 8.7 shows the behaviour of the sea surface and currents in a Kelvin wave. The effect of the Coriolis force is to constrain the wave propagation to the boundary of the channel. The elevation, ξ, is given by:

$$\xi = \xi_0 \exp\left(-\frac{y}{L}\right) \cos(\sigma t - kx) \tag{8.11}$$

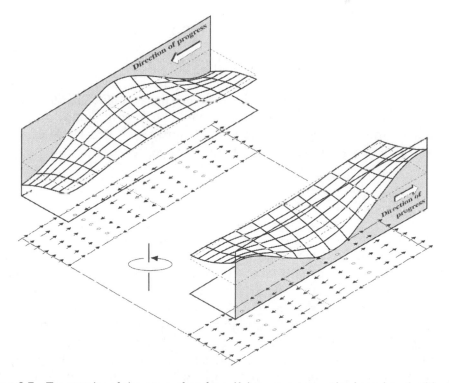

Figure 8.7 Topography of the sea surface for a Kelvin wave in a wide channel in the Northern Hemisphere. Note the wave progresses with the boundary on its right-hand side in the Northern Hemisphere. Reproduced, with permission, from Gill, A.E., 1982, Atmosphere-Ocean Dynamics, Academic Press: page 380, figure 10.3

where $L = \frac{\sigma}{fk} = \frac{\sqrt{gh}}{f}$ and f is the Coriolis parameter (see Section 7.2). σ is the wave frequency and k is the wavenumber. x is the distance along the channel and y is the perpendicular distance from the side of the channel.

At the side of the channel, $y = 0$, the wave has a maximum amplitude ξ_0. The amplitude decays with distance and will become negligible for distances much greater than L. For a long-wave speed of $30\,\mathrm{m\,s^{-1}}$, typical of waves on the continental shelf at $43°$ latitude, $L = 30/10^{-4} = 300\,\mathrm{km}$. Basically, the Kelvin wave propagates as a long wave along the boundary but it is in geostrophic balance normal to the boundary. Therefore, the current in the x direction is in geostrophic balance with the pressure gradient caused by the slope in the sea surface i.e.:

$$fu = -g\frac{\partial \xi}{\partial y} \tag{8.12}$$

where ξ is the height of the sea surface from mean sea level and f is the Coriolis parameter (see Chapter 7).

Substituting for ξ in equation 8.12 gives:

$$u = \sqrt{\frac{g}{h}}\xi \tag{8.13}$$

Therefore, high elevation is associated with a current in the direction of wave propagation and low elevation with a current in the opposite direction.

In a basin closed at one end, a Kelvin wave will propagate around the coast in an anticlockwise direction in the Northern Hemisphere and a clockwise direction in the Southern Hemisphere. Examples of these waves will be discussed with tides and storm surges in later sections of this chapter.

8.4   Internal waves

The previous sections of this chapter have described only waves on the ocean surface but in the ocean and atmosphere there is another type of wave known as an internal wave. Consider two fluids having densities ρ_1 and ρ_2 in hydrostatic equilibrium. If the interface between the two fluids is moved upwards from the equilibrium position, then an element of denser fluid ρ_2 will be subject to a downward buoyancy force proportional to $(\rho_2 - \rho_1)g$. If the two fluids are the ocean and the atmosphere, then ρ_2 is approximately $1025\,\mathrm{kg\,m^{-3}}$ and ρ_1 is $1.25\,\mathrm{kg\,m^{-3}}$. In this case the buoyancy force is, to a very good approximation, proportional to $\rho_2\,g$. The fact that the ocean density is so very much larger than the atmospheric density is the reason why the density of the atmosphere is neglected in formulae for surface waves.

However, for two fluids of similar densities, the $(\rho_2 - \rho_1)$ term, and hence the buoyancy force, will be small. For example, consider a layer of fresh water overlying a layer of ocean water, a situation commonly encountered in estuaries and fjords. The difference in density is $25\,\mathrm{kg\,m^{-3}}$, assuming that the fresh water has a density of $1000\,\mathrm{kg\,m^{-3}}$ and the ocean has a density of $1025\,\mathrm{kg\,m^{-3}}$. The restoring buoyancy force is only 2.5% of the buoyancy force produced by a surface gravity wave and therefore the restoration of the interface to equilibrium will take considerably longer than for a surface wave. Internal waves, therefore, generally have longer periods and move more slowly than their surface counterparts.

When the layer of lighter water overlies a considerably deeper layer of denser water, then the propagation speed of the internal wave is:

$$c = \sqrt{\left(\frac{gh_1(\rho_2 - \rho_1)}{\rho_2}\right)} \qquad (8.14)$$

where h_1 is the depth of the upper layer. In a fjord the surface fresh water layer may be approximately $4\,\mathrm{m}$ deep and therefore, from equation 8.14, the internal wave speed will be $1\,\mathrm{m\,s^{-1}}$. This may be compared with the surface wave speeds in Table 8.1.

Figure 8.8 shows the circulation induced by an internal wave in a two-layer fluid such as a fjord. It is noted that the density interface has much larger amplitude displacements than those which occur at the surface. Therefore the circulation in the upper layer becomes intensified at the crest of the internal wave in the opposite direction to the direction of propagation of the internal wave. A ship in a fjord will not only produce a surface bow wave but also an internal wave, if the surface layer is shallow. Depending on the

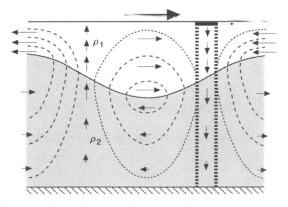

Figure 8.8 Circulations (streamlines are shown as dashes) associated with progressive internal waves on a density interface between a lighter fluid, ρ_1 and a denser fluid, ρ_2. Surface slicks form in the region of surface convergence

speed and length of the boat, the reversed surface circulation at the crest of the internal wave will impede the progress of the boat. This phenomenon is known as 'dead' water because boats caught in these regions will make reduced headway through the water.

A second example of internal waves includes those whose wavelengths are much longer than the total depth of the fluid. Such waves are termed long internal waves. Their wave speed, c, is given by:

$$c = \sqrt{\left(g \left(\frac{h_1 h_2}{h_1 + h_2} \right) \left(\frac{\rho_2 - \rho_1}{\rho_2} \right) \right)} \qquad (8.15)$$

where h_1 and h_2 are the depths of the top and bottom layer, respectively. The ocean may be considered, albeit very simply, as a two-layer fluid with warm thermocline water with density $\rho_1 \sim 1025\,kg\,m^{-3}$ overlying deep, cold abyssal waters with density $\rho_2 \sim 1027\,kg\,m^{-3}$. If $h_1 = 500\,m$ and $h_2 = 3500\,m$, then, from equation 8.15, $c = 2.9\,m\,s^{-1}$. This wave speed can be compared with a long surface gravity wave, such as a tsunami traveling at $200\,m\,s^{-1}$ in $4\,km$ depth of water. For a long internal wave having a period of $1000\,s$, the wavelength will be $2.9\,km$, whilst a surface long gravity wave of a similar period would have a wavelength of $200\,km$.

The amplitude of a typical long internal wave can be estimated in the following way. The hydrostatic pressure variations due to a surface wave of amplitude a_s will be $g\,\rho\,a_s$. The corresponding pressure variations due to an internal wave of amplitude a_1 are $g[(\rho_2 - \rho_1)/\rho_2]a_1$.

If it is assumed that hydrostatic pressure changes are similar for both waves, then:

$$a_1 = \left(\frac{\rho_2}{\rho_2 - \rho_1} \right) a_s \qquad (8.16)$$

From the previous example, $\rho_2/(\rho_2 - \rho_1)$ is 513, and therefore a long surface wave of 20 cm height may produce an internal wave approximately 102 m in height.

In the above examples it has been assumed that the ocean approximates to a two-layer fluid. In reality, the density field changes continuously with depth and therefore internal waves are rather more complicated than previously suggested. In general, the largest amplitude internal waves are located in regions where the vertical density gradient is a maximum, such as the base of the seasonal thermocline, the halocline in the Arctic Ocean or in the main pycnocline.

In regions of continuous density variation, the buoyancy frequency described by equation 2.10 is a useful parameter. The buoyancy frequency corresponds to the highest frequency of the internal waves that can occur. Also, by analogy with equation 2.10, it can be appreciated that the vertical variation in density will produce corresponding variations in the buoyancy frequency. Hence the internal waves will propagate at different speeds and refraction of the wave fronts will occur. The internal waves will, therefore, generally propagate both horizontally and vertically. An internal wave generated in the thermocline will tend to propagate downwards into the deeper ocean.

The mechanisms that produce internal waves are numerous. Internal waves in the seasonal thermocline are often caused by storms, but tides and surface currents may also generate them, particularly close to the shelf break or at the entrance to an estuary. They are found literally everywhere in the ocean and often, because of their large amplitudes, they may cause problems for oceanographers. For example, hydrographic data may be contaminated by internal waves which may result in considerable errors in the determination of geostrophic currents by the dynamic method. In addition, sound waves may be refracted by a group of internal waves which will add noise to the received signal. For submariners, internal waves may cause considerable disturbance both to their boat and their sonar systems.

8.5   Ocean tides

The tides are ubiquitous in the ocean. In the deep ocean, tidal currents ranging between 1 and $10\,\mathrm{cm\,s^{-1}}$ are observed in most current meter records and in the trajectories of neutrally buoyant floats. To the dynamical oceanographer, these tidal fluctuations represent an annoyance and they have to be removed from the record to obtain long-term currents. Measurements by bottom pressure gauges, and by satellite altimeters, have shown for the first time the regular rise and fall of the tides, typically 10–100 cm in range, over large areas of the deep ocean. Over the continental shelves and especially in shallow, semi-enclosed seas, such as the North Sea, the tidal ranges are an order of magnitude larger than those in the deep ocean. Ranges of over 10 m occur in the Severn Estuary, United Kingdom, whilst in the Bay of Fundy, Nova Scotia, the range is 14 m.

The associated tidal currents are usually from 1 to $2\,\mathrm{m\,s^{-1}}$ and are generally larger than wind-induced currents and geostrophic currents. In the English Channel, between the island of Alderney and Cap de la Hague, tidal currents up to $4.5\,\mathrm{m\,s^{-1}}$ occur, whilst near Bodo, north Norway, at the entrance to the Skjerstadt Fjord, currents of $8\,\mathrm{m\,s^{-1}}$ are regularly observed.

The relationship between the tides and the Moon's position has been known since antiquity. Early Roman writers described the twice-monthly spring and neap cycle, and a Persian philosopher, in the second century BC, showed that the difference between the height of consecutive tides, known as the diurnal inequality, varied with the Moon's position north and south of the equator. The first theory of tides was proposed by Newton (1687), following Galileo's ideas on orbital motion. Newton recognized that the orbital motion of the Moon around the Earth, though itself in equilibrium, would produce an imbalance of forces on the Earth's surface which would, in turn, distort the sea surface.

The rotation of the Earth about its own axis will be considered later but will be ignored in the present discussion. Figure 8.9 shows the net forces produced by the Moon at different points on the surface of the Earth. If the Earth and the Moon are considered as point masses, then it can be shown that they will rotate about their common centre of gravity with the outward centrifugal force balanced by the central force of gravitational attraction. However, on a sphere, as opposed to a point mass, this balance of forces is only achieved at the centre of mass of the Earth. At a point Z, the zenith on the Earth's surface, the gravitational attraction is greater than that at the centre of the Earth because of its close proximity to the Moon, whilst at the point N, the nadir, the gravitational attraction is less than at the centre of the Earth. Furthermore, because all points on the Earth describe circles of the same radius as the Earth rotates about the common centre of gravity of the Moon and the Earth, the outward centrifugal force at each point on the Earth's surface is the same. Therefore, at the point Z, the central gravitational force exceeds the centrifugal force and there is a net force directed towards the Moon. At the point N, the centrifugal force exceeds the central gravitational force and there is, therefore, a net force directed away from the Moon.

The net force F per unit mass acting at the point Z is given by the difference between the gravitational force at the centre of the Earth and that at the point Z. The force per unit mass, F_e, at the centre of the Earth is given by:

$$F_e = \frac{Gm_L}{R^2} \tag{8.17}$$

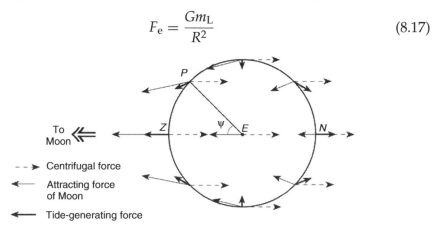

Figure 8.9 Tide-generating force as a resultant of attracting force and centrifugal force along a meridional section through the Earth. Z and N indicate the Moon in zenith or nadir, respectively; E is the centre of the Earth. Reproduced, with permission, from Dietrich, G. et al., 1980, General Oceanography, Wiley Interscience: page 421, figure 180

where G is the gravitational constant, m_L is the mass of the Moon, and R is the distance between the centres of the Moon and the Earth. Since the distance to the point Z is $R - a$, where a is the radius of the Earth, the net force per unit mass at Z is:

$$F = Gm_L \left[\frac{1}{(R-a)^2} - \frac{1}{R^2} \right] \tag{8.18a}$$

$$F = \frac{Gm_L}{R^2} \left[\frac{1}{\left(1 - \dfrac{a}{R}\right)^2} - 1 \right] \tag{8.18b}$$

Since $a/R = 1/60$, then by the binomial theorem:

$$\frac{1}{\left(1 - \dfrac{a}{R}\right)^2} = 1 + \frac{2a}{R} + \dots$$

Hence, neglecting all higher-order terms:

$$F = \frac{Gm_L}{R^3} 2a \tag{8.19}$$

It is noted that the net tidal force depends on the inverse cube of the Earth-Moon distance, whilst the gravitational force depends on the inverse square.

For a general point, P, on the Earth's surface, the tidal force can be resolved into vertical (Z_1) and horizontal (H_1) components. It can be shown that these components are given by:

$$Z_1 = \frac{-2Gm_L a}{R^3} \frac{3}{4} \left[\cos 2\psi + \frac{1}{3} \right] \tag{8.20}$$

$$H_1 = \frac{2Gm_L a}{R^3} \frac{3}{4} [\sin 2\psi] \tag{8.21}$$

The angle ψ is between the point P and the position where the Moon is overhead. The Moon is overhead at the equator in Figure 8.9. Numerical evaluation of Z_1 shows that the component is typically 9×10^{-6} of the Earth's acceleration due to gravity, g. Therefore, the tidal force in the direction of the local vertical is very small compared with g.

Only in the horizontal direction is the tidal-force component comparable with other horizontal forces such as the pressure-gradient force. This is an analogous situation to that described in Section 7.1 where the vertical component of the Coriolis force was neglected and the horizontal component retained. The horizontal components of the tidal-generating force, shown in Figure 8.9, are directed towards the zenith and nadir. At Z and N, the horizontal component vanishes since $\Psi = 0$. This convergence of forces will cause a horizontal acceleration towards the zenith and nadir, and in consequence, a rise in sea level will occur at these two points. A reduction in sea level will occur along the great circle path at $90°$ longitude to the zenith and nadir.

The Moon's orbit varies relative to the Earth's equator and so the latitude of the zenith and nadir will change with time, but the relative distribution of tidal forces will remain similar. In addition, it is noted that the rotation of the Earth does not affect the tide-producing forces and therefore, as the Earth rotates, the tide will appear to rotate around the globe with the Moon. The apparent period of rotation of the Moon around the Earth is 24 hours and 50 min, and therefore the semi-diurnal tide will be delayed by 50 min each solar day. Hence, consecutive high waters will be delayed by 25 min.

In addition to the semi-diurnal tide, variations in the Moon's position will produce other longer-period tides. For example, the distance between the Earth and the Moon varies with a period of 27.55 days, and a larger range of tides will occur when the Earth is closest to the Moon. In addition, the angle between the position of the Moon and the Earth's equatorial plane, known as the declination angle, has a period of 27.32 days and this variation also affects the tidal ranges.

The tidal forces due to the Sun can be obtained in exactly the same way as described for the Moon. The ratio of the tidal force produced by the Sun compared with that of the Moon is given by $(m_S/m_L)(r_L^3/r_S^3)$ from equation 8.19, where m_S and m_L are the masses of the Sun and Moon, respectively, and r_S and r_L are the respective orbital distances from the Earth. The ratio has a value of 0.46 and therefore the lunar tides will always dominate the solar tides. These solar tides will have semi-diurnal and diurnal periods, as well as a semiannual period which is associated with annual variations in the Earth - Sun distance and in the declination angle.

Table 8.2 shows the principal tidal frequencies associated with the orbital variations of both the Moon and the Sun relative to the Earth. These tidal frequencies are known as the astronomical tides. Detailed astronomical

Table 8.2   Principal astronomical tides

Tide	Generating force	Type	Period
M_2	Moon	Semi-diurnal	12 hours 25 minutes
S_2	Sun	Semi-diurnal	12 hours 00 minutes
N_2	Moon	Semi-diurnal	12 hours 40 minutes
K_2	Moon and Sun	Semi-diurnal	11 hours 58 minutes
K_1	Moon and Sun	Diurnal	23 hours 56 minutes
O_1	Moon	Diurnal	25 hours 49 minutes
P_1	Sun	Diurnal	24 hours 04 minutes

calculations of the orbits of the Sun, Earth and Moon have shown up to 396 distinct tidal frequencies can be identified.

Despite the considerable qualitative success of the astronomical theory of tides, the observations of tides, which commenced in the 17th century, show considerable discrepancies with the theory. First, the amplitude of the surface elevation for the semi-diurnal, or M_2, tide is $25 \cos^2 \Psi$ cm according to the astronomical theory. For $\Psi = 45°$ the theoretical amplitude is 12.5 cm, whilst observations of the M_2 tide in the North Atlantic Ocean reveal amplitudes which are an order of magnitude higher. Only in the Mediterranean Sea is the observed tide of the predicted amplitude. Second, the M_2 high tide at a particular place should, in theory, correspond to the passage of the Moon over the meridian but observations show a delay in the time of high water by many hours. The magnitude of this delay varies from place to place. Thus, the 'equilibrium theory' could not explain the propagation of the tide around the globe as shown in Figure 8.10.

In the eighteenth and nineteenth centuries, it was recognized by Laplace and other mathematicians that the tidal wave would be distorted by:

(i)   The horizontal component of the Coriolis force.
(ii) The distribution of the continents.
(iii) The variation in the depth of water as the tide travels around the globe.

Therefore, for a complete theory of tides, it would be necessary to solve the complete set of hydrodynamic equations on a rotating sphere subject to the astronomical tide-generating force. Although the mathematicians Laplace and Hough did find solutions to these equations, they were too simple to describe the complex tidal response of an ocean basin with complicated bottom topography and irregular coastlines.

The gulf between tidal theory and observation prompted the adoption of a different approach to tidal predictions that is widely used today. This approach is known as harmonic analysis. It had long been recognized that, although the response of the ocean to the astronomical tidal forces was complicated, each unique astronomical frequency would produce a corresponding signal in each tidal record, provided that the ocean behaved as a linear system. Thus, a tidal record can be described by a summation of sinusoidal waves having known astronomical frequencies (see Table 8.2) and for N frequencies the sea surface elevation is given by:

$$\xi = \sum_{i=1}^{N} a_i \cos(\sigma_i t - \phi_i) \tag{8.22}$$

where a_i and ϕ_i are the amplitude and phase corresponding to the known astronomical frequency σ_i. In order to calculate a_i and ϕ_i, it is necessary to

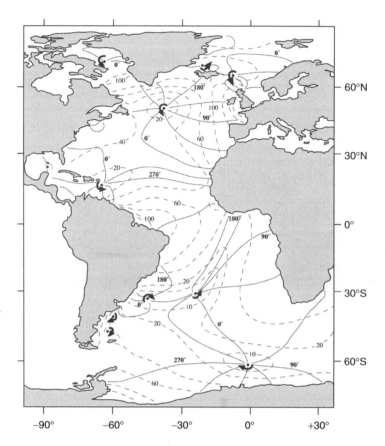

Figure 8.10   Co-tidal map of the M_2 (semi-diurnal lunar) tide for the Atlantic Ocean. Solid lines: co-tidal lines (phase) in degrees. Dashed lines: co-amplitude lines in centimetres. The direction of rotation of the tidal wave is shown by a solid arrow. Reproduced, with permission, from Pugh, D.T., 1987, Tides, Surges and Mean Sea Level, Wiley, page 158 Figure 5.7

have records at least 10 tidal periods long and, in practice, because of the spring - neap cycle, it is usual to take a minimum record of 28 days' duration.

Table 8.3 shows the harmonic analysis of tidal records from Immingham (UK) and Manila (Philippines). It can be seen that, at Immingham, the semi-diurnal, or M_2, tide is dominant, whilst at Manila the diurnal, or K_1, tide is more important. The harmonic analysis shows that, in addition to the dominant components, there are a number of other frequencies present. At

Table 8.3   Amplitudes of constituent tides at Immingham, UK, and Manila, Philippines

	M_2	S_2	N_2	K_2	L_2	O_1	K_1
Immingham (amplitude/cm)	223.2	72.8	44.9	18.3	12.1	16.4	14.6
Manila (amplitude/ cm)	20.3	–	–	–	–	28.3	29.7

Immingham, the S_2. N_2 and K_2 semi-diurnal tides are present and, in all, it is possible to distinguish 23 harmonic components. At Manila, the diurnal tides O_1 and K_1 are larger than the semi-diurnal tide M_2.

Having obtained the phases and amplitudes of the tides for each astronomical frequency, it is then possible to make tidal predictions indefinitely into the future. However, in harmonic analysis, other components appear which do not correspond with astronomical frequencies. First, variations in wind stress, atmospheric pressure and ocean temperature all cause fluctuations in sea level from daily to annual periods. Second, long-term changes in sea level may be present in a tidal record as the result of climatic variations and the relative movement of the land. Third, the hypothesis that the ocean is a linear system is not valid in shallow water, i.e. water less than 100 m deep where large-amplitude tidal currents occur. Non-linear interactions between the tidal components produce higher-frequency harmonics. For example, the M_4 and M_6 tides are harmonics of the M_2 semi-diurnal tide and they have periods of 6 hours 12 min and 4 hours 4 min, respectively. In the central English Channel, the M_4 tide has an amplitude as large as the M_2 component and it produces a double high water at a number of ports. Indeed, at Southampton, the M_2 and M_4 interaction give a prolonged stand of high water. These tides are called shallow water tides or overtides.

From the harmonic analysis, it is a straightforward process to obtain the range of spring and neap tides. For a station such as Immingham, where S_2 and M_2 are the dominant constituents, the range of the spring tide, when the tidal forces of the Sun and the Moon are in phase, is given by $2(a_{M2} + a_{s2})$. The range of the neap tide at such a station is $2(a_{M2} - a_{s2})$ and, at this time, the Sun and the Moon are 90° out of phase. For Immingham (Table 8.3), the spring range is 5.92 m and the neap range is 3.01 m.

Harmonic amplitudes and phases for a given frequency can be used to obtain a geographical representation of the tide in an ocean basin or a sea. The lines of constant phase are known as co-tidal lines and they correspond to all points which are at the same tidal stage relative to the Moon's position and with respect to the Greenwich or Universal meridian. For example, the time of high water will occur at the same time along a co-tidal line. The lines of constant amplitude, or co-range, correspond to points having the same amplitude.

Figure 8.10 shows co-tidal and co-amplitude lines in the Atlantic Ocean, predicted by a model, for the M_2 tide. It can be seen that the M_2 tide rotates in an anticlockwise direction about a point in the centre of the North Atlantic Ocean, known as the amphidromic point. At this point the tidal range is a minimum. Typical amplitudes in the Atlantic Ocean range from 50 cm in the centre up to 150 cm on the eastern Atlantic coast. The highest tidal amplitudes tend to be found close to the continental coasts. Generally, amphidromic points occur in the central parts of ocean basins and this is why mid-ocean islands often have small tidal ranges. In the North Sea two amphidromic points exist. One is located in the southern North Sea and the other is just off the southern Norwegian coast. The M_2 tides rotate in an anticlockwise direction around the North Sea. Measurements made using tide gauges, bottom pressure recorders in the deep ocean and satellite altimeters have mapped the tides in most of the ocean basins.

As briefly discussed earlier, the equilibrium theory of tides cannot account for the amphidromic points and the observed ranges in sea surface elevation. This arises because it is not possible for the sea surface to come into equilibrium with the tide-generating forces on the short time scale of the major tides. An ocean forced by tidal frequencies will produce long gravity waves whose frequencies will be similar to the Coriolis frequency. Such long gravity waves will, therefore, be considerably modified by the horizontal component of the Coriolis force. In a semi-enclosed basin these waves will propagate as Kelvin waves (described in Section 8.3), and they will be reflected and rotated by the boundary. This rotation will be in an anticlockwise direction in the Northern Hemisphere and this corresponds, in a qualitative manner, to the behaviour of the M_2 tide in the North Atlantic Ocean.

During the nineteenth and twentieth centuries, many mathematicians studied the behaviour of long waves in a variety of idealized basins of uniform depth by analytical methods. More recently, however, direct numerical solutions of the linear hydrodynamic equations for the world's oceans, including realistic bottom topography and coastlines, have been obtained by forcing the models with astronomical tidal forces. These models have shown that good agreement with observations can be obtained, as had originally been suggested by Laplace (1778). However, for the tides on the continental shelves, there exist non-linear interactions. The variations in bottom topography are also important in these regions and frictional processes play a crucial role. For all the above reasons, more complicated models are required to give reasonable predictions of tidal elevations and currents on the continental shelves.

Despite the success of the observations and tidal models, there are some tidal problems which have not been solved. One of these problems is accounting for all the dissipation of the tidal energy generated by the Earth's interaction with the Moon and the Sun. This will be discussed in Chapter 9.

8.6   Storm surges

The harmonic method can give predictions of sea level at a port to an accuracy of a few centimeters for a decade ahead. However, systematic departures from tidal predictions do occur as the result of variations in surface pressure and wind stress. Generally, these departures are of the order of a few centimetres and they go largely unnoticed. Occasionally, however, they may be of the order of 1 m and, in combination with a high spring tide, can produce considerable flooding of low-lying coastal regions.

On 31 January and 1 February 1953, one such storm surge in the North Sea reached a height of between 2 and 4 m above the predicted sea level and inundated large areas of eastern England and Holland, with a loss of 1700 lives. Surges also occur in other parts of the world where there are partially enclosed seas which are relatively shallow. The northern Adriatic Sea, including the Venetian region, is vulnerable to surges caused by intense winter depressions in the western Mediterranean Sea. Tropical cyclones in the Bay of Bengal have been known to produce surges of between 4 and 7 m along the coast of Bangladesh. These latter surges are particularly devastating because of the low relief of the Ganges delta and they have been known to flood areas up to 100 km inland.

Although positive surges in sea level are of importance to those on land, negative surges can be equally important to the master of a ship. For instance, a negative surge of 1–2 m will cause difficulties for a large oil tanker navigating a shallow entrance to a port and, in exceptional cases, it may lead to a ship floundering on a sandbank.

A 1 hPa reduction in surface pressure will cause a 1 cm rise in sea level, assuming that the water is in hydrostatic equilibrium. An intense depression, similar to that associated with the 1953 storm surge in the North Sea (Figure 8.11), having a central pressure of 970 hPa will produce a rise in sea level of approximately 42 cm or 0.42 m assuming a mean surface pressure of 1012 hPa. This rise in sea level is about one order of magnitude less than the observed value and it is therefore apparent that most of the surge must be accounted for by the direct effect of wind stress on the sea surface.

Consider now a steady wind stress, τ, blowing against a coast. In equilibrium, the wind stress force must balance the pressure gradient caused by the slope in the sea surface:

$$g\frac{\partial \xi}{\partial x} = \frac{\tau}{\rho h} \tag{8.23}$$

where h is the depth of the sea and $\frac{\partial \xi}{\partial x}$ is the sea surface slope. For a given wind stress, the slope in the sea level is inversely proportional to the depth of the sea. Thus, it is only for shallow seas that a major change in sea level is expected.

A steady wind of 22 m s^{-1} or 43 knots, will produce a stress of approximately 1.16 N m^{-2} using the bulk aerodynamic formula (equation 7.40). For the North Sea, $h \sim 40$ m and therefore, from equation 8.23, the sea surface slope is 2.9×10^{-6}. The North Sea is about 600 km in length and therefore the increased height of the sea surface in the southern North Sea, produced by the southward wind stress, would be approximately 1.7 m. Hence, it can be

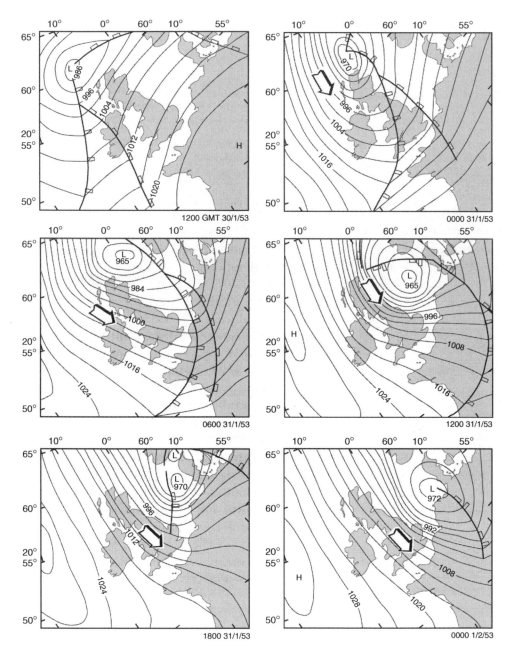

Figure 8.11　The North Sea floods of 1953. Meteorological charts for the period 12.00 hours GMT on 30 January to 00.00 hours GMT on 1 February 1953 (contour interval 4 mb). Reproduced, with permission, from Flather, R.A., 1984, Quarterly Journal of the Royal Meteorological Society, 110(465): page 594, figure 1

seen that a surface-wind stress similar to that which produced the 1953 storm surge can account for a much higher sea surface elevation than pressure factors alone. The total estimated rise in sea level due to both wind stress and pressure is approximately 2.4 m, which is lower than that observed along the English and Dutch coasts in 1953. The reasons for this discrepancy are related to the effect of the focusing of the surge by estuaries, the very shallow depth of water along the southern North Sea and time-dependent dynamics.

An additional feature of North Sea surges is that they propagate in an anti-clockwise direction around the coasts. Figure 8.12 shows the surge elevation on 31 January and 1 February 1953 for the North Sea. The surge was first observed at Aberdeen at 14.00 hours on 31 January 1953 with a height of 0.6 m. It subsequently traveled down the east coast of the UK and arrived at Chatham 11 hours later, when its height was 2.5 m. It then followed the Belgian and Dutch coasts where it reached a maximum height of 3.2 m.

Its propagation was very similar to that of the semi-diurnal tide, which also travels anticlockwise about the North Sea. This surge propagation is related to the Coriolis force, which has been neglected in equation 8.23. First, the Coriolis force causes a rotation of the wind-driven current to the right in the Northern Hemisphere and thus a southward wind stress induces a transport of water towards the eastern coast of Britain which, in turn, induces a rise in sea level. This rise in sea level is then transmitted along the coast by a Kelvin wave, which is trapped to the coast by the Coriolis force (see Section 8.3). The speed of the Kelvin wave for the North Sea, assuming a depth of 40 m, is calculated to be $20 \, \text{m s}^{-1}$ (in an anticlockwise direction). The coastal distance between Aberdeen and Chatham is 800 km, and therefore the expected time delay, for the theoretical speed of $20 \, \text{m s}^{-1}$, is 11.1 hours, which is remarkably close to the observed value. Thus, it has been shown that wind stress, surface pressure and the Coriolis force can account for many features of North Sea storm surges.

The southward Kelvin wave propagation makes it possible to predict storm surges on the east coast of England some 6–12 hours in advance from the observation of tidal residuals (i.e. the difference between the actual sea level and the sea level predicted from tide tables) on the north-east coast of Scotland. This technique was used routinely for sea-level prediction by the Meteorological Office subsequent to the floods of 1953. However, numerical models of the shelf seas coupled with weather prediction models have replaced the statistical methods and are now used to obtain accurate predictions for all parts of the North Sea. Similar techniques have also been used in the Adriatic Sea to give predictions of sea level for up to 12 hours ahead. All forecasts are limited by the predictability of the position and intensity of the storm in question. For regions such as the Bay of Bengal, accurate sea-level predictions require very detailed forecasts of the paths of tropical cyclones. A recent example of the improving capability to forecast

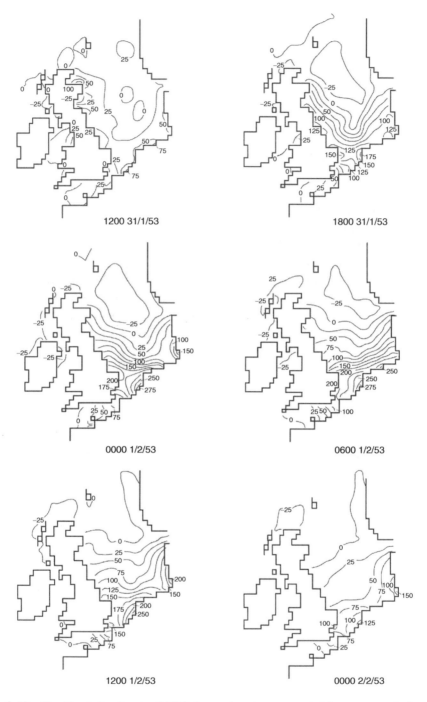

Figure 8.12 The North Sea floods of 1953. Surge elevations computed from a numerical model for the period 12.00 hours GMT on 31 January to 00.00 hours GMT on 2 February (contour interval 25 cm). Reproduced, with permission, from Flather, R.A., 1984, Quarterly Journal of the Royal Meteorological Society, 110(465): pages 605–608, figure 5

storm surges is the accurate forecast 72 hours ahead of the arrival of Hurricane Katrina at the Louisiana coast of the Gulf of Mexico during August 2005, when parts of the city of New Orleans were flooded.

8.7   Atmospheric waves and tides

So far in this chapter all the attention has been on waves within the ocean and there has been little discussion about waves in the atmosphere. Chapter 7 showed there are large scale Rossby waves in the atmosphere and highlighted their importance for the location of mid-latitude weather systems. In this section other atmospheric waves are discussed, including the atmospheric tide.

Surface pressure observations in Indonesia, during the early 20th century, have shown a regular 12-hour cycle of 1–2 hPa as shown in Figure 8.13. This atmospheric solar tide has many causes including the absorption of solar radiation in the upper atmosphere, the release of latent heat in deep convective clouds and gravitational forces. The atmosphere is influenced by the gravitational forces of the Moon and the Sun, but their influences are small. The main influence on the atmosphere is the daily cycle of solar heating rather than gravitational tidal generating forces and the atmospheric tide tends to have regular periods of 12 hours and 24 hours, as compared to the semi-diurnal and diurnal Moon tides (12.4 hours and 24.8 hours) of the ocean. As has already been shown, the daily cycle of solar radiation has a major influence on the mesosphere and lower thermosphere (see Figure 3.7) where pressure is less than 0.1% of the surface pressure. The thermal inertia here is small and the absorption of solar radiation, particularly by ozone, is important (see Section 3.4).

The major difference between the ocean and atmosphere is the large vertical variation of density in the atmosphere (see Section 2.4). The density (ρ) of the atmosphere is given by $\rho = \rho_0 \exp(-z/H)$ where H is the scale height and z is the height above mean sea level (see equation 2. 4). By the conservation of energy, pressure variations will be amplified with height in the atmosphere and therefore the atmospheric tides, which are rather modest in the lower atmosphere, will have large amplitudes in the stratosphere and mesosphere; producing large variations of temperature, pressure and winds in the upper atmosphere. These atmospheric tides propagate vertically and therefore have similarities to the internal waves in the ocean (discussed in Section 8.4). They are a global phenomena (see Figure 8.13), and have their largest signal in the tropical latitudes.

Atmospheric tides are a special class of phenomena called atmospheric gravity waves. The buoyancy frequency in the atmosphere is typically $10^{-2}\,\mathrm{sec}^{-1}$ or a period about 10 minutes (see Section 2.6). This frequency

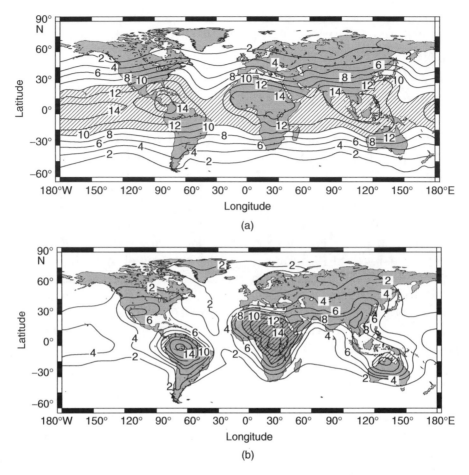

Figure 8.13 Contours of Semi-diurnal (a) and diurnal (b) components of atmospheric tides measured by surface atmospheric pressure. Units: 1 dPa = 10 Pa. Note the tidal pressure variations are much smaller than surface pressure variations in extra-tropical latitudes. Reproduced, with permission, from Dai A. and Wang J., 1999, Journal of Atmospheric Science, 56: pages 3887–3888, figures 15 and 16

can be compared with a typically buoyancy frequency of 10^{-4} sec^{-1} or a period of many hours in the deep ocean. The atmospheric gravity waves move vertically upward in the atmosphere and they carry momentum and energy from the troposphere into the upper atmosphere (i.e. the stratosphere, mesosphere and thermosphere).

These waves are also produced in the troposphere by deep convective storms, which can extend into the stratosphere and mesosphere; by jet streams in the upper troposphere and by flow over mountains. They have a vertical scale of 10–30 kilometers.

The waves can 'break', similar to ocean waves, and this causes them to dissipate. They are not only important at the high frequency time scales (minutes to hours), but also at the low frequency time scales associated with Rossby waves (see Chapter 7). Planetary scale Rossby waves generated in the troposphere by mountain ranges, for example by the Rockies or the Andes, can transmit energy into the stratosphere and above. These waves are responsible for 'sudden warmings' in the stratosphere of 20 K or more, and cause a major disruption of the winds surrounding the polar vortex.

In the equatorial atmosphere, planetary Kelvin waves are generated. These waves are similar to coastal Kelvin waves (see Section 8.3) but they move east-west along the equator. They propagate upwards into the stratosphere and can be absorbed by the mean flow, exchanging the wave momentum for momentum of the mean flow, and this causes an acceleration of the mean zonal wind circulation. This mechanism gives rise to the quasi-biennial oscillation of the stratospheric winds with a period of about 28 months (range of 20–36 months). Atmospheric waves are therefore important not only for understanding high frequency changes in the atmosphere but also for longer period changes in the wind circulation at all levels of the atmosphere. With changes in atmospheric composition (e.g. ozone, and carbon dioxide) due to anthropogenic sources, the vertical transmission properties of the atmosphere are expected to change which in turn will alter the wind circulation particularly in the higher atmosphere.

9

Energy Transfer in the Ocean-Atmosphere System

9.1 Modes of energy in the ocean – atmosphere system

Table 9.1 shows the magnitude of the major energy sources and sinks in the Earth-ocean-atmosphere system. It shows that the major fraction of the incident solar energy is used in the evaporation of water into the atmosphere whilst the kinetic energy of the winds, dissipated into heat by friction, is a relatively minor fraction of the solar flux. The changes in CO_2 and other greenhouse gases, due to human activities, has caused a change in the radiation at the Earth's surface of a similar magnitude to that dissipated by the global wind circulation. The other energy sources are at least one order of magnitude smaller than the dissipation due to the winds.

The direct conversion of visible solar radiation into carbohydrate by photosynthesis is seen to be a relatively inefficient process when compared with the total flux of radiation. Its estimated value, which is based on values of photosynthesis on land and ocean, is probably too large because primary productivity in the oceans is limited by a number of bio-chemical processes. It is interesting to note that world energy production, though small, is now a non-negligible fraction of photosynthetic production. In addition, the potential for biomass as a source of fuel is very limited.

It can be seen from Table 9.1 that the world energy production would have to increase one-hundredfold before it matched the dissipation of kinetic energy by the winds. Geothermal heat, produced by radioactive decay in the Earth's interior, is very small on the world-wide scale, even though it can be locally large in volcanic regions both on the Earth's surface and at ocean ridges. Dissipation of energy by ocean tides is virtually insignificant on a global scale. However, the majority of the dissipation occurs on the continental shelf, especially in the Bay of Fundy, the Bering Sea, the Sea of Okhotsk, the NW European shelf and the Patagonian Shelf where energy fluxes are of the order of 10^3 times larger than the world-wide average value.

The Atmosphere and Ocean: A Physical Introduction, Third Edition. Neil C. Wells.
© 2012 John Wiley & Sons, Ltd. Published 2012 by John Wiley & Sons, Ltd.

Table 9.1   Magnitude of major energy sources and sinks in the Earth-ocean-atmosphere system: Energy flux/$W\,m^{-2}$

Solar radiation	342
Directly absorbed by atmosphere (0.20×342)	68
Latent heat	78
Rate of kinetic energy dissipation in atmosphere and Earth's surface.	2.3
Anthropogenic radiative forcing (2005)	1.6
Photosynthesis	0.03 – 0.25
Geothermal heat flux	0.09
World energy production (2008)	0.031
Solar reflection from full Moon	0.014
Ocean tides	0.003

In this discussion of the energy budget of the ocean and atmosphere, it is possible to neglect all of the sources of heat except the incident flux of solar energy. Reference to Figure 3.10 shows that, apart from the 31% of the incident solar flux which is reflected back into space, the majority of the incident solar radiation is absorbed by the atmosphere and the ocean. However only 20% of the solar radiation is directly absorbed by the atmosphere and the remaining energy is converted into latent heat of condensation and fusion, turbulent heat conduction and long-wave radiation at the surface, to be subsequently released or absorbed into the atmosphere.

From the first law of thermodynamics, the energy absorbed can reside as either internal (heat) energy or it can be used to do work against the environment, appearing as potential or kinetic energy (see Section 2.5). Consider a volume of fluid which is heated uniformly. The temperature, and therefore the internal energy, of the fluid will increase. At the same time, the fluid will expand and do work against gravity. The potential energy of the fluid will therefore increase. The internal energy per unit mass (IE) is C_vT, where C_v is the specific heat at constant volume and T is the absolute temperature (K). The potential energy per unit mass (PE) is gz, where g is the acceleration due to gravity and z is the height above mean sea level. For a column of fluid of unit cross-section, the mass is ρdz, where ρ is the fluid density, and therefore:

$$IE = \int_0^h C_vT\rho dz \tag{9.1}$$

and

$$PE = \int_0^h \rho gz dz \tag{9.2}$$

where h is the depth of the fluid.

From the hydrostatic equation, $p = \rho g z$ and, for the atmosphere, $p = \rho R T$, and thus:

$$PE = \int\limits_{0}^{h} RT\rho\, dz \qquad (9.3)$$

The ratio of the potential energy to the internal energy for the atmosphere is therefore R/C_v or $2/5$. Thus, for each unit of heat absorbed in the atmosphere, $5/7$ of the heat will go into internal energy and $2/7$ will go into potential energy. If it is assumed that 70% of the incident solar radiation is absorbed both directly and indirectly by the atmosphere, then the fraction of solar radiation going into potential energy is $0.7 \times 2/7$ or 0.2 of the incident solar flux.

Most of the solar radiation (99%) in the ocean is absorbed in the upper $100\,\text{m}$. The ocean has a small compressibility and therefore virtually all of the absorbed solar radiation will appear as internal energy and only 0.01%, a negligible fraction, will go directly into potential energy. However, this heat is mixed and transported by ocean currents, such as Ekman currents, downwards into the deep ocean. These processes will cause heating and therefore changes in potential energy to depths of at least the main thermocline and eventually to the abyssal depths.

Returning once more to the atmosphere, it is noted that, for the wind circulation to be in a steady-state balance, the energy dissipated by the wind must be balanced by an equal energy input which must, in turn, be derived from the potential energy of the atmosphere. From Table 9.1, the conversion of PE into kinetic energy (KE) is about $2.3\,\text{W}\,\text{m}^{-2}$, and this may be compared with a rate of $68\,\text{W}\,\text{m}^{-2}$ for the production of PE. Therefore, only 3% of the potential energy is used to drive the general circulation of the atmosphere. For the potential energy and internal energy to be in a steady-state balance, the majority of the potential and internal energy must be continuously and directly dissipated by long-wave radiation into space without being involved in the generation of kinetic energy. In thermodynamic terms, the atmosphere heat engine has a very low efficiency.

To understand why the atmosphere is so wasteful of its potential energy, consider two immiscible fluids of different density, ρ_1 and ρ_2 as shown in Figure 9.1a. The fluids are separated by a vertical partition and have equal volume. It can be shown that the potential energy of the system is $(gh^2/2)(\rho_1 + \rho_2)$ where h is the depth of the fluid. If the partition is now removed, the denser fluid will flow under the lighter fluid and the lighter fluid will flow back over the denser fluid, and it will eventually take up the position shown in Figure 9.1b.

During this process, kinetic energy will be generated and it will, in its turn, be dissipated by the internal viscosities of the two fluids. In its final

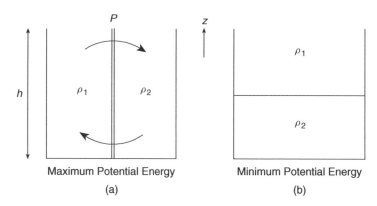

Figure 9.1   Potential energy in a two-layer fluid. (a) Maximum potential energy, before partition removed. (b) Minimum potential energy

state the system still has potential energy but, because the interface is parallel to the geopotential surface, it is not possible to derive any more kinetic energy from the potential energy field. This remaining potential energy is known as unavailable potential energy and it can be shown to be equal to $(gh^2/4)(\rho_2 + 3\rho_1)$. The difference between the initial and final potential energy is $(gh^2/4)(\rho_2 - \rho_1)$. It is this fraction of total potential energy which is converted into kinetic energy and it is known as the available potential energy (APE).

The ratio of kinetic energy to total potential energy is $\frac{1}{2}[(\rho_2 - \rho_1)/(\rho_2 + \rho_1)]$. If the density difference between the two layers of the fluid is small, then the APE will be much less than the initial potential energy.

In the atmosphere, the most satisfactory indicator of APE is the potential temperature because it avoids the adiabatic effect (see Figure 9.2). The horizontal gradient of potential temperature, and therefore APE, is caused by the difference in net radiation between the equator and the poles. The largest slope of the isotherms is in the middle latitudes. Most of the equatorial heating occurs at low levels, whilst the polar cooling occurs at high levels in the troposphere. The potential temperature gradient is therefore between the lower equatorial troposphere and the upper polar troposphere.

In the deep ocean, the horizontal density differences are less than 1% of the mean density and therefore the potential energy is more than 100 times the ocean kinetic energy. The calculation of APE is not straightforward because the flow in the deep ocean basins below 2 km is severely constrained by the mid-ocean ridges and sills (see Section 1.6). However, in the upper ocean the APE can be used. In the upper ocean, an appropriate indicator of APE is the horizontal variation of the potential density. An estimate of APE for the upper ocean is about 28 times larger than the kinetic energy estimate (see Figure 9.3).

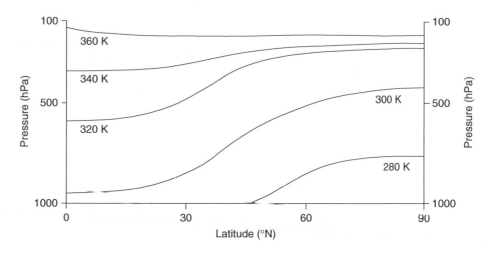

Figure 9.2   Zonal averaged atmospheric potential temperature

The majority of the APE in the ocean basins is in upper 1 km and is associated with the slope of the main pycnocline. This slopes upwards both polewards and equatorwards from the centres of the sub-tropical gyres. The western boundary currents in the sub-tropical gyres are associated with large slopes of the main pycnocline.

In the Atlantic and Pacific Oceans, the largest APE is concentrated on the poleward and western boundaries of the gyres where the main pycnocline rises from its deepest point to the surface in a horizontal distance of 200–300 km. In the ocean basins, below the main pycnocline, the slopes of the density surfaces are smaller and therefore, in common with the above analogue, there is less APE in the deeper layers. The exception is the Antarctic Circumpolar Current system where large slopes in the density surfaces occur across the sub-tropical and Antarctic convergence zones and the deep flows extend to the bottom.

Figure 9.3 shows the energy cycle of the atmosphere and ocean in terms of the KE and APE reservoirs. For the atmosphere, the APE is about four times larger than the KE and the total energy per unit area of the system is 615×10^4 J m^{-2}. If the energy of the atmosphere was not replenished by radiation, then all of its energy would be dissipated on a time scale of $(615 \times 10^4)/2.3 = 2.67\,10^6$ seconds or 31 days, which is very similar to the radiative time scale discussed in Chapter 2.

Quantitative global energy estimates for ocean are not as well determined as those for the atmosphere. Figure 9.3 shows estimates from the literature. There is uncertainty in the values for APE but one estimate is shown for the upper 1.5 km of the ocean. Values for the KE reservoir are better known. It is noted that the APE is many times larger than the KE; in this case is about

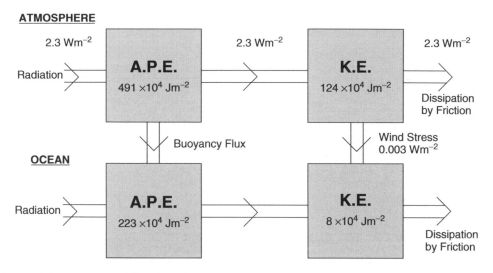

Figure 9.3   Energy diagram of the atmosphere and ocean. Note that the estimates of energy in the ocean are uncertain

a factor of 28. Therefore the APE will store energy and this energy will be released to the ocean to maintain the circulation.

There are a number of sources of energy for the ocean. These are the tidal motions, the direct driving of ocean currents by the wind stress and the surface buoyancy flux. A surface wind of $10\,\mathrm{m\,s^{-1}}$, typical of middle latitudes, will cause an energy transfer of approximately $1\,\mathrm{W\,m^{-2}}$ into the oceans. Of this energy input, some will be dissipated quickly in the surface layer; a fraction will go into the surface waves and swell; some will go into internal waves and the remainder will go into ocean currents. The energy transfer into ocean currents is probably about 0.3% of the work done by the wind stress. The energy input to ocean circulation by the surface buoyancy flux is not as certain. If the buoyancy is added and removed at the same pressure, then it does not generate an ocean circulation. However, in the North Atlantic Ocean positive buoyancy added in low latitudes is mixed slowly to the main thermocline, whilst at high latitudes, strong cooling can cause deep convection to depths of over 1 km. The strong mixing of this cold saline water increases the density of the deep water at high latitudes. The horizontal density difference between low and high latitudes can therefore generate APE. One study has shown that the APE generated by surface cooling in the Northern Hemisphere is less than the energy input by the surface winds. Therefore, the ocean is different from the atmosphere in that the ocean circulation is mainly derived from the kinetic energy input by the winds. Assuming that the dissipation rate is $0.3 \times 10^{-2}\,\mathrm{W\,m^{-2}}$, then the ocean would lose all of its KE in about 300 days if it were not being continually replenished by the surface wind stress. The APE could, however, maintain the circulation for far longer. Based on the values in Figure 9.3, this maintenance could be about 24 years.

9.2   The kinetic energy of the atmosphere and ocean

The kinetic energy can be partitioned into two components:

(i)   The mean kinetic energy (MKE), which is associated with the long-term average circulation of the ocean or atmosphere, is:

$$\text{MKE per unit volume} = \frac{1}{2}\rho\bar{v}^2$$

where $\bar{v}$ is the time mean velocity averaged over a number of years and ρ is the density.

(ii)   The fluctuating or eddy kinetic energy (EKE), which is associated with the time variability of the currents and the winds, is:

$$\text{EKE per unit volume} = \frac{1}{2}\rho\overline{v'^2}$$

where v' is the fluctuation velocity.

The velocity at any given time, $v(t)$, is therefore given by $v(t) = \bar{v} + v'$.

In the atmosphere the MKE is approximately the same magnitude as the EKE. For the observed kinetic energy in the atmosphere, the typical mean velocity and fluctuation velocity would be $12\,\text{m s}^{-1}$.

In the ocean the MKE is similar to the EKE only in the western boundary currents whilst in the ocean interior the EKE may be about 5–10 times larger than the MKE. In the ocean, away from western boundaries, the mean velocity is between 2 and $4\,\text{cm s}^{-1}$ and the corresponding eddy velocity is $10 - 20\,\text{cm s}^{-1}$. The largest fraction of the total KE occurs in the upper ocean, above 1000 m.

In both the atmosphere and the ocean there is a high correlation between the magnitude of the MKE and the EKE. This indicates that energy is being continually exchanged between the two modes. In the atmosphere, the majority of the KE is associated with the locations of the jet streams in the upper troposphere between 30 and 50° latitude. The largest MKE occurs in the sub-tropical jet streams (see Figure 6.11), whilst the transient polar jet streams contribute most to the EKE.

The equivalent high-energy regions for the ocean are the western boundary currents and the equatorial currents but, in all of these regions, the fluctuating energy is often larger than the MKE. Figure 9.4 shows that even the most well-defined 'climatological' currents, such as the Gulf Stream, are regions of considerable variability. In the interior ocean, away from the boundary currents, most of the kinetic energy is in eddies and only a small fraction, usually less than 10%, is found in the mean flow. Generally, both MKE and EKE decrease from the western to the eastern basins of both the Pacific and Atlantic Oceans. In the Antarctic Circumpolar Current, eddies are found along the entire length of the current and they are most intense near to the water-mass boundaries of the sub-tropical and Antarctic convergence zones.

It is interesting to compare the relative contributions of the different scales of atmospheric and ocean circulation systems to the overall budget of kinetic energy. Figure 9.5 shows the kinetic energy of a variety of atmospheric phenomena and Figure 9.6 is a similar diagram for the ocean. It can be seen that the most intense atmospheric systems, such as the tornado and the tropical cyclone, do not make a major contribution to the total kinetic energy of the atmosphere. For example, a sea-breeze system, which may penetrate inland for up to 100 km, has a kinetic energy equivalent to about 10 000 tornadoes. Another example is the extra-tropical cyclone that has a kinetic energy which is between 10 and 100 times larger than that of a tropical cyclone. The former system has a diameter about 10 times as great as a tropical cyclone and so its larger kinetic energy is readily appreciated. Thus, most of the fluctuation energy in the atmosphere is bound up in jet streams, Rossby waves and extra-tropical cyclones in middle latitudes; whilst the other scales of motion, although of local importance, make little contribution to the total

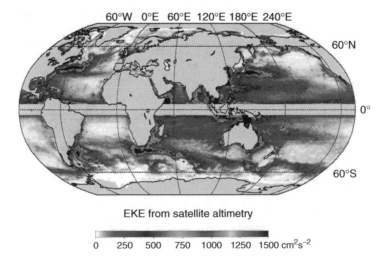

Figure 9.4 EKE in global ocean. Reproduced, with permission, from Barnier, B, Medec, G. *et al.* Ocean Science, 59, page 9, Figure 5b. See plate section for a colour version of this image

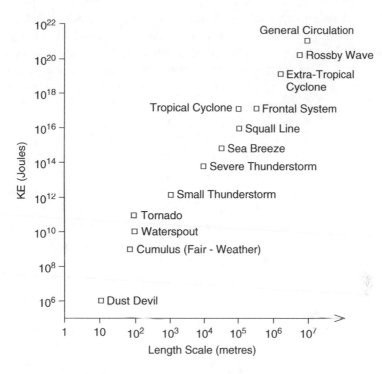

Figure 9.5   Kinetic energy of atmospheric phenomena

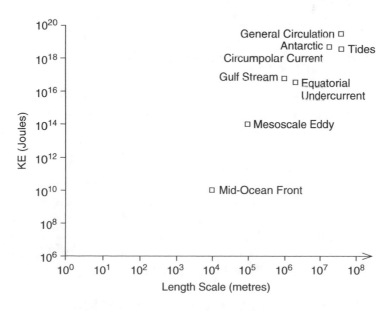

Figure 9.6   Kinetic energy of oceanic phenomena

kinetic energy budget. In lower latitudes the seasonal Asian Monsoon makes a major contribution to the total kinetic energy budget of the atmosphere.

In the ocean, the scale dependence of the circulation systems' contributions is again apparent. The Antarctic Circumpolar Current; the western boundary currents, such as the Gulf Stream and the Kuroshio; the equatorial currents and the tides are the principal contributors, whilst an individual mesoscale eddy, though locally energetic, has an energy which is only about 5×10^{-5} of the total kinetic energy. Thus, 20 000 mesoscale eddies would be required in the ocean at any one time if their contribution was to be equivalent to the large-scale systems. Ocean eddies have typical scales of 50–100 km, and so the requirement would imply that virtually the whole ocean would have to be filled with eddies. Mesoscale circulations have been found in all of the ocean basins and therefore they do make an important contribution to the total kinetic energy of the ocean. Less energetic motions are associated with mid-ocean fronts which may extend horizontally for 1000 km, but the frontal boundary may only be 10 km wide. Deep convection which occurs on scales between 1 and 10 km will make only a small contribution to the total kinetic energy budget.

Surface waves and internal wave energy cannot be ignored in the kinetic energy budget. A wind of $10\,\mathrm{m\,s^{-1}}$ will generate an equivalent surface-wave energy of approximately $2 \times 10^4\,\mathrm{J\,m^{-2}}$. If this energy was typical of the whole ocean, it would be of a similar magnitude to the total kinetic energy of the circulation. It is also noted that internal waves have been shown to have an energy up to $\sim 10^4\,\mathrm{J\,m^{-2}}$, and therefore they also make a significant contribution to the total energy of the ocean circulation.

9.3   Mechanisms of kinetic energy transfer

The simplest method by which kinetic energy is generated, from available potential energy, is by the mechanism in the tank experiment described in Section 9.1. In this experiment the dense fluid runs under the lighter fluid when the barrier is removed, thus releasing kinetic energy. The vertical circulation that is set up in the tank is known as *direct circulation* because it always converts APE into KE. A direct circulation can be visualised in the ocean where the sinking of dense water formed as the result of cooling or evaporation occurs, as in the Mediterranean Sea and the Red Sea (see Figure 5.13). The formation regions of dense water are very localised and occupy a very small proportion of the area of the global ocean. However, these regions are partly responsible for the global thermohaline circulation.

On a non-rotating Earth, the radiative imbalance between low and high latitudes in the atmosphere will drive a direct circulation with potentially warmer air rising at the low latitudes and descending air at higher latitudes. In the observed atmosphere there is a direct circulation known as the Hadley Cell, shown in Figure 6.13. The potentially warm air, driven by the radiative imbalance and by the release of latent heat of condensation, ascends into the ITCZ. The upward branch moves air polewards in the upper troposphere where it cools by long-wave radiation emission into space, and then it descends in the sub-tropical latitudes. The buoyancy source in the lower equatorial troposphere maintains the circulation and thus APE is converted into KE. Therefore it has been shown that direct circulations are associated with the upward motion of less-dense fluid and the downward motion of denser fluid.

Vertical circulations in which the denser fluid rises and the lighter fluid sinks are known as *indirect circulations*. In such cases, the system increases the APE at the expense of the KE. The Ferrel cell, in the middle latitudes of the troposphere (see Figure 6.13) is an example of an indirect, meridional circulation and acts as a sink to the KE of the atmosphere. The loss of KE due to the Ferrel cell is, however, small when compared with the KE gained from the Hadley circulation.

Indirect circulations are responsible for the generation of the large reservoir of APE in the upper ocean. In Section 7.5 it was shown that Ekman currents, produced by the large-scale wind circulation, cause surface convergence in the sub-tropical zones, and divergence on the equator and in higher latitudes. The vertical Ekman circulation pushes warmer water downwards in the sub-tropical gyre and brings cold water up to the surface in the equatorial zone. It thus increases the APE. As has been shown, a surface-wind stress along the equator will produce a longitudinal tilt in the thermocline in both the Atlantic (see Figure 7. 17b) and Pacific Oceans, and this will also produce APE. In equatorial regions, because of the direct response of the thermocline to the wind, the APE reservoir can be replenished relatively quickly within about one year. In higher latitudes the process is rather less efficient and it would take in the region of a couple of decades to produce the observed tilt of the permanent thermocline.

The direct circulation is the simplest mechanism by which kinetic energy can be generated from available potential energy. In the atmosphere or in the ocean there is an additional constraint to the release of PE and that is the rotation of the Earth. Consider what would happen if the tank in Figure 9.1 was rotating when the partition separating the two different density fluids was released. Initially, the behaviour would be similar to the non-rotating case but, after a short time, the Coriolis force would deflect the parcels of fluid perpendicular to the density gradient, as shown in Figure 9.7. In the Northern Hemisphere the dense fluid would be deflected in a westward direction and the lighter fluid would be deflected in an eastward direction. Imagine the tank was an annulus, and therefore the motion was unimpeded perpendicular to

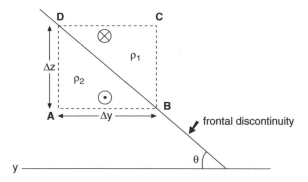

Figure 9.7   Geostrophic equilibrium for a two-layer fluid; y is directed northwards, cross denotes eastward and dot denotes westward geostrophic winds. Note rotation inhibits the adjustment of the fluid to a state of minimum potential energy (see Figure 9.1)

the density gradient, and an equilibrium between the Coriolis force and the horizontal pressure gradient would eventually be achieved. The zonal flows would be geostrophic and the interface would slope as shown in Figure 9.7.

The rotation of the Earth therefore inhibits the conversion of APE into KE. The slope of the interface can be determined from the geostrophic and hydrostatic equations in the following way. If the geostrophic flow of the dense fluid is u_2 and that of the lighter fluid is u_1, then:

$$f\rho_1 u_1 = \frac{p_C - p_D}{\Delta y} \tag{9.4a}$$

and

$$f\rho_2 u_2 = \frac{p_B - p_A}{\Delta y} \tag{9.4b}$$

using the notation of Figure 9.7. From the hydrostatic equation:

$$p_B = p_C + g\rho_1 \Delta z \tag{9.5a}$$

and

$$p_A = p_D + g\rho_2 \Delta z \tag{9.5b}$$

Subtracting 9.5b from 9.5a and dividing through by Δy:

$$\frac{p_B - p_A}{\Delta y} = \frac{p_C - p_D}{\Delta y} - g(\rho_2 - \rho_1)\frac{\Delta z}{\Delta y} \tag{9.6}$$

Substituting for the geostrophic equations (9.4a and b) into equation 9.6 to obtain the slope of the interface:

$$\frac{\Delta z}{\Delta y} = \frac{f(\rho_1 u_1 - \rho_2 u_2)}{g(\rho_2 - \rho_1)} \tag{9.7}$$

As density variations are relatively small compared with velocity variations:

$$\rho_1 u_1 - \rho_2 u_2 \approx \bar{\rho}\,(u_1 - u_2)$$

where $\bar{\rho} = (\rho_1 + \rho_2)/2$, and hence:

$$\frac{\Delta z}{\Delta y} = \frac{\bar{\rho} f\,(u_1 - u_2)}{g\,(\rho_2 - \rho_1)} \tag{9.8}$$

It can be seen that the interface, or front, will have a small slope when the density difference $(\rho_2 - \rho_1)$ is large and the velocity difference $(u_1 - u_2)$ is small. The interface will have a steep slope if the converse is true.

Consider the frontal interface between the northern boundary of the Gulf Stream at approximately 20°C and the Labrador Sea water at about 5°C. Assuming that the Gulf Stream water has an eastward velocity of $0.5\,\mathrm{m\,s^{-1}}$ and that the cold water is quiescent, then from the equation of state:

$$\rho_1 = 1024\,\mathrm{kg\,m^{-3}}$$

$$\rho_2 = 1021\,\mathrm{kg\,m^{-3}}$$

assuming that the temperature dominates the density field. Hence, from equation 9.8:

$$\frac{\Delta z}{\Delta y} = \frac{1022.5 \times 10^{-4} \times 0.5}{9.8 \times 3} \approx 1.7 \times 10^{-3}$$

Thus, the frontal surface will slope upwards towards the north by 170 m in 100 km.

For the atmosphere, equation 9.8 has to be modified by substitution of the ideal gas equation ($p_1/RT_1 = \rho_1$ and $p_2/RT_2 = \rho_2$) and, with the assumption that at the interface $\rho_1 = \rho_2$, the following expression is obtained:

$$\frac{\Delta z}{\Delta y} = \frac{f\bar{T}}{g}\left(\frac{u_1 - u_2}{T_2 - T_1}\right) \tag{9.9}$$

where $\bar{T} = (T_1 + T_2)/2$. For the polar front $u_1 - u_2 = 30\,\mathrm{m\,s^{-1}}$, $T_2 - T_1 \sim 15\,\mathrm{K}$, $\bar{T} = 280\,\mathrm{K}$ and

$$\frac{\Delta z}{\Delta y} = \frac{10^{-4} \times 280 \times 30}{9.8 \times 15} \approx 5.7 \times 10^{-3} \text{ or } \frac{1}{175}$$

Therefore, the frontal surface will have a calculated slope of 5.7 km in a horizontal distance of 1000 km, which is similar to the observed value shown in Figure 9.8.

It has been shown that the effect of the Earth's rotation is to produce regions of discontinuity of density in the ocean and of temperature in the

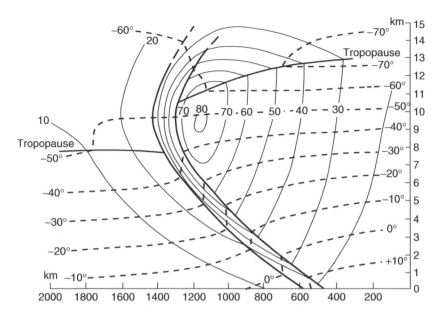

Figure 9.8   Schematic isotherms (dashed lines, °C) and isotachs (thin solid lines, m s⁻¹) in the polar front zone. Heavy lines are tropopauses and boundaries of frontal layer. Note the wind maximum (jet stream) occurs close to the tropopause and on the warm side of the polar front. Reproduced, with permission, from Palmen, E. and Newton, C.W., 1969, Atmospheric Circulation Systems, Academic Press: page 176, figure 7.4

atmosphere. These frontal regions also demarcate the boundaries of different air masses and different water masses. However, these frontal regions are rarely in equilibrium, as described by equations 9.8 and 9.9. Such a balance of forces would not allow for the transfer of heat and other properties across the frontal surface.

As an extreme example, if the atmospheric transfer of heat across the 50° latitude circle was stopped, then the atmosphere polewards of 50°N would cool by 100 K in 100 days. This increase in the horizontal gradient of temperature would in turn cause an increase in the upper-level winds relative to the low-level winds and, at a certain point, the wind shear between the upper and lower troposphere would reach a critical value and the flow would become hydrodynamically unstable. This is known as 'baroclinic instability', and Figure 9.9a shows the initial deformation of a frontal surface as the result of this instability. The warm air (ρ_1) moves polewards and upwards, overriding the cold air (ρ_2), and the cold air moves equatorwards and downwards, undercutting the warm air.

If the warm air follows the frontal surface, as shown in Figure 9.9b, an external supply of energy will be required to lift the warm air above the cold air. However, if the warm air follows trajectory AA', it will find itself in a

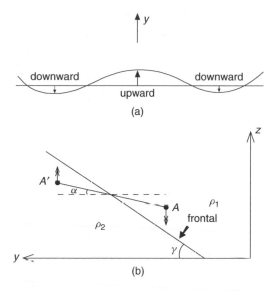

Figure 9.9 (a) Deformation of a frontal surface – plan. (b) Unstable perturbations across a frontal discontinuity. Parcels of air following trajectories AA' will be unstable with respect to their environment, which will cause a conversion of potential energy (PE) to kinetic energy (KE). This will occur if $\alpha < \gamma$

colder environment, and it will accelerate polewards and upwards without any energy input. Similarly, a parcel of cold air following path A'A will find itself in a warmer environment, and it will accelerate equatorwards and downwards. This is very similar to the behaviour of air parcels in a vertically unstable environment where the air parcels continue to accelerate away from their initial positions when once displaced (see Section 2.6). All of the parcels in the present example which have a trajectory between the horizontal and the frontal surface will gain KE from the APE of the front. This kinetic energy will produce a slantwise circulation across the isotherms, and therefore this process is sometimes known as 'slope convection'.

Figure 9.10 shows a more detailed picture of the trajectories in an extra-tropical cyclone. The warm tongue can be seen to move upwards and polewards whilst polar air, developing along the polar front in the rear of the system, can be seen to descend from high levels towards the surface, thus releasing potential energy for the development of the extra-tropical cyclone. Latent-heat release in the region of ascent will provide an additional source of kinetic energy for the system. Eventually, the tongue of warm air will be completely undercut by polar air and the system is then said to be occluded. At this stage most of the APE has been converted to KE and the extra-tropical cyclone will begin to decay as surface friction dissipates its kinetic energy. A typical extra-tropical cyclone will grow to its maximum intensity in one to three days and will decay in about five to six days. In some cases the

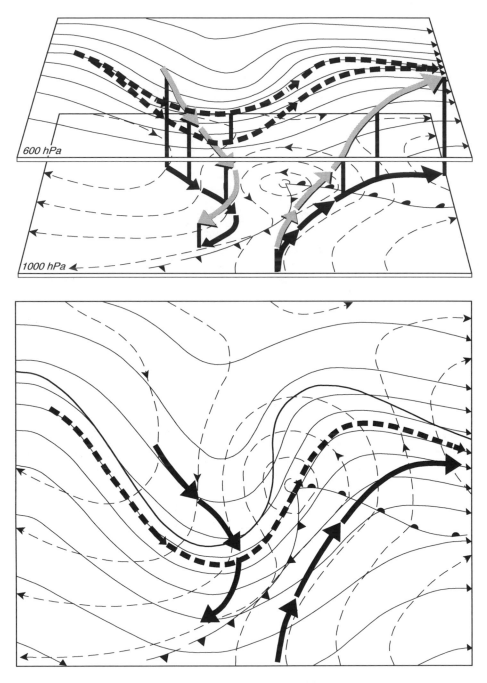

Figure 9.10 Perspective view of a 1000 hPa cyclone and 600 hPa contour pattern. Heavier arrows indicate three-dimensional trajectories in the main ascending and descending branches; thin arrows, their projection on to 1000 hPa or 600 hPa surfaces. Reproduced, with permission, from Palmen, F. and Newton, C.W., 1969, Atmospheric Circulation Systems, Academic Press: page 310, figure 10.20

development can be rapid, of the order of a few hours, and this makes accurate prediction of these systems difficult.

Baroclinic instability is thus occurring at all times in the middle latitudes and it acts to transfer heat between low and high latitudes. However, this process does show a strong seasonal behaviour. In summer, with relatively weak horizontal gradients of temperature, the APE in the atmosphere is smaller and the extra-tropical cyclones are generally weaker. By contrast, in winter larger horizontal temperature gradients are produced not only between the pole and the equator, but also between the continents and the ocean. These large sources of APE produce intense cyclonic systems which transfer heat both polewards and towards the cold continental land masses. These zones of large horizontal temperature gradient and strong vertical wind shear are known as baroclinic zones. In the Northern Hemisphere they are located on the eastern seaboards of the North American and Asian continents in winter, where very cold continental air occurs adjacent to the very warm Gulf Stream and Kuroshio Currents. The baroclinic zones are located at the western end of the extratropical storm track, in the North Atlantic and North Pacific Oceans.

In conclusion, it is seen that vertical instability in low latitudes and baroclinic instability in middle and high latitudes are the major processes by which energy is transferred in the atmosphere.

In the ocean, baroclinic instability is also responsible for the transfer of heat and other properties, such as salinity and nutrients, across frontal surfaces. Figure 9.11(a) shows the development of a Gulf Stream meander and the eventual formation of isolated eddies of cold and warm water on either side of the front. The growth rates of these meanders are slower than their atmospheric counterparts. It takes between one week and one month to produce Gulf Stream rings but not all the meanders grow into eddies. The 'cold' Gulf Stream rings move southwards and westwards into the Sargasso Sea and they may be tracked for up to two years before they finally dissipate. In contrast, the 'warm' Gulf Stream rings, situated on the northern side of the Gulf Stream, tend either to be dissipated on the continental slope by friction or to be reabsorbed into the Gulf Stream system.

Mesoscale ocean eddies are also important for the transfer of heat in the higher latitudes of the Southern Hemisphere. The Antarctic Circumpolar Current is the only zonal current which flows unimpeded around the globe and it therefore has some similarities to the zonal mid-latitude atmospheric flows. Eddies are an important mechanism for transferring heat and salt across the zonal frontal boundaries. The energy for the eddies comes from the slope of the isopycnal surfaces between 40 and 60°S. Studies have shown that these Southern Hemisphere eddies may account for a poleward heat transfer of 0.4×10^{15} W.

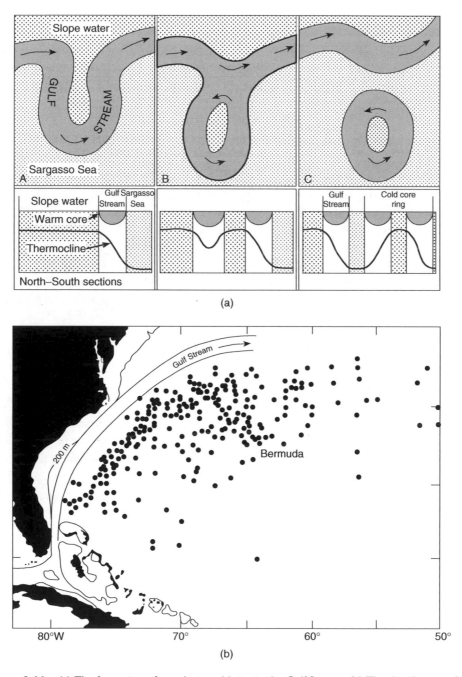

(a)

(b)

Figure 9.11 (a) The formation of a cyclonic cold ring in the Gulf Stream. (b) The distribution of Gulf Stream cold rings from 1932 to 1976. Reproduced, with permission, from Richardson, P.L., 1983, In Eddies in Marine Science, ed. Robinson, A.R., Springer-Verlag: pages 20 and 30, figures 1 and 6

Thus it has been shown that eddies produced by baroclinic instability in frontal zones, in both the atmosphere and ocean, make an important contribution to the transport of heat and freshwater around the planet.

9.4   General circulation of the atmosphere

The atmosphere and the ocean are both highly complex systems which fluctuate on a large variety of time and space scales. Both systems are dissipative and there is, therefore, a continuous flux of energy through the systems which maintains their circulations. Because of the dissipation of energy, the memories of the ocean and atmosphere are limited. In the atmosphere the KE can be dissipated in about five days if not replenished from the APE reservoir, whilst in the ocean the KE can be dissipated within 300 days. These limited dynamical memories imply that the predictability of the behaviour of, say, an extra-tropical cyclone or an ocean eddy is limited by the dissipation time scale. It is, therefore, quite impossible to describe the behaviour of the global atmosphere or ocean in terms of the influence of an individual eddy circulation. The only possible approach is to evaluate the statistical influence of these highly variable eddies on the behaviour of the general circulation.

To describe the behaviour of the general circulation of the atmosphere, it is necessary to concentrate attention on two quantities:

(i)   The energy of the atmosphere and, in particular, the transfer of energy between the equator and the pole.

(ii) The momentum balance of the atmosphere. This is important because it gives an insight into the reasons for the observed distribution of wind, as shown in Figure 6.11, and also why the sub-tropical jet streams are located at 30° latitude. It also yields information about the role of extra-tropical cyclones in maintaining the climatological wind distribution.

A poleward flux of energy is required by the ocean - atmosphere system to counteract the net radiation imbalance between the poles and the equator. The atmospheric energy, E, can be written as:

$$E = KE + IE + PE + LE \tag{9.10}$$

where KE is the kinetic energy; IE is the internal energy; PE is the potential energy and LE is the latent heat of condensation of the atmospheric water vapour. The KE of the atmosphere is small compared with the other energy terms and it is generally ignored in the total energy budget. The IE and PE are usually grouped together as the potential heat flux (see Figure 9.13).

The change in the energy of the atmosphere with time, dE/dt, is given by:

$$\frac{dE}{dt} = Q - D \tag{9.11}$$

where Q is the energy source and D is the energy dissipation by frictional processes. The diabatic heating term, Q, is composed of the net radiation heating, the sensible heat flux from the surface and the latent heat released by condensation.

For the atmosphere as a whole the dissipation, D, must balance the net heat input, Q. However, between the equator and the pole, equation 9.11 does not balance at every point and therefore a transfer of energy is required to maintain the distribution of heat sources and sinks. Figure 9.12 shows the meridional distribution of Q in the Northern Hemisphere winter. Everywhere in the troposphere the atmosphere is cooled by long-wave radiation at a rate of between 1 and $2\,\mathrm{K\,day^{-1}}$. Sensible heat flux from the surface heats the lower 1–3 km of the troposphere, especially in the winter hemisphere, and latent-heat release is responsible for heating all of the troposphere in the equatorial zone and for heating the lower and middle troposphere at higher latitudes. The net heating distribution shows the whole equatorial zone and the surface boundary layer as net heat sources, whilst the remaining regions are net heat sinks.

To maintain this distribution, energy has to be transferred both polewards and upwards by the general circulation. The horizontal flux of energy, F_A, across a latitude circle is given by:

$$F_A = \frac{2\pi a \cos\phi}{g} \int_0^{p_0} [\overline{vE}]\,dp \tag{9.12}$$

where v is the northward velocity, E is the energy and $2\pi a \cos\phi$ is the distance along a latitude circle. The bar denotes a time average and the square brackets denote a zonal average.

The mean northward transport of energy can be partitioned into two terms:

(i) $\bar{v}\bar{E}$ - the transport of mean energy by the time averaged meridional circulation such as the Hadley and Ferrel cells.
(ii) $\overline{v'E'}$ - the transport by the fluctuating component of the meridional circulation, or the eddy flux. This term includes fluctuations caused by the day-to-day variations in the position of the Rossby waves in the upper troposphere and by synoptic-scale motions in the lower troposphere.

Figure 9.13 shows the contributions of the potential heat flux and the latent heat flux to the total heat transport. It can be seen that the mean meridional

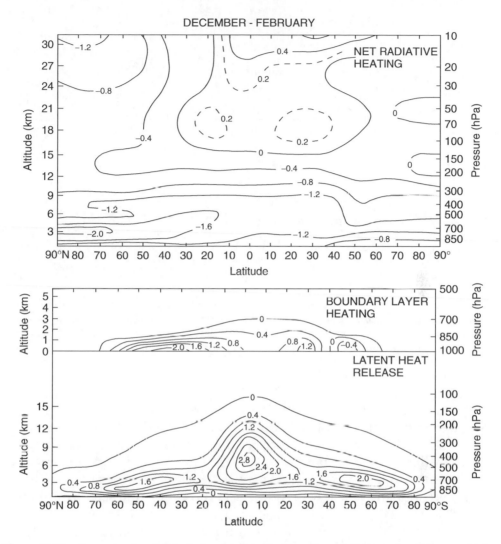

Figure 9.12 Components of diabatic heating for the atmosphere for December-February. Units: degrees per day. Reproduced, with permission, from Newell, R.E. *et al.*, 1969, The Global Circulation of the Atmosphere, ed. Corby, GA., Royal Meteorological Society: page 63, figure 7

circulation is very effective at transferring heat at low latitudes but it is less efficient in middle and high latitudes. The Ferrel cell in the middle latitudes actually transfers heat towards the equator against the temperature gradient. At middle and higher latitudes, all of the poleward heat flux is achieved by the eddy circulations. These eddy circulations have not only to transfer the heat demanded by the distribution of heat sources and sinks but they also have to transfer an extra amount to counterbalance the equatorward transport of heat by the mean circulation in middle latitudes.

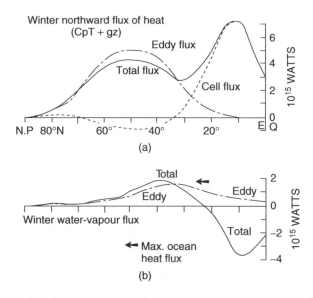

Figure 9.13   (a) Northward heat flux, and (b) water-vapour flux in Northern Hemisphere winter. Arrow shows latitude and magnitude of strongest northward heat flux by ocean currents (an annual average). Reproduced, with permission, from Newton, C.W., 1969, In The Global Circulation of the Atmosphere, ed. Corby, G.A., Royal Meteorological Society: page 138, figure 1

The latent heat flux shows a different behaviour from that of the potential heat flux. At low latitudes, the Hadley cell transports water vapour towards the equator where it is required because of the net deficit of water in the rising branch of the Hadley cell, as described in Section 5.4. At higher latitudes, the eddy flux of water vapour, like the eddy heat flux, is much larger than the mean meridional circulation. However, because the majority of the water vapour is confined to the lowest 5 km, the Ferrel circulation actually produces a latent heat flux in the same direction as the eddy transport. In addition, the major contribution of the eddy latent heat flux is made by the synoptic-scale motions in the lower atmosphere. The total transport of energy by the atmosphere necessary to balance the distribution of heat sources and sinks can be obtained by the summation of the potential heat flux and the latent heat flux.

In conclusion, it has been shown that the variability of the atmosphere is not purely random but that it acts in a systematic manner to transfer energy from low to high latitudes, and eddy circulations play an important role in this process. A similar approach is also necessary to explain the momentum balance of the atmosphere.

The total angular momentum of the Earth, including the atmosphere and the ocean, is constant. Any changes in the angular momentum of one component of the system must be balanced by a corresponding change in the angular momentum of another of the system's components. The atmosphere exchanges angular momentum with the solid Earth by the frictional torque that it exerts on the Earth's surface, including the ocean surface. If this net torque were in the direction of rotation, then the Earth would increase its angular momentum, and hence its rotation rate, at the expense of the atmosphere's angular momentum. Correspondingly, a net torque in the opposite direction would slow the Earth's rotation rate and the atmosphere would gain angular momentum. Although small variations in the angular velocity of the Earth do occur (the length of a day may change by about 1 millisecond), it can be assumed that, over a period of a few years, the angular velocity is constant. Hence, the total angular momentum of the atmosphere should be in balance when it is averaged over several seasonal cycles. The ocean circulation can produce bottom torques on the sea floor, and therefore could change the angular momentum balance. The atmosphere wind circulation however drives a large part of the ocean circulation and therefore it is accounted for by the surface wind stress and will be ignored in the following discussion.

The net torque on the Earth's surface does vary with latitude. At low latitudes the winds have a slower rotation rate than the Earth and therefore the surface torque acts in the opposite direction to the Earth's rotation. The solid Earth will lose angular momentum and the atmosphere will gain eastward angular momentum at these low latitudes. In the middle latitudes the surface winds have a higher rotation rate than the Earth and they produce a torque in the same direction as the Earth's rotation. In this case, the atmosphere will lose and the solid Earth will gain angular momentum. There will, therefore, be a sink of atmospheric angular momentum in the middle latitudes and a source in low latitudes. In order to maintain this distribution of sources and sinks, the meridional component of the atmospheric circulation must transport angular momentum from low to middle latitudes. Over the polar regions the surface winds are westwards and these areas therefore act as a small source of angular momentum which will also be transferred to middle latitudes.

The angular momentum per unit mass of a parcel of air is given by:

$$(u + \Omega a \cos \phi)a \cos \phi \tag{9.13}$$

where u is the zonal component of the wind; Ω is the rotation rate of the Earth; a is the radius of the Earth and ϕ is the latitude.

The first term in the above equation is the relative angular momentum of the zonal wind and the second term is the angular momentum of the solid Earth. Furthermore, the annual average of the northward flux of angular momentum, M_ϕ, integrated over the depth of the atmosphere to balance the surface torque τ_o, can be shown to be:

$$M_\phi = \frac{2\pi a^2 \cos\phi}{g} \int_0^{p_0} [\overline{uv}] + \Omega[\overline{v}] a \cos^2\phi \, dp = \frac{2\pi a^2 \cos\phi}{g} \int_0^{p_0} \tau_o a \cos\phi \, dp \quad (9.14)$$

The overbar indicates a time average and the square brackets indicate a zonal average. v is the meridional component of the wind.

It will now be shown that the total transport of angular momentum consists only of the first term in equation 9.14. Consider a column of the atmosphere extending around a latitude circle. A northward mass transfer into the column at one level must be balanced by an equal southward mass transfer out of the column at a different level in order to avoid a net accumulation of mass in the column. Hence, the total mass transport over the whole atmospheric column is zero and integration of the zonal average meridional velocity $[\overline{v}]$ over the column is also zero. Therefore:

$$M_\phi = \frac{2\pi a^2 \cos\phi}{g} \int_0^{p_0} [\overline{uv}] dp \quad (9.15)$$

Now the term $[\overline{uv}]$ is the time-averaged correlation between the zonal wind and the meridional wind. It can be partitioned into two components in a similar manner to the potential heat and latent heat water vapour flux. The first component is the contribution of the mean circulation, $[\overline{u}]$ and $[\overline{v}]$, and the second is the contribution of the eddy circulation $[\overline{u'v'}]$ where u' and v' are the departures from the mean wind.

The northward transport of angular momentum from low to middle latitudes shown in Figure 9.14, which is required to maintain the surface torque balance, is dominated by the eddy flux $[\overline{u'v'}]$ and it is only in equatorial latitudes where the Hadley cell contributes an equivalent mean transport. The Hadley cell produces a positive contribution to the $[\overline{u}][\overline{v}]$ term, since the poleward branch of the cell (i.e. $[\overline{v}] > 0$) is associated with an eastward wind (i.e. $[\overline{u}] > 0$), whilst the low-level equatorward flow (i.e. $[\overline{v}] < 0$) has westward winds for which $[\overline{u}] < 0$. Therefore, the product $[\overline{u}][\overline{v}]$ is positive in both cases and the Hadley cell is responsible for a poleward transport of angular momentum.

The Ferrel cell produces a small equatorward transport of angular momentum and therefore, as with the potential heat flux, the eddy momentum flux is the only mechanism available to produce the required poleward momentum flux. The maximum eddy momentum flux occurs in the upper troposphere a few degrees equatorwards of the sub-tropical jet streams. It is this convergence of the eddy momentum flux which provides the source of angular momentum to maintain the jet streams.

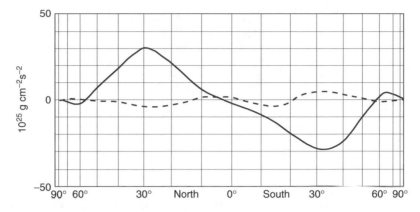

Figure 9.14 The transport of angular momentum by eddies (solid curve) and by the mean meridional circulation (dashed curve). Northward transport is positive. Reproduced, with permission, from Lorenz, E.N., 1967, The Nature and Theory of the General Circulation of the Atmosphere, World Meteorological Organization: page 82, figure 40

Although the meridional cells are not the major contributors to the meridional flux of angular momentum, they are important in the vertical exchange of momentum. The Earth's angular momentum, represented by the 'Ω' term in equation 9.14, is a maximum at the equator and decreases towards the poles. The ascending branch of the Hadley cell has a larger angular momentum than the descending branch at a higher latitude. The Hadley cell will therefore produce a net upward flux of eastward momentum. The Ferrel cell acts in the opposite direction and brings eastward momentum down to the surface. Vertical eddy fluxes, associated with large convection cells in low latitudes and synoptic-scale systems in higher latitudes, will also contribute to the vertical flux of angular momentum.

It is seen, therefore, that both eddy fluxes and the mean meridional air circulations contribute to the horizontal and vertical transfer of angular momentum necessary to maintain the observed distribution of surface wind. The eastward winds at middle latitudes and westward trade winds at low latitudes are a necessary consequence of the conservation of angular momentum. Furthermore, because the surface torque is given by $\tau_0 \cos \phi$, where τ_0 is the surface wind stress and ϕ is the latitude, the eastward winds at middle latitudes must generally be stronger than the westward winds at low latitudes in order to maintain the surface torque balance.

9.5   General circulation of the ocean

The general circulation of the ocean is very different from that of the atmosphere. First, the atmosphere is warmed in the lower troposphere of the tropics and is cooled in the upper troposphere of the polar regions by radiation. The heating occurs at higher pressure than cooling and this drives a strong global circulation. In contrast, the ocean is warmed and cooled close to the surface, and therefore the heating and cooling occurs at a similar pressure, and the direct thermally driven circulation, the thermohaline circulation, is relatively weak.

Second, the wind stress on the ocean is the major source of energy (see Table 9.1 and Figure 9.3) for the ocean circulation. Surface Ekman currents converge and diverge, producing vertical motions which push downwards and pull upwards the pycnocline (see Section 7.5 and Figure 7.15). The horizontal density gradients, produced by the vertical displacement of the pycnocline, are associated with the large horizontal gyres in the ocean basins. These flows are strong at the western boundaries of the basin and are generally confined to upper 1 km of the ocean. In the Southern Ocean, the wind stress drives a deep eastward ocean current which extends to the sea floor, known as the Antarctic Circumpolar Current. This one current accounts for a large proportion of the K.E. in the global ocean. The wind stress also produces surface waves and internal waves which contribute to the vertical mixing of energy downwards into the ocean. This mixing together with wind driven currents move heat away from the surface into the permanent thermocline, at depths of up to 1 km, and into the abyssal ocean. The presence of the thermocline at this depth indicates there has to be upwelling of deeper water to balance the vertical mixing. Without this upwelling the thermocline would mix eventually to the sea floor. The thermocline is present in all the ocean basins and therefore this upwelling is a global process.

To conserve the total mass of the ocean upwelling in one region has to be compensated by downwelling elsewhere. The downwelling areas are localized to small areas of the sub-polar regions whilst the upwelling regions cover most of the ocean and are relatively weak. A large proportion of downwelling occurs in the northern part of the North Atlantic Ocean, in particular the Labrador Sea and Greenland Seas. These are regions where sinking of cold water masses occurs as the result of strong surface cooling and a weak vertical stratification. The water originating from the Greenland Sea sinks to 1 - 3 km in the North Atlantic to produce North Atlantic Deep Water. This is the largest volume of deep water in the world ocean and it subsequently spreads into all of the major ocean basins, except the Arctic Ocean. This water mass is not the densest water mass in the ocean. The densest water is Antarctic Bottom Water, which is produced in the Weddell Sea and is found in all the ocean basins. However, it is formed in smaller quantities than the North Atlantic Deep Water and therefore has less of an influence on the global thermohaline circulation introduced in Chapter 6 and shown in Figure 9.15.

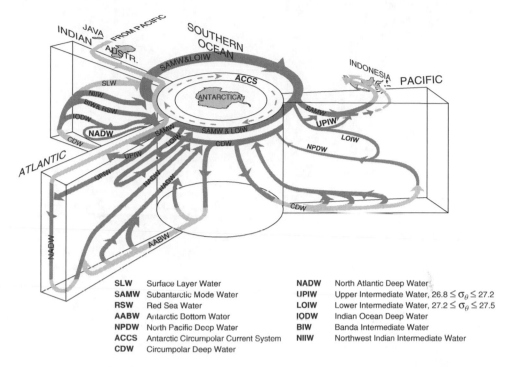

SLW	Surface Layer Water		**NADW**	North Atlantic Deep Water
SAMW	Subantarctic Mode Water		**UPIW**	Upper Intermediate Water, $26.8 \leq \sigma_\theta \leq 27.2$
RSW	Red Sea Water		**LOIW**	Lower Intermediate Water, $27.2 \leq \sigma_\theta \leq 27.5$
AABW	Antarctic Bottom Water		**IODW**	Indian Ocean Deep Water
NPDW	North Pacific Deep Water		**BIW**	Banda Intermediate Water
ACCS	Antarctic Circumpolar Current System		**NIIW**	Northwest Indian Intermediate Water
CDW	Circumpolar Deep Water			

Figure 9.15 Global thermo-haline circulation. The water masses are shown on constant potential density surfaces denoted by σ_θ. This is sea water density -1000, and is evaluated using potential temperature rather than in-situ temperature. Reproduced from Siedler, figure 1.2.7, Schmitz, and Gordon A.L., 1996. See plate section for a colour version of this image

The global thermohaline is related to the surface wind stress and the tides, which mix water between the surface and the abyssal depths, and strong surface cooling in the North Atlantic and around Antarctica.

The relationship between the wind driven circulation and the thermohaline circulation will now be discussed. The wind driven circulation in an ocean basin is driven by the convergence and divergence of Ekman currents. In the regions of convergence, for example in a subtropical gyre, the pycnocline reaches its deepest extent between 500 m and 1000 m. This drives a horizontal geostrophic circulation with a strong western boundary current.

Figure 9.16 shows a schematic diagram of the wind driven circulation in a rectangular ocean basin. The downward motion in the subtropical gyre is compensated by an upward motion in the subpolar gyre and all the circulation is restricted to the upper ocean. These wind driven circulations are relatively strong compared with the thermohaline circulation.

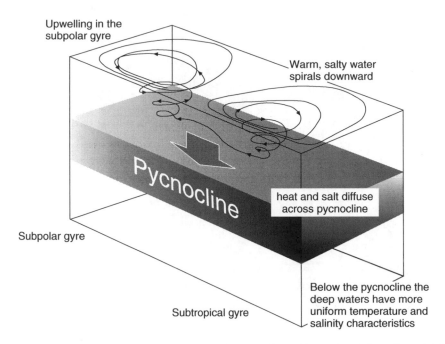

Figure 9.16 Wind driven circulation in an simple ocean basin. Reproduced from Bean, M. S., 1997 University of Southampton: Ph.D thesis

The thermohaline circulation is a meridional overturning circulation that acts to transport warm water polewards in the upper 1 km and cold water equatorwards in the deeper layers. In the subtropical gyre the western boundary current will be enhanced by the addition of this circulation and it creates a warm flow in the upper ocean from the subtropical gyre towards the subpolar gyre (see Figure 9.17). This warmer water will flow around the subpolar gyre but it will lose heat by air-sea exchange, particularly in winter. The subsequent buoyancy changes will enhance the wind driven flow in the sub-polar gyre. Deep water will be produced in the centre of the subpolar gyre and will help to maintain the thermohaline circulation.

This simple description shows how the wind driven circulation and thermohaline circulation interact to produce the observed circulation. It is most applicable to the Atlantic Ocean, where the Gulf Stream extension forms the source of the North Atlantic current which extends into the subpolar gyre and further north to the Norwegian and Barents Seas and the Arctic basin. In the North Pacific the thermohaline circulation contribution is relatively weak and therefore the overall circulation is dominated by the surface wind circulation (see Figure 9.16).

The Gulf Stream extension is one of the most dynamically active areas in the global ocean and it is where mesoscale eddies strongly interact with

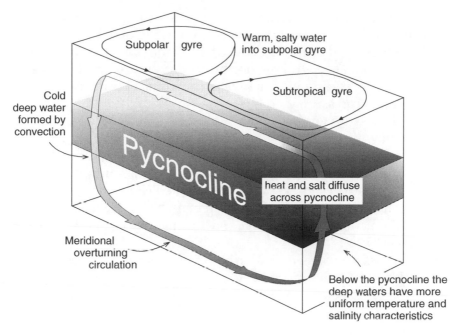

Figure 9.17   As Figure 9.16 but with the addition of a thermo-haline circulation driven by buoyancy exchange (heat and freshwater) with atmosphere

the mean flow. Studies on the momentum balance of these intense current systems have shown that their eddy circulations are very effective at pumping angular momentum from the surface layers into the deep ocean. The addition of momentum to these deeper regions produces well-organised, quasi-permanent deep gyre circulations under the Gulf Stream which are larger than the individual mesoscale eddies. These eddies also cause horizontal transfers of momentum and which produce larger mean horizontal circulations either side of the Gulf Stream.

The circulation in the Southern Ocean is dominated by the Antarctic Circumpolar Current (ACC). Unlike the Northern Hemisphere ocean, there is a circumpolar flow whose dynamics have some similarities to those in the atmosphere. In particular, the mesoscale eddies play a crucial role in the circulation. Consider the ACC as a flow in a circumpolar flat bottomed channel, driven by eastward winds. The eastward winds in the Southern Hemisphere will drive a northward flow in the wind driven Ekman layer. This northward flow will converge at the northern wall of the channel and diverge at the southern wall, which will cause the sea level to rise at the northern wall and fall at the southern wall. The surface pressure gradient will be southward and therefore the surface geostrophic flow will be eastward (see Figure 7.5b). The northward transport of water will converge at the northern wall, sink downwards and return southwards in the deeper water below. This return flow can occur only in the bottom frictional layer of the channel. The reason for this lies with the geostrophic principle. The eastward flow is in a geostrophic balance with the north-south pressure gradient (see Figure 7.5) with low pressure around Antarctica. To obtain a northward geostrophic flow, an east-west pressure gradient is needed. If the channel is circumpolar the east-west pressure gradient is always zero, there cannot be a southward geostrophic flow in the channel and hence the flow can only return in the bottom frictional layer. In similarity with the atmosphere, the ACC contains frontal regions that are hydrodyamically unstable (baroclinic instability), and this instability produces mesoscale eddies, which can transport momentum, heat and freshwater in the meridional direction across the eastward current. Both these circulations transfer heat and freshwater around the planet (see Figures 5.6 and 5.11).

Therefore it has been shown that, in both the ocean and the atmosphere, the transient eddies not only contribute to the exchange of heat and other tracers but are also instrumental in maintaining the observed mean circulation.

10

Mathematical Modelling of the Ocean and Atmosphere

10.1 Introduction

A meteorologist or an oceanographer is at a distinct disadvantage compared with scientists in many other disciplines for he or she has only one laboratory and that is either the atmosphere or the ocean. Experimental programmes carefully designed to study a particular scale of motion, such as a tornado or a mesoscale eddy, are themselves exposed to the vagaries of the atmosphere and ocean. Often the measurements obtained are incomplete and definitive conclusions cannot be drawn. Sometimes atmospheric or ocean conditions can lead to equipment failure. For example, a well-designed current meter array may be rendered useless by the loss of moorings in a strong current or a storm. Despite improvements in the reliability of technology, the lack of control over experimental conditions is a unique factor in the sciences of meteorology and oceanography.

This basic experimental difficulty has encouraged the use of laboratory models of both the atmosphere and the ocean. These models are of two types. The first type is a physical analogue model. For instance, an analogue of the atmospheric general circulation can be obtained by rotating an annulus containing a fluid on a turntable. The temperature gradient between the pole and the equator is simulated by heating the outer rim of the annulus and by cooling its inner rim. The experiment facilitates the study of the behaviour of the flow under controlled conditions. The experimental scientist is able to change the rotation rate and temperature gradient at will and is able to repeat the experiment many times to check the validity of the results.

The second type of model is the mathematical model. These models are generally based on the numerical solution of a set of differential equations which approximate closely to the equations of motion on a rotating Earth (i.e. equations 7.14 and 7.15). These models can be used to study a variety of

The Atmosphere and Ocean: A Physical Introduction, Third Edition. Neil C. Wells.
© 2012 John Wiley & Sons, Ltd. Published 2012 by John Wiley & Sons, Ltd.

scales of motion, from the general circulation of the atmosphere and ocean to the detail of a thunderstorm or an ocean eddy. Again, the experimenter has the advantage that the parameters in the model can be changed at will in order to gain a deeper insight into the mechanisms of the phenomena under study. Furthermore, the modeller has a complete set of calculated data from which, for example, the energy and momentum balance of the atmosphere or ocean can be deduced.

Mathematical modelling is a very necessary tool for the prediction of the behaviour of individual components of the Earth system, such as the atmosphere and the ocean. Both atmospheric and oceanic models are used for operational purposes: that is they are used every day for predicting the future state of the environment. Weather forecasts are produced by national meteorological centres using atmospheric models. These models attempt to predict the future state of the atmosphere a few days ahead. For example the variables that they predict include the winds; temperature; humidity and pressure from close to the Earth's surface to a height of about 100 km above the Earth's surface over the whole globe.

Operational ocean models are used on an ocean basin scale for the North Atlantic to predict the behaviour of the Gulf Stream weeks ahead. These models are used by national governments, coastguards and navies. For example, in air-sea rescue, it may be many hours or even days before the remains of a ship or plane are found, in which time the survivors and debris will have drifted hundreds of kilometres from the point where the ship sank or the plane crashed. Ocean models can help in providing guidance for the rescue and recovery.

Climate models, which typically model both the atmosphere and the ocean, are used to predict the future behaviour of the global climate from decades to centuries ahead. These predictions are used by governments to plan for the future of our cities and towns. For example, London, U.K. is one of many cities that are vulnerable to coastal flooding and therefore sea level rise is a key prediction from climate models.

In this chapter the basis of scientific modelling will be explained and its applications to the ocean, atmosphere and climate are discussed.

10.2   Scientific modelling: A simple model of the surface layer of the ocean

Consider the following example. A model is needed to predict the sea surface temperature of the ocean. Many complex factors influence the surface layer of the ocean and some of these factors are shown in Figure 10.1. These processes

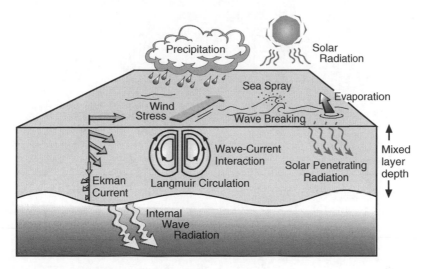

Figure 10.1 A schematic of the physical processes influencing the surface ocean mixed layer in the Northern Hemisphere. See plate section for a colour version of this image

include the vertical heat fluxes at the surface and the bottom of the surface mixed layer; freshwater fluxes and the momentum fluxes. The freshwater fluxes change the salinity and the momentum fluxes, due to winds and waves, would change the surface wind drift currents and the vertical mixing. Additionally, there are processes associated with ocean currents and gradients of temperature and salinity that have to be considered. The complexity of these physical processes makes the modelling of the surface layer a daunting prospect. However, in the construction of a model the simplest and most basic approach is often the most successful. For the model of the temperature of surface layer the most important factors are the heat fluxes into and out of the layer. The freshwater fluxes will only have an indirect influence by changing the salinity and the vertical stratification, while the momentum fluxes will control the vertical mixing and therefore the depth of the surface mixed layer. For the present these two factors will be excluded from the model. The simplest model is therefore a surface layer of constant depth with surface heat fluxes into the surface layer (Figure 10.2). The flux of heat and momentum at the bottom of the surface layer will also be ignored. Equation 10.1 shows the temperature change with time of the layer of constant depth, h, is proportional to the net surface flux Q_T. The constant of proportionality C is the heat capacity of the layer.

$$C\frac{dT}{dt} = Q_T \qquad (10.1)$$

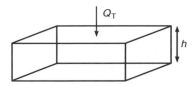

Figure 10.2   A model of surface ocean mixed layer depth h subject to a surface heat flux Q_T

where

$$C = \rho C_p h$$

$$Q_T = \text{net surface heat flux}$$

$$dT/dt = \text{ temperature change with time}$$

$$h = \text{depth of surface layer}$$

$$\rho = \text{water density}$$

$$C_p = \text{specific heat at constant pressure}$$

To solve equation 10.1 we need to know the temperature T at $t = 0$ which we shall call T^0. Furthermore we need to know Q_T for all t from $t = 0$. For example the temperature T may be required at the end of the month, and therefore Q_T would be needed from the beginning of the month ($t = 0$) to the end of the month ($t = t_1$). Rearranging equation 10.1 and integrating over time we get:

$$T^{(t_1)} = T^{(0)} + \int_0^{t_1} \frac{Q_T}{C} dt \qquad (10.2)$$

Equation 10.2 is the solution of equation 10.1 but to provide a prediction of temperature at the end of the month we must know the surface flux Q_T and the heat capacity C of the layer for the whole month. These could be measured every minute, hour or every day, but here only the integral Q_T/C over the month is necessary.

Let us assume we have measured the surface heat flux Q_T every day over the month, say from a research ship, but we are not sure of the depth of the surface layer, h. We guess a value of 15 m, based on a single profile taken the previous year and assume it remains constant over the month. Based on this value we calculate C, assuming that density and specific heat also remain constant. Equation 10.2 can be integrated repeatedly for each day provided $T^{(0)}$ is measured at the beginning of the first day of the month.

Equation 10.3 shows the first and second steps in the calculation. $F^{(1)}$ and $F^{(2)}$ are the integrals over day 1 and day 2 respectively, whilst $T^{(1)}$ and $T^{(2)}$ are the temperatures at the end of day 1 and day 2 respectively.

$$\text{Step 1} \quad T^{(1)} = T^{(0)} + F^{(1)} \quad \text{where } F^{(1)} = \int_{t=0}^{t=1} \frac{Q_T}{C} dt$$

$$(10.3)$$

$$\text{Step 2} \quad T^{(2)} = T^{(1)} + F^{(2)} \quad \text{where } F^{(2)} = \int_{t=1}^{t=2} \frac{Q_T}{C} dt$$

A time series of the predicted values of T are produced for the month. Figure 10.3b shows the predicted temperature at the end of each day against the measured temperature over the month (Figure 10.3a).

The question then arises of how these predicted values compare with the measured values over the month? In this case the predicted temperatures are much higher than the measured values. To improve the prediction we can choose a deeper depth, say 20 m, and repeat the calculation. This time the predictions are closer to the observations, as shown in Figure 10.3c.

In this case we are changing a parameter of the model, h, to provide a better fit to observations. This tuning process allows us to calibrate the model with observations.

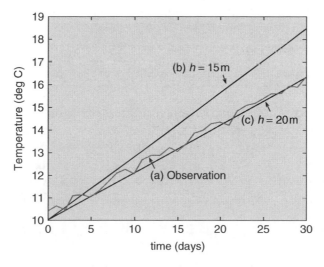

Figure 10.3   The temperature variation for the surface mixed layer over 1 month (a) observed (b) model $h = 15$ m and (c) model $h = 20$ m

Once the model has been calibrated with observations, it can then be used to predict the temperature variation for another month. The procedure is similar to the previous one but with one important distinction. The parameter h remains at the value previously chosen in the calibration procedure (i.e. 20 m). Comparison of the model with observations for the second month provides an objective test of the model. If the comparison is good then the model is a suitable model for prediction and our simple assumptions are valid. In the event that the results are poor the initial assumptions have to be re-examined carefully. For example, there may be large changes in the layer depth during the month and these variations would have to be included to improve the model.

The model that is calibrated, objectively tested against another data set and shown to be good can then be used for prediction. A further question arises to the meaning of the word 'good' which is subjective and dependent on the expert making the assessment of the model. An objective assessment can be made by defining a suitable statistical measure of the difference between the model and observations. A common measure is the root mean square error or standard error.

The standard error $= \frac{s}{n-1}$ where $s = \frac{1}{n}\sqrt{\sum_{i=1}^{n}(T \text{ model} - T \text{ observed})^2}$ and n is number of observations. In Figure 10.3 the standard error is 0.75°C for $h = 15$ m but 0.18°C for h $= 20$ m, and thus the objective measure supports our subjective assessment.

Additional physical processes may be added to the model, and they can be objectively assessed. Only if the errors show a reduction will the new model replace the original model.

10.3   A dynamical model of the ocean surface layer

Scientific modelling uses mathematical equations to describe our complex world. To further understand this subject we will consider a dynamical example of the ocean surface layer. A simple hypothesis is the rate of change of speed, u, of a surface parcel of water is directly proportional to the wind force, X, acting on the parcel over time, t. This is an example of Newton's 2nd Law of Motion where the acceleration is proportional to the force per unit mass.

This hypothesis can be expressed as a mathematical equation:

$$\frac{du}{dt} = X \tag{10.4}$$

If we say the speed is $u = 0$ at $t = 0$ and u_1 at a time in the future t_1, then the solution for equation 10.4 is:

$$u = X t \tag{10.5}$$

Since the hypothesis states a linear relationship between force and speed, the speed of the parcel will increase continuously with time. However, our physical intuition suggests this is not correct because ocean currents do not get faster and faster with time. Let us add an additional term to this equation, $-Ku$, where u is the speed and K is a constant coefficient. This term has a negative sign and represents a frictional force proportional to the speed, u. This frictional force will oppose the wind force.

$$\frac{du}{dt} = X - Ku \tag{10.6}$$

The solution for equation 10.6 is given by:

$$u = \frac{X}{K}\left[1 - e^{-Kt}\right] \tag{10.7}$$

The addition of the frictional term means that after a long time u will tend to a constant value of $u = X/K$.

Equations 10.6 and 10.7 represent a model or a mathematical representation for the speed of a parcel of water when subjected to a wind force, X, and a frictional force given by Ku. Equation 10.7 is a much more reasonable solution than equation 10.5 because the speed reaches a constant value after sufficient time. This agrees with our physical intuition and observation of ocean currents.

The elements of modelling represented by this example are as follows:

(i) We commence with a hypothesis that the rate of change of water speed is proportional to a wind force. To solve equation 10.4 we need an **initial** value for u at $t = 0$. In the above example we assumed $u = 0$ at $t = 0$. This is an **initial condition** and all predictive ocean and atmosphere models require initial conditions to solve the equations. For example we may have measured the current speed at $t = 0$ to be u_0 in which case

$$u = u_0 + X t \tag{10.8}$$

(ii) These initial conditions allow a mathematical solution to be obtained and evaluated but we have to use physical intuition and observations to check that the model produces reasonable behaviour and values. In the above case our physical intuition or knowledge of Newton's 2nd Law of Motion could have guided us. However, the most useful factor for obtaining reasonable current speeds is our knowledge of ocean currents from observations. Therefore, we need observations to check the model is producing a reasonable representation of reality. This is called **validating** the model.

(iii) Once a model has been validated against observations and our physical intuition, it may then be used to predict the speed of the current subject to the given wind force, X, and the initial condition.

This is the scientific process of modelling and is illustrated in Figure 10.4. The example given is rather simple in that only one initial condition is necessary to solve the equation. In atmosphere and ocean models the flow will vary spatially as well as with time and therefore the flow field

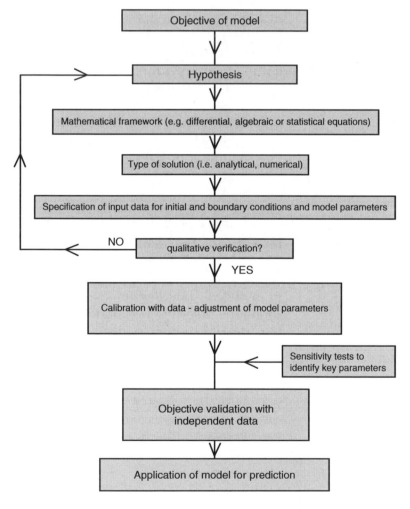

Figure 10.4   The scientific process of modelling

at t = 0 will need to be specified as a set of initial conditions. In the above example values of the wind force need to be given not only at $t = 0$ but also at subsequent times. The ocean has boundaries at the surface with the atmosphere and at the ocean bottom and values at these boundaries have to be given. This is known as a **boundary condition** and is necessary to solve the equations. A further example of a boundary condition is at the ocean floor. We could say that there is no flow through the sea floor and therefore we can state the normal component flow is zero.

To be able to use a mathematical model we need to provide initial and boundary conditions which mainly come from observations.

10.4   Numerical solutions of mathematical models

The equations described in the previous section have simple analytical solutions i.e. the solution of equation 10.7 is an exact solution of the initial differential equation (equation 10.6) and it is valid for all values between $t = 0$ and $t = t'$. The equations used to describe the motion of the atmosphere and ocean are much more complex than those described above, as they include pressure gradients; frictional forces; forces due to rotation of the Earth; gravitational forces due to Earth and tidal forces due the interaction of the Moon and Sun with the Earth. The complexity of these equations means that they cannot be solved by analytical methods and do not generally have mathematically exact solutions.

These complex equations can, however, be solved by numerical techniques. Numerical solutions are different from exact analytical solutions in that they are approximations to the real solution. These numerical methods have been used by mathematicians for over a century for calculating solutions to equations. They are based on finite differential calculus rather than continuous differential calculus. In finite differential calculus we set up a line of points along the time, t, axis, with every point separated by a time interval, Δt, so that the time at each point is given by

$$t = (n - 1)\Delta t \text{ where } n = 1, 2, 3 \ldots$$

At each point on the t axis we define u^n which is the value of the speed u at point n (see Figure 10.5). This method will be applied to equation 10.6, assuming that X and K have known values which do not change with time.

We first approximate the first term in equation 10.6 by a finite difference:

$$\frac{u^{n+1} - u^n}{\Delta t} = X - Ku^n \tag{10.9}$$

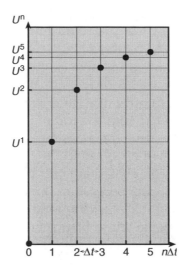

Figure 10.5   Finite difference solution (U^n) for $n=1,2,3,4,5$ and time step Δt

where u^{n+1} is the predicted value of u, u^n is a known value and Δt is the time interval.

Rearranging this equation

$$u^{n+1} = u^n + \Delta t \left(X - K u^n \right) \tag{10.10}$$

For our initial condition let us assume that at $t = 0$, $u = 0$. Using finite-differences this corresponds to $n = 1$ and $u^{n=1} = 0$

Hence equation 10.10 gives:

$$u^{n=2} = \Delta t \left(X \right)$$

$u^{n=2}$ is the predicted value at $t = \Delta t$. This is the value of u one time step later.

The next step is to calculate $u^{n=3}$

From equation 10.10

$$u^{n=3} = u^{n=2} + \Delta t \left(X - K u^{n=2} \right) \tag{10.11}$$

Therefore by repeated use of equation 10.10, values of u are calculated along the time axis at the grid points. The values are only at the grid points unlike the analytical solution which can be evaluated at any point along the time axis.

This example illustrates that from the initial condition $u = 0$ at $t = 0$ we can step ahead in time intervals Δt and produce a set of values of u along the

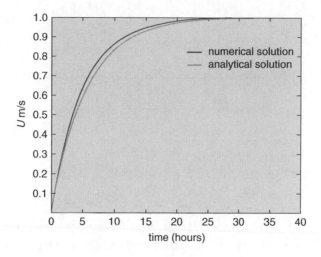

Figure 10.6   Analytical solution compared with numerical solution from Figure 10.5

time axis t. The numerical solution will produce values of u which should be similar to values of u calculated from equation 10.7 (see Figure 10.6). They will not be exactly the same values because numerical solutions always have an error associated with them. This error arises because we are approximating the exact gradient of a curve by finite values as shown in Figure 10.5. In finite differential calculus we have many methods for calculating the numerical solution to equation 10.6, which have different advantages and disadvantages. Many methods tend to have a restriction on the size of the time step Δt. This restriction is necessary because the very small errors arising from the finite differences can grow quickly in time and overwhelm the true solution.

However, with careful choice of the numerical method one can obtain good solutions, which closely approximate the exact solution of equation 10.6 as shown in Figure 10.6.

10.5   Numerical solutions for momentum on a rotating Earth

The general equations for momentum on a rotating Earth (see Chapter 7) are rather more complicated than those in the previous example. However, the same method can be applied.

In these equations there are now spatial gradients of quantities as well as temporal gradients, for example the horizontal pressure gradient shown in equations 7.14 and 7.15.

To represent the horizontal pressure gradient, $-\frac{1}{\rho_0}\frac{dp}{dx}$, we set up a line of points along the x axis a fixed distance Δx apart. The position x is defined as $x = (i-1)\Delta x$ where $i = 1, 2, 3 \ldots$

The horizontal pressure gradient at the grid point i is represented in finite differences by:

$$-\frac{1}{\rho_0}\left(\frac{p_{i+1} - p_{i-1}}{2\Delta x}\right)$$

where p_i is the pressure at grid point i.

More generally the atmosphere or ocean can be represented by a set of grid points in the x, y and z directions (see Figure 10.7) which form a lattice structure.

These numerical equations can be solved in theory for an infinite time into the future, but we shall see there are limitations to the numerical solution of the momentum equations since the solutions are mathematically chaotic.

It was mentioned previously that to solve these equations one needed initial conditions. In our simple example this was the value of u at $t = 0$

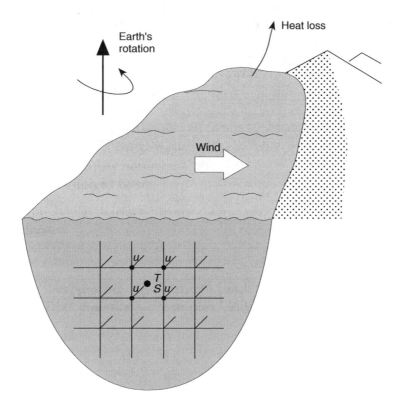

Figure 10.7   3D lattice of points x, y, z for an ocean circulation model

but in the ocean or atmosphere model we would need the values of u at all the grid points on the lattice in Figure 10.7. Furthermore, the velocity has three components given by u, v, w in the directions x, y and z respectively and therefore one needs a large number of values to prescribe the initial conditions for velocity. A global atmosphere model from the sea surface to a height of say 100 km, may have up to 90 levels in the vertical and 180 by 360 grid points in the east and north direction for a horizontal grid spacing of 100 km. Therefore the total number of grid points is about 5.8 million. In addition to the velocity it is necessary to specify the temperature; pressure; density and water concentration (vapour, liquid, and solid) at these grid points, to completely specify the initial state of the atmosphere.

It has been found that no matter how accurate the initial conditions are, tiny perturbations in the initial conditions will cause the solutions to diverge from each other over a few days. Even if the initial conditions were changed by a very small amount, say one in a trillion ($1{:}10^{12}$), the solutions will still diverge from each other. This sensitivity to initial conditions is known as mathematical chaos.

For numerical weather prediction initial conditions are obtained from observations of the state of the atmosphere at a particular time. It is not possible to measure the atmosphere at every point and every time but, because of the large scales of many atmospheric features, a very good representation of the flow can be obtained for specific times each day, normally every 6 hours, at a horizontal resolution of 100 km and with 90 levels in the vertical. To provide measurements for the state of the atmosphere at $t = 0$, a whole array of observing systems are used which include profiling balloons, satellite observations and aircraft observations (see Chapter 6).

These global numerical weather forecasts rely on standardised reporting methods and telecommunication systems operating around the clock. It is a major international effort which is orchestrated by the World Meteorological Organisation, but continuously maintained by the individual nations. Using this data, weather predictions with good skill can be obtained for a few days ahead up to about 10 days. Skill is a measure of how close the weather prediction is to the observed weather. For example it could be the inverse of the root mean square error of the sea level pressure (Figure 10.8) and therefore, as the error became smaller, the skill in predicting the surface pressure would increase. The skill generally decreases with time and therefore a forecast 24 hours ahead will have a higher skill than a forecast for 48 hours ahead.

10.6   Atmospheric and climate general circulation models

First consider a General Circulation Model (GCM) of the atmosphere (see Box 10.1). This type of model is based on a set of equations which describe

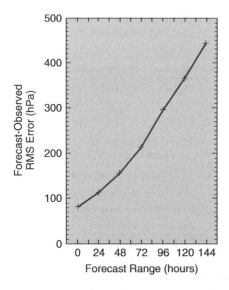

Figure 10.8 An example of the growth of the root mean square error with time in numerical weather prediction models. The predicted and observed variable is sea level pressure (hPa) over the Northern Hemisphere. Supplied by Met Office, Exeter, UK

the momentum of the air and the transport and mixing of temperature; water vapour; water droplets; ice crystals and other chemical tracers.

Box 10.1 An atmosphere general circulation model

Predictive equations

Momentum equation

Rate of change of horizontal momentum = Transport of momentum across faces of box
+ Coriolis acceleration
+ Horizontal pressure gradient
+ Friction and small scale processes

Temperature equation

Rate of change of temperature in box = Transport of temperature across faces of box
+ Mixing of temperature between adjacent boxes
+ Adiabatic term
+ Diabatic term

The adiabatic term is the temperature change due to adiabatic cooling or warming mainly associated with vertical motion.

Diabatic terms represent the temperature changes due to radiation, latent heat due to condensation or evaporation and sensible heat from ocean and land surfaces.

Water vapour equation

Rate of change of water vapour in box $=$ Transport of water vapour across faces of box $+$ Mixing of water vapour between box and adjacent boxes $+$ Gain/Loss Terms

The loss terms are due to condensation and precipitation, and the gain terms result from by evaporation from clouds and rain, or from surfaces of oceans and continents

Diagnostic equations

Equation of state

Gas equation for moist air which relates pressure to air density and virtual temperature.

Hydrostatic equation

This equation relates the decrease of pressure with height to the air density.
The equation is valid for horizontal scales of motion larger than 20 km.

Continuity equation

The continuity equation expresses the conservation of mass for a box, and it is required to determine the vertical wind speed and the change in surface pressure.

The conservation of mass states the mass of a box does not change with time. The mass coming into through the vertical sides is balanced by mass going out of the horizontal sides of the box. The mass coming through the sides is the product of the velocity normal to the side of the box and its density. The density is obtained from the equation of state above and is known at each face of the box. If we know the horizontal velocity through the vertical sides of the box, and we know the vertical velocity at the bottom of the box, the vertical velocity at the top face of the box can be determined.

If the box is adjacent to the ocean or land surface the vertical mass flux and hence vertical velocity of the air at the lowest face of the box will be zero. With this boundary condition and the equation of continuity the vertical velocity can be determined in all the atmospheric boxes.

The momentum equations, together with conservation of mass, provide the dynamical part of the model. These equations are the equations of motion and therefore give predictions of the three dimensional velocity field throughout the global atmosphere (see Chapter 7 and Box 6.1).

The principal forces that drive air flows are the pressure gradients both in the horizontal and vertical direction. In the vertical direction the balance of forces is described by the hydrostatic balance equation (see Chapter 2) which is accurate for large scales of motion in the atmosphere. This equation can be used to obtain the pressure at any point in the atmosphere, provided the density at every point is known. To obtain the density we need to use the gas equation for a moist atmosphere (see Chapter 4). Therefore provided values of humidity and temperature are available throughout the atmosphere, the pressure gradients can be determined.

The other important set of forces is that of the rotational forces and in particular the Coriolis force (see Chapter 7). The Coriolis force only requires the velocity and the latitude to be known and therefore is determined easily. Other forces include frictional forces which depend on the gradients of the velocity and tidal forces which can be prescribed as a known external force similar to gravity.

The temperature and humidity in the atmosphere are transported around the atmosphere by the winds and time dependent equations for the advection and mixing for temperature and humidity can be solved in a similar way to those for momentum described in Section 10.4. The temperature is also dependent on the solar and planetary radiation. These two components of radiation are described by the radiation transfer equations similar to the ones discussed in Chapter 3. This is a complex set of equations which require in principle all the concentrations of all the gaseous emitters and absorbers of radiation; the atmospheric aerosol, which is mostly the solid particles, and the liquid droplets which include the stable haze droplets as well as the water and ice cloud particles. In earlier models these emitters and absorbers of radiation were given set values and the absorption and emission equations were directly solved. The temperature changes due to radiation emission and absorption can then be added to temperature changes caused by the wind to predict the temperature.

Humidity, in similarity with temperature, is also advected and mixed by the wind circulation. In addition, the processes of condensation in clouds and precipitation from the clouds need to be included. For example, in regions of upward motion caused by the convergence of the surface winds (see Chapter 6), cloud formation will tend to occur. Mathematical equations for these processes can be included in our global model.

Finally, to complete the description, the Earth's surface needs to be included. There are three major surfaces present: the air-sea interface, the land surface and the ice surface. The simplest way to model the air-sea interface is to

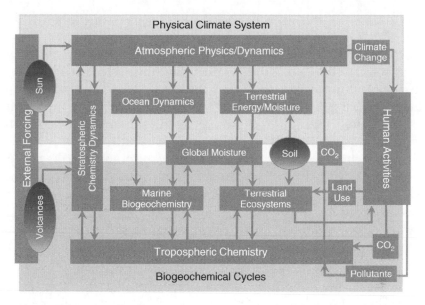

Figure 10.9 Components of an IPCC climate model. Reproduced, with permission, from Glantz, M.H. and Krenz, J.H. 1992 Human components of the system in Climate System Modelling. Ed: Trenberth K.E. Cambridge University Press, Plate 2. See plate section for a colour version of this image

predict the sea surface temperature by using a model of the surface mixed layer, similar to that described in Section 10.2, but it would also need to include the prescribed heat transport by ocean currents. Over the land a model of the soil temperature and moisture would need to be included as well as information on the surface properties, for example the type of vegetation and its albedo. Over an ice sheet or glacier the albedo is an important property as it will determine the rate of heat absorption by the ice and possible melting of the ice (Figure 10.9).

Such models can be used for many different purposes from weather forecasting to climate prediction.

For example, the global atmosphere model can be used for climate prediction up to 100 years ahead. The 100-year time scale has been decided by the Intergovernmental Panel on Climate Change (IPCC) as a reasonable time scale for climate prediction. The variability in the ocean and atmosphere at the interannual and decadal time scale may mask the climate change signal (see Chapter 11).

This type of climate model includes a representation of the ocean and the land surface, which is very important for climate prediction models, because the ocean and land surface vary more slowly than the overlying atmosphere.

To investigate the response of the climate to changes in greenhouse gases over the next century, a model can be initialised to the observed atmosphere in the year 2000. Greenhouse emissions into the atmosphere, based on global population growth and assumptions on agriculture and industrial production, are then estimated from the present to 2100. The concentrations of these gases can then be defined at all grid points over the 100-year period and the global climate predicted up to 2100.

Experiments have been made with models from climate centres around the globe. The models, though quite similar to the model described, may have different methods for describing the physics of clouds, radiation and surface processes, and therefore the predictions are different. Furthermore small changes to the physics in a single model can produce a different prediction. This set of models gives a number of predictions for the same greenhouse gas concentration, and this enables climate scientists to not only predict the mean temperature in 2100 but also give a measure of the spread of the mean temperatures for 2100, i.e. the uncertainty in the prediction.

10.6.1 Parameterisation

In General Circulation Models (GCM) physical processes which occur on length scales smaller than the model grid scale are replaced by a representation of the process known as a parameterisation. Processes which are parameterized in GCM's include cloud formation, radiative transfer and the effect of the Earth's surface on the momentum, temperature and moisture of the lower atmosphere.

Solar and planetary radiation affect the temperature of the atmosphere. These two components of radiation are modelled by the radiative transfer equations similar to those described in Chapter 3. This complex set of equations requires, in principle, knowledge of the emitters and absorbers of radiation including all gases; liquid droplets; atmospheric aerosols; water and ice cloud particles. In early models radiative transfer was parameterised by assuming fixed amounts of these emitters and absorbers. However, more recently they have been directly linked to microphysical parameterisations which calculate the distribution of cloud particles. Temperature changes due to emission and absorption are then added to temperature changes caused by advection and mixing to predict the temperature.

The effect of the Earth's surface on the momentum, moisture and heat in the surface layer can also be parameterised. As previously mentioned, there are three major surfaces present: the air-sea interface, the land surface and ice surface. In the following discussion, the air-sea interface is used as an example. The simplest way to model the air-sea interface is to predict the sea surface temperature by using a model of the surface mixed layer. This would be similar to that described in Section 10.2 but it also includes the prescribed heat transport by ocean currents. A parameterisation of the air-sea interface would also include a representation of transfer of momentum and moisture between the surface winds and surface waves. The bulk aerodynamic formulas discussed in Chapter 5 are parameterisations of the complex transfers of momentum, heat and moisture at the interface.

10.6.2   *Data assimilation*

Typically a meteorological agency may receive around 1/2 million observations of atmospheric conditions around the globe every day. However, even this number of observations does not contain enough information to know about atmospheric conditions at all heights above the Earth's surface. For example, there are some inaccessible regions on land where few observations are available. To fill these gaps we must combine the observations with forecasts in a process known as data assimilation. In data assimilation we aim to find the current state of the atmosphere by minimising the difference between the observations and the forecast. Both the forecast and observations are weighted by their accuracy to achieve an appropriate balance between the uncertainty in the observations and the forecast. Once, the best estimate for the current state of the atmosphere, known as the analysis, has been determined it is used as the initial conditions for the next forecast run.

10.7   Global ocean models

Global ocean models have many similarities to atmospheric models. Ocean circulation models are often designed to predict the global pattern of ocean currents and their associated heat and freshwater transports (see Section 9.5).

The elements of a typical model will now be described (see Box 10.2). The dynamical equations describing the ocean circulation are very similar to those in the atmosphere. They consist of the two horizontal equations of motion (momentum equations) and an equation of mass continuity, which provide predictions of the three components of velocity. The vertical pressure gradient can be obtained by the hydrostatic equation from knowledge of the ocean density. The ocean density is provided by the equation of state of sea water which depends on temperature, salinity and pressure. To obtain the pressure at a particular depth the hydrostatic equation is integrated from the required depth to the sea surface. The height of the sea surface will vary with time and position and these variations have to be taken into account. They can be predicted from the equations of motion, providing the wind stress and surface atmospheric pressure are included as additional forces. These forces can be predicted by a weather forecasting model, a climate model or directly from global observations. Hence the dynamical ocean model has many conceptual similarities to the dynamical atmosphere model.

Box 10.2 An ocean general circulation model

Momentum equation

$$\begin{matrix} \text{Rate of change} \\ \text{of horizontal} \\ \text{momentum} \end{matrix} = \begin{matrix} \text{Transport of momentum across faces of box} \\ +\,\text{Coriolis acceleration} \\ +\,\text{Horizontal pressure gradient} \\ +\,\text{Friction and small scale processes} \end{matrix}$$

Temperature equation

$$\begin{matrix} \text{Rate of change of} \\ \text{temperature in box} \end{matrix} = \begin{matrix} \text{Transport of temperature across faces of box} \\ +\,\text{Mixing of temperature between} \\ \text{box and adjacent boxes} \\ +\,\text{Solar heating} \end{matrix}$$

Heat fluxes at the ocean surface, rivers and glaciers appear as boundary conditions.

Salinity equation

$$\begin{matrix} \text{Rate of change of} \\ \text{salinity in box} \end{matrix} = \begin{matrix} \text{Transport of salinity across faces of box} \\ +\,\text{Mixing of salinity between} \\ \text{box and adjacent boxes} \end{matrix}$$

The fresh water fluxes though the ocean surface, from rivers and glaciers, appear as boundary conditions.

Equation of state

Density is dependent on pressure, temperature and salinity and is described by the equation of state. This equation is complex and is solved by a computer programme.

Hydrostatic equation

The vertical variation of pressure is related to density. The equation is valid for horizontal scales greater than 10 km.

Conservation of mass or continuity equation

The continuity equation expresses the conservation of mass for a box, and it is required to determine the vertical component of velocity.

The conservation of mass states the mass of a box does not change with time. The mass coming into through the vertical sides is balanced by mass going out of the horizontal sides of the box. The mass coming through the sides is the product of the velocity normal to the side of the box and its density. The density is obtained from the equation of state above and is known at each face of the box. If we know the horizontal velocity through the vertical sides of the box, and we know the vertical velocity at the bottom of the box, the vertical velocity at the top face of the box can be determined.

If the box is adjacent to a horizontal ocean bottom the vertical mass flux and hence vertical velocity at the lowest face of the box will be zero. With this boundary condition and the equation of continuity the vertical velocity can be determined in all the ocean boxes.

To predict the temperature and salinity another set of equations are required. These time dependent equations consist of the advection terms and the mixing terms. The three advection terms describe the transport of temperature by the two horizontal components of ocean current and the third vertical component of the ocean current. Another three diffusion terms describe the mixing of temperature in the horizontal and vertical directions. The radiative equations can be added to the upper ocean but they are usually simpler than those for the atmosphere, because the longwave or planetary radiation is absorbed in top 1 mm and most of the solar radiation is absorbed in the top 100 m of the ocean.

An equation for salinity, similar to the temperature equation, can be added to the model. The equations can be solved in time on a three dimensional grid using numerical methods described in Section 10.4.

Operational ocean prediction models have similarities to their atmosphere counterparts in weather forecasting. They require observations over the three dimensional grid at $t = 0$. They may have 20 to 60 levels in the vertical extending over the total ocean depth. Their horizontal resolution is typically 1/10 deg. latitude by 1/10 deg. longitude. The higher resolution is needed to resolve the mesoscale ocean eddies which provide most of the ocean variability on time scales from weeks to months. The models can extend either over the global ocean or a single ocean basin, depending on the model application.

Ocean observations are needed to provide the initial conditions for the predictions. Satellite measurements of sea surface temperature and ocean colour can give information about the surface structures, for example ocean eddies and fronts. Indirectly the kinematics of surface ocean flows can also be determined by following parcels of similar temperature or ocean colour. Drifting buoys and anchored buoys can provide this information on surface flows directly. Ocean flows will vary with depth and therefore observations are needed throughout the water column. Ocean profiling floats known as Argo floats can measure temperature and salinity from the surface down to 2 km every 10 days and they drift with the ocean current at depths of typically 1 km. They are thinly spaced with roughly one float in each 300 km by 300 km square over the globe. Ocean eddies, which have horizontal scales from 10 km–200 km, will therefore be poorly sampled by the Argo floats. However, the satellite altimeter, which measures the variations of the sea surface height with at least 1/2 deg. x 1/2 deg. resolution every 10 days can provide valuable observations on the larger ocean eddies in the deep ocean. Sea level variations produce a pressure signal which can be measured at the ocean floor. A measurement of the height of the sea surface, together with a density profile from the sea surface to the ocean bottom, allows the calculation of pressure at any depth in the ocean by use of the hydrostatic equation. All of these methods help to provide the initial conditions for operational ocean prediction models.

In present climate models the ocean is represented simply as a surface mixed layer model (see Section 10.2) where the horizontal heat transports are represented indirectly. It is possible to replace this simple model with a global general ocean circulation model (OGCM) which allows for ocean circulation to change as well as the atmospheric circulation. These ocean models have a lower horizontal resolution, typically 100 km, than those used in operational ocean models because they are integrated from seasonal to millennial time scales and are therefore very computationally demanding.

Coupled atmosphere ocean models have the capacity to describe the complexities of large scale air-sea interaction. For example they are used to predict El Niño Southern Oscillation phenomena (see Section 11.3). They are being used to produce experimental weather forecasts for a season ahead by many weather and climate centres. An example of the prediction of the sea surface temperature in the El Niño region located in the eastern equatorial Pacific Ocean is shown in Figure 10.10.

Atmosphere, ocean and climate models provide an additional tool with which to understand the complexities of our fluid environment and to forecast changes in that environment. A plentiful supply of reliable and accurate observations, from in situ measurements and satellites, is needed to provide the necessary initial conditions for these models over the globe. This is further discussed in Section 10.8.

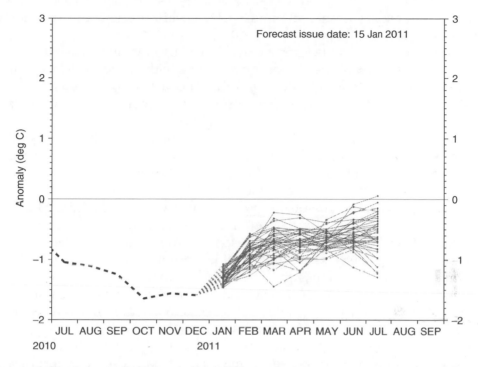

Figure 10.10 An example of the prediction of sea surface temperature in the Eastern Equatorial Pacific Ocean (El Niño region) using a coupled ocean-atmosphere-land model. This model contains some of the components of an IPCC climate model, and is suitable for prediction a few months ahead. Very small changes are introduced to the initial conditions ($t = 0$) and therefore many forecasts can be made. The spread of the forecast plume from December 2010 gives a measure of the forecast uncertainty. Supplied by ECMWF, Reading, UK

10.8   Observations of the ocean and atmosphere

Both meteorology and oceanography are observational sciences and therefore the growth in understanding of how these complex systems work has been largely dependent on improvements in the methods of measurement. For example, the establishment of the synoptic network of upper air stations using the radiosonde in the 1940s and 1950s provided, for the first time, daily information on the pressure and wind fields in the troposphere and lower stratosphere. These observations led to an improved understanding of the development of weather systems which, in turn, laid the foundations for numerical weather predictions. Similarly in oceanography, it has been seen that the neutrally buoyant float provided the first important evidence for mesoscale eddies in the deep ocean, as well as some of the first observations of the abyssal western boundary currents. These phenomena had not been previously observed because of the coarse horizontal resolution of hydrographic data.

Despite the considerable improvements in instrumentation, both meteorologists and oceanographers today are confronted with questions which, if they are to be answered satisfactorily, require additional methods of measurement to those discussed previously. One important question involves the role of the ocean in seasonal and interannual variations in climate.

In Section 11.3 it is shown that the El Niño-Southern Oscillation (ENSO) is the most prominent signal in the atmosphere-ocean system on interannual time scales. To be able to provide forecasts of ENSO, tropical ocean models are coupled to global models of the atmosphere. These models, however, require regular measurements of the atmosphere and of the tropical oceans in order to provide the initial conditions from which the forecasts can be made. Fortunately, we have a reasonable measurement system for the troposphere and lower stratosphere which provides the initial conditions for weather forecasts across the globe. For the ocean, a system called the Tropical Pacific Ocean Observing System provides regular observations from Papua New Guinea to the Americas. This system involves moored and free-floating buoys, volunteer observing ships and a network of tide gauges. The ocean observations include surface wind; heat and freshwater fluxes; sea surface temperature and sea level, as well as the vertical profiles of temperature, salinity and currents down to the depth of the equatorial thermocline.

Similar systems for the tropical Atlantic and Indian Oceans have been deployed and are providing regular observations. The measurements are relayed from the instruments by satellite to climate centres around the world to provide the data required by the coupled ocean-atmosphere models for ENSO predictions. This is the first step in the provision of regular and high-quality measurements of the surface layers of the ocean.

Scientists are currently designing a global climate observing system, which will have an ocean component called the global ocean observing system. The scientific understanding gained from the recent international research programmes, such as the World Ocean Circulation Experiment (WOCE) and the Tropical Ocean Global Atmosphere (TOGA) experiments in the 1990's, is being used to design a system from which climate forecasts can be made. The Tropical Pacific Ocean Observing System illustrates the requirement to provide a methodological approach to measurements if we are to use all our knowledge to address the important questions discussed in Section 12.5.

An extremely important part of a global observation system is the use of satellite platforms to observe the Earth. The development of satellite instruments to measure the properties of the atmosphere and the underlying ocean and land surfaces has been very impressive in recent decades. Weather prediction has been greatly enhanced by the use of radiometers which can measure the reflective energy in the visible or the emitted radiation in the thermal infra-red and microwave bands from clouds. This enables them to pinpoint tropical and extra-tropical cyclones as well as frontal cloud bands in remote regions of the world. The infra-red radiometers provide measurements of the vertical structure of temperature in the stratosphere and troposphere.

First, we will discuss the satellite orbits and then the measurements which can be made. A satellite can be placed into a geostationary orbit over the equator or it can be placed in a polar orbit. The geostationary satellite has a nadir view over the same position on the equator but the view is limited to a maximum latitude of 60°. A set of five geostationary meteorological satellites can provide continuous monitoring over most of the globe. A disadvantage is the height of the geostationary satellite above the Earth's surface, about 36 000 km, which limits the horizontal resolution of the radiometers. In contrast, the polar orbiting satellite can provide an improved global coverage with a lower orbit of between 800 and 1200 km above the Earth's surface. The improved spatial resolution is bought at the expense of the temporal resolution. For example, an altimeter mounted on a polar satellite measures sea level but covers the same part of the globe only once every 10 days.

The selection of the oceanographic and meteorological parameters which can be determined from the present generation of instruments is shown in Table 10.1. The instruments measure

(i)   The passive emission of radiation from a wavelength in the infra-red and microwave bands.
(ii) The reflection of sunlight.
(iii) The reflection of a radar or lidar signal.

The measurement of atmospheric temperature, cloud-top temperature, the sea surface or land surface temperature relies on the emission of thermal infra-red and microwave radiation. By using multi-spectral sensors, more discrimination in either the vertical structure of the atmosphere or in the surface properties can be obtained.

Until recently, most meteorological satellite systems were passive. They measured the radiation emitted in the infra-red and microwave regions of the spectrum and they also monitored reflected solar radiation, as shown in Table 10.1. Passive sensors in the visible are able to delineate regions of clouds, ice boundaries and the colour of the land and sea surfaces during daylight hours. However, thermal infra-red sensors, which measure the radiation temperature of the Earth's surface or cloud tops, are able to delineate variations both during the day and the night. The thermal emission from the sea surface can be used to obtain the sea surface temperature using known values of the emissivity. This sea surface temperature is representative of the

Table 10.1   Selection of atmospheric and oceanic parameters measured by satellite

Environmental parameter	Satellite sensor	Accuracy	Method
Tropospheric wind	Visible/infra-red	$\pm 4\,\mathrm{m\,s^{-1}}$; $\pm 20°$	Cloud displacement from geostationary satellite
Surface winds	Scatterometer (radar)	$\pm 2\,\mathrm{m\,s^{-1}}$: $\pm 15°$	Scattering of radar beam by sea surface
Waves	Altimeter (radar)	± 0.3 m	Shape of reflected pulse related to significant wave height
Sea surface elevation	Altimeter (radar)	± 0.02 m	Time delay of radar pulse
Vertical atmospheric temperature profile	Infra-red/passive microwave	± 2 K	Multichannel radiometer samples emitted thermal radiation from different levels of atmosphere
Sea surface temperature	Infra-red/passive microwave	± 0.3 K	Thermal radiation from sea surface Corrections for atmospheric contamination Infra-red limited to cloud-free areas
Solar radiation at sea surface	Visible/near infra-red	$\pm 10\,\mathrm{W\,m^{-2}}$	
Precipitation	Visible/infra-red	–	Measures cloud-top temperature empirically related to precipitation
	Passive microwave		Measures liquid water content of atmospheric column
Ice cover	Visible/infra-red Passive microwave	–	High resolution in visible and infra-red Low resolution, but all weather, in microwave

upper $100\,\mu\mathrm{m}$, and it is therefore different from the bulk temperature typical of the upper 1–3 m measured by ships. However, except under low wind conditions, the differences are usually less than 1°C.

Unfortunately for sea surface temperature measurements, the atmosphere also emits radiation at infra-red wavelengths due to the presence of carbon dioxide, ozone and water vapour, and therefore the signals have to be corrected for atmospheric effects. Furthermore, clouds will absorb all thermal radiation emitted from the sea surface and therefore the technique is limited to cloud-free regions. Microwave energy is also emitted from the sea surface and, at certain wavelengths, it can be transmitted through non-precipitating clouds and used to obtain sea surface temperature measurements. The thermal emission of radiation by the atmosphere has been used to obtain the vertical distribution of temperature in the atmosphere. The principle of the instrument used is that different wavelengths of infra-red and microwave radiation originate at different levels in the atmosphere. The radiance at each wavelength can therefore be related to the temperature at the appropriate level. The vertical resolution is not as good as that obtained from a radiosonde, being about 5 km, but the horizontal resolution is much better, being typically 100 km for infra-red observation and 250 km for passive microwave measurements.

Passive microwave sensors have been used to determine the total water vapour and the total liquid water content of the atmospheric column from simultaneous measurements of the water-vapour absorption line at 1.36 cm wavelength and the liquid water absorption band at 0.97 cm wavelength. From such measurements, global maps of water vapour and liquid water content have been produced. These maps would have been unobtainable with the same resolution from conventional observations. Absorption of microwave radiation at a wavelength of 1.55 cm is dependent on the number and size of the water droplets in the atmospheric column and this fact has been used to deduce precipitation rates. It is interesting that this technique is most suited to ocean areas where the emission of contaminating microwave radiation from the ocean surface is low compared with the emission from land surfaces.

The variations in surface emissivity in the microwave spectrum have been used to detect the distribution of sea ice. The higher microwave emissivity of sea ice compared with sea water allows identification of the boundary between sea ice and open water even in cloudy conditions. Weekly variations in sea-ice cover can thus be monitored by the satellite network. In addition, it is known that the emissivity of ice more than one year old is lower than that of newly-formed ice and therefore the age distribution of the ice can be mapped.

In the field of oceanography, the development of active microwave satellite instruments has produced quantitative information about the ocean surface. The satellites carry three different types of radar:

(i)   The first radar, known as the scatterometer, measures the back-scattered radiation from the sea surface. The intensity of the back-scattered radiation is proportional to the surface-wind stress, whilst the direction of the scattering is related to the direction of the surface stress. Although the theory of the scattering at this wavelength is not well understood, the technique can provide accurate wind measurements over the world's oceans.

(ii) Second, a synthetic aperture radar, SAR, provides radar images of the sea surface. These images give information about the surface wave spectrum and the direction of the waves. Surprisingly, the radar images also detect internal waves in the ocean, as well as variations in bottom topography in water which is many tens of metres deep. Since SAR wavelengths are absorbed in the upper centimetre of the ocean, this effect is an indirect influence of bottom topography on the surface wave field.

(iii) The third radar is a high-precision radar altimeter which is able to measure the distance between the ocean surface and the satellite to a precision of ±2 cm. Variations in the geometric distance between the sea surface and satellite are related to variations in the satellite orbit, variations in the surface geoid and the ocean surface slope associated with ocean currents. The first effect can be eliminated by accurate position-fixing systems but distinguishing geoid variations from dynamical sea-level variations is very difficult unless independent measurements of the geoid or sea level are available. The determination of the geoid to the high horizontal precision necessary for the determination of ocean eddies and narrow ocean currents remain a fundamental difficulty at the global scale. However, it has been found that repeated passes of the altimeter over a given ocean area can distinguish time-dependent changes in the sea surface elevation (see Figure 7.7).

The satellites are able to detect time variations of mesoscale eddies in the world ocean, as shown in Figure 7.7, and it is now feasible to obtain eddy maps of the ocean similar to the now familiar atmospheric weather maps. In the equatorial ocean, sea-level variations are associated with displacement in the shallow equatorial thermocline. High sea levels are generally associated with a deep thermocline and vice versa. Thus, altimeter measurements may be used to obtain routine observations of the thermocline displacement as well as the heat content in the upper ocean. This technique is invaluable for measuring seasonal and interannual variations in the equatorial ocean associated with ENSO and other large-scale phenomena. With the improved precision of the altimeter, to ±2 cm, and with high-quality determinations of the surface geoid, it is feasible to determine surface geostrophic currents from the satellite measurements. The surface geostrophic current $v_g(0)$ can be calculated from equation 10.12

$$v_g(0) = \frac{g}{f}\frac{\partial \xi}{\partial x} + \frac{1}{f\rho_0}\frac{\partial p_0}{\partial x} \tag{10.12}$$

Where ξ is the height of the sea surface, p_0 is surface pressure, f is Coriolis parameter (Section 7.2), g is the acceleration due to gravity, and ρ_0 is the ocean density.

From meteorological surface pressure measurements, the last term in equation 10.12 can be determined and hence v_g at the surface is obtained.

Moreover, from the scatterometer wind-stress data, the wind-drift current, v_e, can be calculated. Thus the total surface current $v_g(0) + v_e$ can be obtained.

We have seen that the satellites can measure sea surface temperature and sea level over the globe. However, because of the lack of transparency of sea water to electromagnetic radiation, the satellites are limited in what they can tell us about the deep ocean. For the interior ocean, new technologies and measurement methods are being developed. The relative ease with which sound can travel through the oceans could be used to measure global ocean temperature. We can recall from Chapter 2 that the velocity of sound in sea water is strongly influenced by temperature and pressure. These two properties conspire to produce a minimum velocity at depths of about 1000 m, which is known as the sound or SOFAR channel. Low-frequency sound (70 Hz) can travel in the sound channel for thousands of kilometres with little attenuation. An experiment in 1991 transmitted sound from Heard Island in the Southern Ocean to receivers as far away as the Pacific and Atlantic coasts of North America, a distance of 16 000 km.

The travel time is dependent on the temperature of the ocean through which the sound travels; hence changes in travel time from one year to another will be a measure of the average temperature change between the source and receivers. The Acoustic Thermometry of Ocean Climate (ATOC) was planned for the North Pacific Ocean, with sound sources at Hawaii and California and receivers around the Pacific rim. The array was expected to test whether thermal anomalies that have been seen from repeated hydrographic stations can be detected and their evolution monitored. The use of acoustics for the ocean basin and global ocean measurements caused major public concern for marine mammals and this research has been discontinued.

Because of the costs of measurements from ships and the inherent problems in sampling of the ocean, both spatially and temporally, there is considerable interest in the use of autonomous instruments. The development of the Argo programme since 1999, which relies on a global distribution of over 3000 profiling floats, is one such example (see Chapter 6).

Another new development uses an ocean glider. The objective of this project is to provide autonomous hydrographic sections across ocean basins. The key elements of the ocean glider are a very low drag body, which has a small buoyancy system. In similarity with Argo floats, it can surface to telemeter its position and measurements to a satellite. The Argo and glider technologies provide measurements to a maximum depth of 2000 m for Argo and 1000 m for gliders. Other automonous vehicles can now sample depths to 5000 m and possibly 6000 m. The Autosub based at National Oceanography Centre, Southampton, U.K. is an example of such a vehicle. These vehicles therefore have the potential to access about 99% of the ocean floor.

Though the measurements of temperature, salinity and pressure are the main objectives of these autonomous vehicles and floats, there are also biological and chemical sensors that can be attached. These sensors can measure the optical properties of the water, to determine chlorophyll and other pigments derived from phytoplankton, and the chemical properties such as dissolved carbon dioxide, oxygen and nitrates.

These developments in instrumentation will not on their own provide the answers to the many questions being asked of ocean scientists, meteorologists and climatologists, because of the complexity of the Earth's system. To be able to discern the global and regional changes in the ocean, we require mathematical models of the world ocean into which these observations may be added. These predictions can be considerably improved by the addition of observations. The data assimilation process allows observations to be smoothly introduced into the model as it is integrated forward in time. The data sets obtained are dynamically consistent and provide an improvement on both the original observations and the model.

The global ocean models are in continuous development to improve their ability to handle the complex topography of the ocean basins and coastlines; to transport and mix properties within the ocean and to produce improved transports of heat and fresh water between ocean basins. This development goes hand-in-hand with the expansion of data collection. By the end of the World Ocean Circulation Experiment (1990–97) more measurements had been taken than in all the previous experiments.

Global ocean models rely on powerful parallel computer systems to obtain the necessary computational power to run them over many decades at a resolution sufficient to resolve the most energetic turbulent eddies in the ocean. Lower resolution ocean models, which parameterize the eddies, can be included in general climate models and used to predict climate changes for decades to centuries ahead.

We should not neglect the problems of how these ocean models and atmospheric models are coupled together. The interchange of properties between the atmosphere and ocean occurs in turbulent boundary layers (see Chapter 5). The processes by which this occurs are not well understood at present, though measurements have considerably improved with the advent of oceanographic satellites. These measurements of surface stress, surface waves and sea surface temperature provide useful information which can help to validate models of the ocean boundary layer.

The modelling of pack ice remains a difficult problem, residing as it does between the atmospheric and ocean boundary layers. The ice interface is far from uniform in thickness or extent. Improvements in the modelling of sea ice will require better understanding of the processes of the atmospheric boundary over the ice. The modelling of sea ice has to include the heterogeneity of ice from the thin ice at the edge of the pack to the solid pack with its associated pressure ridges and ice rafting.

It is not difficult to appreciate that there is considerable scope to improve our knowledge and understanding of the ocean-atmosphere system. Though it is difficult to predict the particular development of systems as complex as the ocean and the atmosphere, we can see a common thread in the study. Improved understanding comes from a combination of observations and scientific principles which, in turn, provide the basis of the mathematical models.

11

Atmosphere-Ocean Interaction

11.1 Air-sea interaction: An introduction

11.1.1 Air-mass transformation over the ocean

The modification of a continental air mass as it moves over the ocean is one of the more important processes necessary to the understanding of the interaction between the ocean and the atmosphere. In order to illustrate this process, two examples of air-mass transformation will be considered.

The first example is the transformation of the trade winds on their journey from the sub-tropical land masses to the ITCZ. Figure 11.1 shows a schematic diagram of some of the salient features of the trade-wind system between the north-west African coast and the equatorial Atlantic Ocean. Dry, warm surface air from the continent is rapidly transformed by contact with the cooler surface water in a shallow boundary layer. The moisture content in the boundary layer is rapidly increased by evaporation from the ocean, whilst sensible heat is lost from the air to the ocean because of the reversal of the vertical temperature gradient. In addition, the air is cooled by the emission of long-wave radiation into space. The surface air therefore cools rapidly to the sea surface temperature and, because of the addition of water vapour and the net cooling, low stratus clouds form in the boundary layer. At this stage the boundary layer may be relatively shallow, having a depth of a few hundreds of metres.

Above the maritime boundary layer the air is potentially warmer and very dry because of the descent of air from high levels in the troposphere associated with subsidence in the subtropical anticyclones. Therefore, at the boundary between the two air masses, a strong trade-wind inversion is formed which tends to suppress vertical motion and mixing. As the maritime air moves over progressively higher sea surface temperatures, the latent heat flux increases and the sensible heat flux becomes upwards from the ocean to the atmosphere. The warming of the air mass results in a break-up of

The Atmosphere and Ocean: A Physical Introduction, Third Edition. Neil C. Wells.
© 2012 John Wiley & Sons, Ltd. Published 2012 by John Wiley & Sons, Ltd.

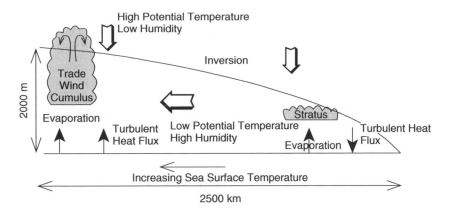

Figure 11.1   Schematic diagram of the modification of the trade-wind boundary layer over the ocean

the stratus cloud and the development of trade-wind cumulus clouds. The shallow cumulus convection mixes the moist surface air with the dry air above the inversion. This results in a deeper boundary layer. The convection also brings drier air towards the surface which, in turn, increases the evaporation from the ocean into the boundary layer. In the western tropical ocean basins the boundary layer reaches a depth of 2–3 km and evaporation is larger than in the eastern ocean basins. In these boundary layers, convection is deep enough to produce showers and this warms the cloud layer by latent heat release. Energy budgets of the trade-wind boundary layer have shown that the downward heat flux through the inversion is as large as the upward surface flux of sensible heat. They have also shown that virtually all of the water evaporated into the boundary layer is exported horizontally out of the trade-wind zone and is subsequently released in the intertropical convergence zones. A typical trade-wind velocity is $6\,\mathrm{m\,s^{-1}}$ and therefore the surface air mass, providing that it remains in the boundary layer, will traverse a distance of 2500 km between the source region and the equatorial region in about five days.

The second example of air-mass transformation is more typical of air-sea interaction in middle latitudes. During winter, very cold, dry air from the northern Asian and North American land masses periodically surges out over the relatively warm waters of the western North Pacific and North Atlantic Oceans. These cold outbreaks occur in the north-westerly flow in the rear of an extra-tropical cyclone and they may last for several days. Figure 11.2 shows the transformation of a cold, dry stable air mass as it flows out over the East China Sea and the warm Kuroshio Current. Initially, the continental air mass may have a temperature as low as $-20°C$, and a very low specific humidity of less than $1\,\mathrm{g\,kg^{-1}}$. The sea surface temperature, in contrast, increases from about 5°C near the coast to 20°C over the Kuroshio Current. As the air mass

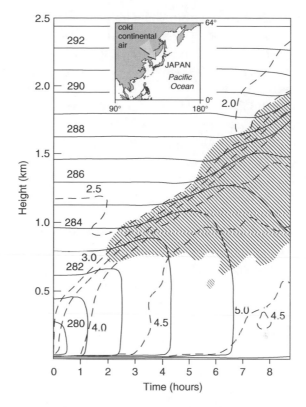

Figure 11.2 The evolution of a moist convectively mixed layer during a cold outbreak over the Kuroshio current. Solid lines indicate potential temperature, K, dashed lines specific humidity, g kg $^{-1}$, and stippled area is cloud layer. Reproduced, with permission, from WMO, 1981, Garp Publication Series, 24: page 167, Figure 6.1

first moves out across the East China Sea, the unstable boundary layer will be rapidly warmed by an upward flux of sensible heat. Although the air is dry, it is initially very cold and is therefore unable to absorb much moisture. As the boundary layer warms and deepens, the surface latent heat flux becomes comparable with the sensible heat flux. In contrast to the trade-wind boundary layer, which deepens only slowly, these cold air outbreaks cause rapid changes in the depth of the boundary layer. Over a horizontal scale of 300 km, changes of 1–2 km are observed.

During the Air-Mass Transformation Experiment (AMTEX) in 1974 and 1975, total surface heat fluxes of 700–800 W m^{-2} were observed over two separate four-day periods. This heat flux, which is half the solar constant (i.e. 1360 W m^{-2}), would heat a column of air 2 km deep at a rate of 30 K day^{-1} and would lead to vertical instability of the boundary layer. The growth of convective cloud is limited by the height of the inversion which marks the boundary between the modified air mass below and the continental air mass above. Near the coast the boundary layer depth is small, and therefore the convective cloud is shallow, being usually stratocumulus and of small horizontal dimensions. However, as the boundary layer grows, both the depth and horizontal scale of the convective cloud increases and it can grow sufficiently to produce showers.

The regions of large air-mass transformation are also regions of frequent cyclonic development, as the result of the larger horizontal gradient of surface temperature. The region south of Japan is the most active region of cyclone generation in the world during the winter season. The average heat loss of the ocean over the Kuroshio Current in February is $350 \, \text{W m}^{-2}$, which is sufficient to cool the upper 100 m layer of the ocean by 2.2°C in one month.

The two previous examples have shown that the extraction of heat from the ocean is strongly dependent on the temperature and humidity characteristics of the overlying air mass relative to the sea surface temperature. In an extra-tropical cyclone, two or three different air masses may cross a given area in a day and produce large variations in surface heat fluxes. Figure 11.3 shows the surface fluxes of latent and sensible heat during a period of relatively weak air-sea interaction in summer in the north-eastern Atlantic Ocean. The cold front marks the boundary between a cooler, drier air mass to the north-west and a warm air mass to the south-east. In the warm sector, both latent heat and sensible heat fluxes are close to zero, and the air mass is therefore receiving little or no heat from the ocean. This is typical for warm, humid air masses moving over progressively lower sea surface temperatures. In the rear of the cold front, the latent heat flux increases to a maximum of $110 \, \text{W m}^{-2}$ about 300 km from the front, whilst the sensible heat flux is much less, being in the range from 10 to $20 \, \text{W m}^{-2}$. This contrasts strongly with the cold outbreaks over the Kuroshio Current where sensible heat fluxes are of a similar magnitude to latent heat fluxes. It will also be noticed that, in the rear of the cold front, there are variations in latent heat flux caused by horizontal variations in the sea surface temperature of 1°C.

11.1.2   *Response of the atmosphere to sea surface temperature*

All scales of atmospheric motion are involved in the transfer of energy between the ocean and atmosphere, and therefore most atmospheric phe-nomena will respond, directly or indirectly, to the sea surface temperature. On the small scale, an increase in sea surface temperature may change the sta-bility of the overlying air mass which may, in turn, change the boundary layer cloud from a stable stratus type to an unstable convective type. Similarly, the

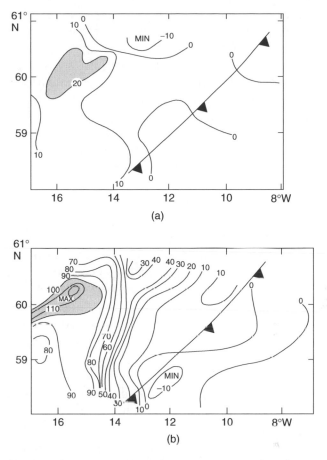

Figure 11.3   (a) Horizontal variation of the surface turbulent heat flux over the ocean during the passage of a cold front (Q_H: W m^{-2}) at 01.00 hours GMT on 31 August 1978. Values of more than 20 W m^{-2} are shown hatched. (b) Horizontal variation of the surface latent heat flux (Q_E: W m^{-2}) at 01.00 hours GMT on 31 August 1978. Values of more than 100 W m^{-2} are shown hatched. Reproduced, with permission, from Guymer, T.H. ci al., 1983, Philosophical Transactions of the Royal Society of London, A308: page 265, figures 10 and 11

distribution of sea fog may be affected by a relatively small change in the sea surface temperature.

Mesoscale atmospheric circulations may also be influenced by the sea surface temperature. The intensity of a sea-breeze circulation depends on the difference in temperature between land and sea to generate the available potential energy which sustains the circulation. Mesoscale convection over the ocean will depend on the air-sea temperature difference to produce latent and sensible heat fluxes to sustain the vertical circulation. In turn, the vertical circulation will bring colder and drier air to the boundary layer, and this will enhance the vertical transfer of heat into the circulation.

The tropical cyclone is an example of the self-sustaining circulation described above. The generation areas of the tropical cyclone are usually limited to the regions of the tropics where sea surface temperatures are above 26°C. Evaporation rates are determined principally by the surface wind and by the sea surface temperature. It is also noted that the vapour pressure has a non-linear dependence (see Figure 4.9), and thus high evaporation rates occur over the tropical ocean. An embryonic low-pressure system will produce a convergence of water vapour in the maritime boundary layer which will release latent heat when uplifted. As the horizontal and vertical circulation gains kinetic energy from the latent-heat release, the surface evaporation rate and the surface convergence of water vapour will both increase, and the circulation will intensify. Provided that there is a copious supply of water vapour, the storm will increase in intensity at an exponential rate until the loss of kinetic energy by surface friction balances the kinetic energy input. A mature tropical cyclone can produce surface latent heat fluxes of $1000 \, \text{W m}^{-2}$ and sensible heat fluxes of $500 \, \text{W m}^{-2}$. These are approximately twice the surface heat flux observed in cold outbreaks over the Kuroshio Current.

The tropical cyclone is a truly oceanic phenomenon because when it strikes land its surface moisture source is severely reduced and this, in combination with higher surface friction, will cause it to dissipate quickly. Occasionally, tropical cyclones will propagate into higher latitudes where they may trigger the development of an extra-tropical cyclone. Over the higher-latitude oceans, the extra-tropical cyclone may increase its energy by ventilating the marine boundary layer with drier air which has subsided from the upper troposphere in the rear of the cold front. The dry air is able to pick up heat and moisture which can both be recirculated into the ascending region of the cyclone. They may be released subsequently as available potential energy and kinetic energy.

The large-scale circulation of the atmosphere is strongly influenced by the distributions of the continents and oceans. However, before discussion of these influences, it is useful to consider first a hypothetical planet without any land surface known as an aqua-planet. The distribution of net radiation discussed in Chapter 1 and atmospheric circulations discussed in Chapter 9 would not be very different to that observed today. The Hadley cells in both hemispheres, with their associated trade winds would still be found, along with Rossby waves and their associated synoptic scale weather systems in middle latitudes. The eastward atmospheric flow in middle latitudes would have baroclinic instabilities, producing transient synoptic scale weather along latitude circles on the aqua-planet. The atmospheric circulation over the circumpolar Southern Ocean between 55°S and 60°S is perhaps the closest to the hypothetical aqua-planet.

Computer models have shown that the climatological atmospheric flow is determined by both large mountain barriers and by the distribution of heat sources and sinks. The upper tropospheric flow is most sensitive to the distribution of topography, whilst the flow in the lower troposphere is more influenced by the pattern of continental and oceanic heating. However, the presence of mountain barriers, such as the Himalayas, impedes the transport of moisture in the lower troposphere and thus indirectly influences the distribution of heat sources.

Large meridional topographic barriers, such as the Andes and the Rockies, cause the formation of a long Rossby wave in the upper troposphere, which produces a trough downstream of the barrier extending for many thousands of kilometres (see Figure 11.4). This influence is embedded within the transient systems and can be seen most clearly when the flow is averaged over a number of years.

The continents and ocean are also regions of heat sources and heat sinks driven by the seasonal cycle of solar radiation. During winter, the North Atlantic and North Pacific Oceans provide an average heat source into the atmosphere of $100 \, \text{W} \, \text{m}^{-2}$. The largest heating is located over the Kuroshio and Gulf Stream Currents. In contrast, the continental land masses in winter are heat sinks, because the heat loss by long-wave radiation exceeds the input

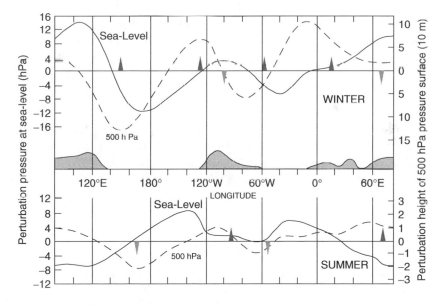

Figure 11.4 Zonal profiles of normal perturbation pressure at sea level (solid line) and normal perturbation height of the 500 hPa surface (dashed line). Continental elevations are indicated by a shaded area. Positions of observed relative heat sources and sinks are indicated by spires pointed upwards and downwards, respectively. Upper part: winter profiles averaged over 20° latitude centred at 45°N. Lower part: summer profiles averaged over 20° latitude centred at 50°N. Reproduced, with permission, from Smagorinsky, J., 1953, Quarterly Journal of the Royal Meteorological Society, 79(341): page 362, figure 7. See plate section for a colour version of this image

by solar radiation. The cooling over the continent and heating over the ocean produces a large horizontal gradient of temperature in the lower atmosphere known as a baroclinic zone (see Chapter 9). This zone is associated with baroclinic instability and the formation of intense extra-tropical cyclones. These cyclones can be tracked by satellites and surface buoys over the oceans, by measuring the position of the storm centre and plotting its position with time. These storm tracks can be traced over the oceans from the region of the Gulf Stream, where the storms are generated, to the North East Atlantic and Northern Europe, where they finally dissipate. Storm tracks are found at similar latitudes in the North Pacific and in the Southern Ocean. The position of the storm tracks may change from one season to another and from one year to another. For example, in the winter time North Atlantic the storm tracks may move southward from their normal position and be directed towards the Mediterranean instead of northern Europe. The positions of the storm tracks may be linked to climate indices, for example the North Atlantic Oscillation (see Section 11.3).

In tropical latitudes the differential heating between the Asian land mass and the relatively cool Indian Ocean is responsible for the energy of the south-west monsoon circulation. These monsoon circulations are associated with a large-scale, inter-hemispheric transfer of latent heat from the southern Indian Ocean and Arabian Sea which, when released over the Indian sub-continent, strengthens the monsoon circulation. The vertical flux of latent and sensible heat from the Arabian Sea during the onset of the south-west monsoon in June has a typical value of $200\,\text{W}\,\text{m}^{-2}$. This large heat loss produces a rapid cooling of the ocean mixed layer of $2°\text{C}$ in one month and a deepening of the oceanic mixed layer from $40\,\text{m}$ in early June to $100\,\text{m}$ in early July.

The above examples illustrate some of the complex interconnections which exist between the ocean and atmosphere. The ocean is able to store vast quantities of heat during the inactive season and then release it into the atmosphere, when required, to enhance and to ultimately drive the atmosphere on a variety of scales, from the mesoscale and synoptic-scale circulations to the large-scale monsoon circulations.

11.2   Seasonal anomalies of the ocean-land-atmosphere system

The seasonal cycle is the largest change in the ocean-land-atmosphere system that is experienced by humankind. The monsoon of Africa and Asia is the response of the system to the seasonal variation of the solar radiation. The position and strength of extra-tropical systems, though more variable than the monsoon systems, nonetheless respond to the seasonal cycle, with more intense storm systems in winter than in summer.

The seasonal cycle of solar radiation at top of the atmosphere, changes little from one year to another (see Chapter 1), though at the Earth's surface we experience variations in weather from one winter to another, or one summer to another. These deviations, or anomalies, from the normal seasonal cycle occur because of interactions between the surface layer of the ocean, the land surface and the atmosphere.

Before considering the possible mechanisms of these interactions, an example of the anomalous behaviour of the atmosphere and ocean in the North Atlantic Ocean between January and June 1972 will be considered.

Figure 11.5 shows both the atmosphere and sea surface temperature anomalies for June 1972. Both patterns are representative of the chosen period between late winter and early summer. The anomaly patterns are obtained from the differences between the observed pattern and the long-term climatic normals for June. The most notable features are:

(i)   The negative temperature anomalies, up to 3°C in the north - eastern North Atlantic, and the positive temperature anomalies, up to 5°C in the western and central Atlantic at 40°N, as well as smaller negative temperature anomalies at lower latitudes.

(ii)   An anomalous cyclonic circulation over the British Isles with a cold north-westerly flow over the northern North Atlantic and an anomalous anticyclonic circulation over the central Atlantic Ocean. The cold north-westerly winds of this year produced a cold wet spring and early summer in north-western Europe, and were associated with the largest volume of ice recorded in the Labrador Current since 1912, when the RMS *Titanic* sank. In June 1972 an extremely active hurricane, called Agnes, followed an unusual northerly path along the east coast of the USA into the Great Lakes region. It is thought that the anomalous southerly wind off the east coast of the USA and the high sea surface temperatures prevailing there influenced the behaviour of Hurricane Agnes.

From Figure 11.5 it can also be seen that the patterns of sea surface temperature anomalies and atmospheric circulation have similar horizontal scales and that they are clearly related. The cold northerly atmospheric flow is associated with negative sea surface temperature anomalies, whilst the warm southerly flow to the west of the anticyclone occurs over the positive sea surface temperature anomaly. Thus, the atmospheric flow is tending to maintain the sea surface temperature anomaly patterns by extracting more heat from the northern North Atlantic Ocean than is normal and by extracting less heat than usual from the western Atlantic. In turn, the large heat flux from the ocean in the northern North Atlantic Ocean would provide an anomalous source of heating which would tend to maintain a low-pressure system downstream of the heat source. Conversely, the reduced heat

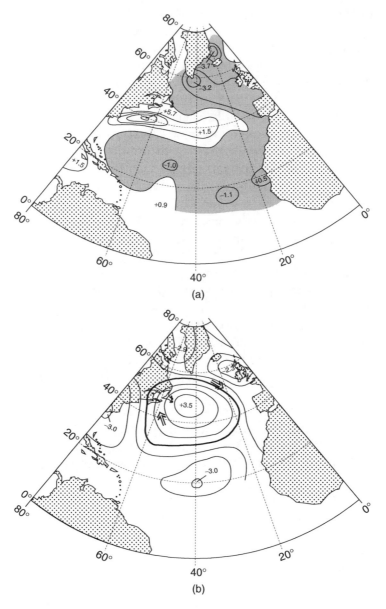

(a)

(b)

Figure 11.5 (a) Sea surface temperature anomalies (°C) for 1–26 June 1972. (b) Atmospheric circulation anomalies at 700 hPa height for June 1972. Intervals are in units of one standard deviation; maxima and minima are labeled. Zero contour is the heavy line. The arrows indicate the direction of the anomalous atmospheric circulation. Reproduced, with permission, from Namias, J., 1973, Quarterly Journal of the Royal Meteorological Society, 99(421): page 509 and 517, figures 2 and 10

flux in the western Atlantic would produce an anomalous heat sink which would maintain an anticyclone in the central Atlantic. These arguments are not necessarily conclusive because large-scale interactions with the tropical atmosphere and the Pacific Ocean may also be responsible for the anomaly pattern, but they do show that large-scale air-sea interactions can produce a persistent, anomalous pattern which may be reinforced by heat exchange between the ocean and the atmosphere. The pattern of sea surface temperature in the North Atlantic shown in Figure 11.5 is known as the North Atlantic tri-pole pattern and is the most important pattern of seasonal variability in this ocean.

In the above example the ocean is behaving as a slave to the atmosphere but, because of its heat storage capacity, it is able to maintain anomalous patterns of heating and cooling. Without the ocean the atmosphere would lose all of its memory of the thermal anomalies within a period of a few weeks. The longevity of the above type of air-sea interaction has aroused much interest in the possibility of using these patterns for long-range seasonal forecasting. However, progress in this area has been limited. Recent studies have shown that the predictability of these air-sea interaction events is limited by the predictability of the atmosphere, which is generally less than two weeks. However, once an air-sea interaction event has developed, the ocean will tend to maintain the pattern for a period of one to a few months. Thus, the atmosphere initiates the anomaly, whilst the ocean tends to make it persist. Ultimately, the seasonal cycle limits the longevity of these air-sea events by changing the distribution of heat sources and sinks. This eventually results in a breakdown of the anomaly pattern.

An interesting example of the effect of the seasonal cycle on air-sea interaction events has been described by Namias (1971). In the northeastern Pacific Ocean, it has been found that warm sea surface temperature anomalies, which develop in the summer months, tend to persist through the autumn and into winter. During the autumn, the extra-tropical cyclones intensify in response to the increase in the poleward temperature gradient and the differential heating between ocean and continent. The presence of a warm sea surface temperature anomaly increases the heat available, via latent heat and sensible heat fluxes. Thus, the energies of the storms are also increased and the cyclones tend to become more intense than is normal during the autumn and winter months. This type of air-sea interaction is, in the end, self-destructive because, once the anomalous supply of heat in the ocean mixed layer has been exhausted, the anomalous atmospheric circulation can no longer be maintained. However, it does illustrate the point that the amount of heat stored in the ocean mixed layer during the summer heating season is an important factor for the prediction of air-sea interaction events in the subsequent autumn and winter seasons. A summer-time sea surface temperature anomaly of $1°C$, extending to a depth of $50\,m$, contains approximately $2 \times 10^8\,J\,m^{-2}$ of excess energy. If this energy is released over the subsequent three months it will provide an anomalous heat source of about $25\,\mathrm{W\,m^{-2}}$, compared with the normal autumnal heat loss of $80\,\mathrm{W\,m^{-2}}$ in the north-eastern Pacific.

A phenomena similar to that in North Pacific Ocean, has been discovered in the North Atlantic Ocean. Anomalous sea surface temperature patterns in the North Atlantic during early summer are related to the atmospheric circulation of the following winter, and this relationship can provide a prediction with a reasonable level of skill for seasonal forecasts over Northern Europe.

There is also evidence that positive sea surface temperature anomalies in the Arabian Sea, which develop in the heating season prior to the onset of the summer south-west monsoon, may provide an additional source of latent heat to enhance the monsoonal circulation described in Section 10.1. In the tropical ocean, relatively small sea surface temperature anomalies of approximately $1°C$ can produce large anomalies in evaporation because of the non-linearity of the water-vapour pressure with temperature.

In all of the above examples, the interaction of the atmosphere with the ocean has been confined to the mixed layer which occupies the upper 50–$200\,\mathrm{m}$ of the ocean. As seen in Section 5.2, most of the seasonal storage of heat is confined to this layer, and therefore many short-term or seasonal anomalies can be explained by variations in the depth and temperature of the mixed layer. However, other interactions may also be significant. The ocean transport of heat shows a seasonal variation both in tropical latitudes and middle latitudes, as illustrated by Figure 5.5. In middle latitudes the variation in seasonal transport is relatively small in comparison with the storage term, but in the tropics it is larger than the storage term. These seasonal changes in the ocean heat transport at low latitudes are related to the seasonal variations in the trade winds, with strong trade winds enhancing the poleward transport of heat. Model studies have shown that this seasonal response of the ocean is mainly confined to the direct wind-stress-driven flow in the mixed layer, via the Ekman balance. Hence, strong westward trade winds will produce a poleward transport of warm surface water and cooler sub-surface water returns towards the equator for mass balance. More recent studies have indicated that the tropical ocean varies significantly on seasonal time scales due to large scale ocean Rossby and Kelvin waves, probably being forced by atmospheric variability on time scales of days to months.

11.3   Interannual fluctuations in the ocean-atmosphere system

The air-sea interaction events discussed in the previous section occur on monthly or seasonal time scales. They are generally connected with the short-term variability of the atmosphere and the seasonal forcing by solar radiation of the atmosphere-ocean system. On the interannual time scale (variations from year to year) there is no large external forcing on the Earth (see Chapter 1), and therefore climate variations arise mainly from internal interactions such as those within the ocean-atmosphere system.

The largest global signal on the interannual time scales is that of the El Niño-Southern Oscillation (ENSO). The El Niño is the ocean component whilst the Southern Oscillation is the atmospheric component of the coupled interaction. The ENSO events are irregular events occurring between two and seven years, with a mean frequency of about four years. The interaction involves the surface and thermocline waters of the tropical Pacific Ocean with the overlying circulation in the global troposphere. The positive phase of ENSO is the El Niño, whilst the negative phase is known as La Niña. The El Niño is a warming of the equatorial Pacific Ocean, of 2–5°C that persists for a number of seasons.

The classical El Niño is the warming of the waters adjacent to the coasts of Ecuador and Peru, which is accompanied by torrential rain and a major reduction in the productivity of coastal waters. The warming usually occurs in December or January, hence the name El Niño or Christ Child. In normal years, the coast has cool surface waters due to strong upwelling of thermocline waters to the surface. The thermocline waters originate from the cool northward Peru Current. This upwelling water, which is rich in nutrients, absorbs large quantities of solar radiation in the surface layers. These conditions are ideal for the growth of phytoplankton, which in turn sustains the productive anchovy fishery. When El Niño occurs, the cool waters of the Peru Current are replaced by warm equatorial waters which are low in nutrients. This leads to a very drastic reduction in the productivity of these waters. Consequently, El Niño has a major influence on the fishing economies of Peru and Ecuador.

More comprehensive measurements of sea surface temperature have shown the warming spreads eastwards from the international dateline to the South American coast (see Figure 11.6) and up to 20°S. This is an area of the ocean which is equivalent to 10% of the Earth's surface. Hence, a rise in the global surface temperature of a few tenths of a degree can occur during an El Niño event. For comparison, this is similar to the longer-term rise in the global surface temperature during the last 150 years and therefore has to be taken into account in the analysis of global warming (see Chapter 12).

We will now consider the influence this warming will have on the overlying atmosphere. The additional thermal energy which is stored in the Pacific Ocean during a strong El Niño event is about 3×10^{22} J, which is equivalent to about eight times the total available potential energy and kinetic energy of the atmosphere. Furthermore, if all this extra heat was converted into available potential energy at the observed rate of 2.5 W m^{-2}, then it would be sufficient to maintain the atmospheric circulation for about nine months – a time scale

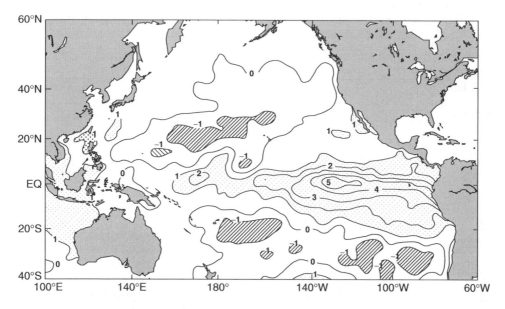

Figure 11.6   Sea surface temperature anomalies (°C) in the tropical Pacific Ocean during an El Niño event (December 1982)

which is slightly shorter than the El Niño event. Therefore, this large-scale warming of the tropical Pacific Ocean will have a profound influence on the global atmosphere.

The Southern Oscillation is the major response of the troposphere to the anomalous warming in the Pacific Ocean. It was first described by Sir Gilbert Walker, the Director General of the Indian Meteorological Office, early in the 20th century. He described a large-scale surface pressure pattern which had its dominant signal over the Indian-Pacific Ocean (see Figure 11.7). This pressure signal can be defined in terms of the seasonal pressure difference between Indonesia and Tahiti in the south-east Pacific.

The climatological surface pressure gradient between Indonesia and Tahiti forces a steady south-east trade-wind flow from Tahiti towards Indonesia. During El Niño events, the surface pressure over Indonesia is higher than normal, whilst pressure is lower over the south-east Pacific Ocean, which causes a reduction in the pressure gradient between the two regions and a weakening of the south-east trade winds. The difference in surface pressure between Indonesia and Tahiti is known as the Southern Oscillation index, and in this case it is anomalously low (negative index). During La Niña with a cooler than normal east Pacific, the pressure is lower than normal over Indonesia and higher than normal in the south-east Pacific and this drives stronger than normal south-east trade winds and the Southern Oscillation index is high (positive index).

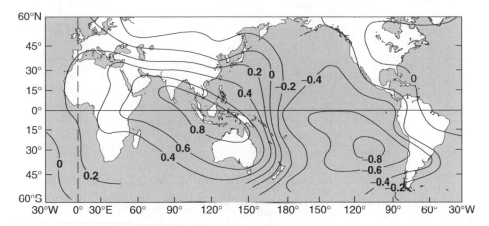

Figure 11.7   The correlation of monthly mean sea-level pressure with that of Jakarta, Indonesia. The correlation is large and negative in the South Pacific, and large and positive over India, Indonesia and Australia. This pattern defines the Southern Oscillation. An index of the Southern Oscillation can be defined as the difference in pressure between Darwin (north Australia) and Tahiti in the South Pacific. Reproduced, with permission, from Philander, G., 1983, Nature, 302: page 297, figure 3

This pressure signal is associated with major changes in the tropospheric wind circulation and, in particular, the position of the inter-tropical convergence zones. These convergence zones are regions of heavy rainfall due to the uplift of warm moisture-laden air brought in by the ocean trade winds in the lower troposphere. The cooling due to the uplift of air releases large amounts of latent heat of condensation which gives additional buoyancy to the air and further uplift.

Figure 11.8 shows the principal features of the tropospheric circulation over the tropical Pacific Ocean for

(i)   normal or cold conditions in the east Pacific (La Niña), and
(ii)   during an El Niño event.

In normal conditions the main tropical convergence zones are the ITCZ, which extend from Indonesia in the west to Central America in the east between 5 and 10°N, and the South Pacific Convergence Zone (SPCZ), which extends from northern Australia south-eastwards into the central South Pacific. In order to balance the mass of air brought into the convergence zones by the trade winds, there has to be a compensating outflow of air in the upper troposphere. The outflow has a low water content due to its low temperature and pressure in the upper troposphere, and the removal of water from the atmosphere by rainfall. This outflow of air is directed towards the south-east Pacific to complete the circulation cell. The north-south component

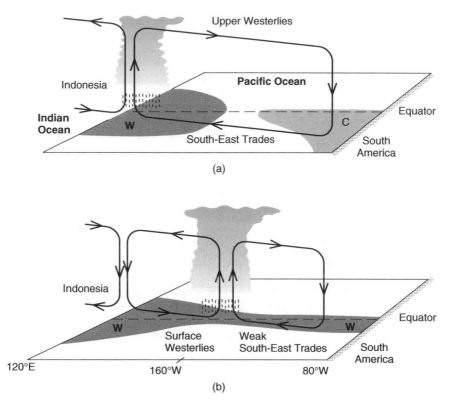

Figure 11.8   The tropical atmospheric circulation over the Pacific Ocean during (a) normal conditions, (b) El Niño conditions. During normal conditions the surface pressure is low over Australia and Indonesia (high rainfall), and high over the south-east Pacific, so the surface trade-wind circulation is strong and the Southern Oscillation index (P_{DARWIN} - P_{TAHITI}, where P is the surface pressure) is high. During El Niño conditions, the pressure is higher over Australia and Indonesia (low rainfall), and lower in the south-east Pacific; consequently, the trade-wind circulation is weaker and the Southern Oscillation index is low (W, warm; C, cold). Reproduced, with permission, from Wells, N.C., Gould, W.J. and Kemp, A.E.S., 1996, In Oceanography – An Illustrated Guide, ed. Summerhayes, C.P. and Thorpe, S.A.T. , Manson Publishing: page 54, figure 3.22

of the circulation is known as the Hadley Circulation, whilst the east - west component is known as the Walker Circulation. The upper tropospheric air which converges towards the south-east Pacific descends to the lower troposphere. Because of adiabatic warming as the air descends, this region is associated with clear skies and low rainfall. This region, which is sandwiched between the convergence zones, is known as the 'dry zone'.

In the ocean the trade winds drive cooler water equatorwards in the Peru Current and cause upwelling along the coasts of South America and the equator. This produces a large cool tongue of water in the eastern tropical Pacific Ocean. When the winds are stronger than normal the water can be unusually cold, and this is known as La Niña.

During an El Niño event, the cold tongue is replaced by warmer water from the central Pacific Ocean and from north of the equator. In an extreme El Niño event, such as occurred in 1982–83, the warming is sufficient to completely remove the cold tongue. The weakening of the trade-wind circulation is associated with major displacements of the tropical convergence zones from their normal position. In particular, the ITCZ tends to move south to the equator, whilst the SPCZ will move further east. Over Indonesia and northern Australia, the rainfall is lower than normal and drought conditions may prevail. Conversely, the islands in the central and eastern Pacific, which are normally in the dry zone, will experience heavy rainfall associated with the displacement of the convergence zones.

Though the major ENSO events are predominant in the tropical Pacific Ocean basin, there are influences well beyond this region. The change in the surface tropical pressure associated with the Southern Oscillation has a global influence on the general atmospheric circulation. Its effects cause changes in the extra-tropical circulation, as well as changes in the monsoonal circulations over the Asian and African continents. Indeed, Sir Gilbert Walker's interest in surface pressure patterns was for determining the reasons for the variations in monsoon rainfall from one year to the next over the Indian subcontinent. It was found that there is a tendency for the Indian monsoon to be weaker than normal during El Niño years.

During El Niño events, the winter temperatures in western North America are warmer than normal. This is the result of a change in the atmospheric circulation in the middle latitudes of the North Pacific, which in turn is driven by the anomalous warming in the tropical atmosphere over the Pacific Ocean. These influences, which go well beyond the location of ENSO, are known as 'teleconnections', and they are used to aid seasonal predictions in regions around the world.

11.3.1   Mechanism of the ENSO

It is known that the principal physical processes associated with ENSO are related to the tropospheric wind circulation, the sea surface temperature and the heat stored above the tropical thermocline in the Pacific basin.

During El Niño there is a major redistribution in the heat content in the Pacific Ocean with a transfer of heat towards the eastern Pacific. This heat transfer is the result of the changing patterns of equatorial currents brought about, for example, by the weakening of the south-east trade winds in the eastern Pacific Ocean. Local processes also have a role to play; for example, a reduction in winds will reduce the local upwelling of cooler thermocline waters, which in turn leads to an increase in sea surface temperature. Hence, the wind circulation in the lower troposphere has a crucial role in determining the ocean circulation and redistribution of heat in the ocean.

The sea surface temperature and heat content have a major role to play in the positioning of the major tropical convergence zones, as we have seen in the tropical Pacific Ocean. These convergence zones have a tendency to be located in regions of maximum sea surface temperature. Thus, changes in the location of maximum sea surface temperature will move the convergence zones and alter the surface wind circulation. This is known as a positive feedback loop between the wind circulation and the sea surface temperature. There are two stable states of the feedback loop. One state is a cold east Pacific (La Niña) which is associated with strong south-east trade winds and the other state is a warm east Pacific with weaker south-east winds, the familiar El Niño.

Though we are still some way off a complete understanding of the processes which cause the ENSO, there has been considerable progress in the measurement of the phenomena. In recent years there has been the deployment of an oceanographic and meteorological observing system over the tropical Pacific Ocean. This array of instrumentation is providing real-time information on the currents; sea level; surface and thermocline temperatures, and surface winds. These observations can then be used in coupled ocean-atmosphere models to provide forecasts of the El Niño many months in advance. These measurements not only provide the initial conditions for the models, but they also provide verification of the models' ability to forecast El Niño, which in turn will lead to an improved understanding of the phenomena.

In addition to ENSO there are other important phenomena in both the Indian Ocean and the tropical Atlantic known as the Indian Ocean Dipole and tropical Atlantic Dipole respectively. In similarity to ENSO, they involve an interaction between the atmospheric wind circulation and the upper layer of the tropical ocean, and account for variability on seasonal to interannual time scales.

11.4 Decadal variations in the ocean-atmosphere system

Decadal variability occurs on a time scale from 10 years to less than 100 years. It is perhaps the least understood part of the ocean-atmosphere system. Some examples of the better studied decadal phenomena are presented here.

On the decadal time scale, teleconnections between the rainfall in the Sahel (a region which borders the southern fringe of the Sahara Desert in Africa) and the sea surface temperature in the tropical Atlantic Ocean have been discovered. The Sahel region suffered a number of devastating droughts in the 1970s and 1980s, whilst in the 1950s it experienced reasonably wet seasons. The difference between the drier and wetter periods is accompanied by major changes in the sea surface temperature in the tropical oceans. Subsequent studies, using global atmospheric models, have shown that the positions of the main rainfall zones in the tropics are sensitive to anomalies in sea surface temperature, which has been demonstrated for the shorter period of ENSO in Section 10.3. In the drier years of the Sahel, the West African monsoon does not penetrate as far north as in the wet years. The lack of northern penetration of the monsoon is associated with warmer than normal water in the south and equatorial Atlantic, Indian and Pacific Oceans, and cooler than normal water in the northern oceans. This example shows the global scale of this ocean-atmosphere interaction.

A second example of decadal variability is that associated with one ocean, the northern North Atlantic during the 1960s and 1970s, as shown in Figure 11.9. During this period, surface waters became fresher (the salinity decreased) over a large region. The pool of fresh water, known as the Great Salinity Anomaly (GSA), moved gently around the sub-polar gyre from Greenland to Scotland in a period of over one decade. This freshwater cap prevented the deep overturning of water during periods of strong cooling in late winter. In turn, there was a reduction in the production of North Atlantic Deep Water, which has been shown to be the main water mass in the abyssal depths of 1000 to 2000 m of the world ocean and is associated with the deeper arm of the thermohaline circulation. How did the GSA develop? Some evidence suggests that it was caused by a large export of fresh water, associated with pack ice from the Arctic Ocean moving into the Greenland Sea and then into the North Atlantic Ocean. The GSA indicates that small changes in the climate system, in this case fresh water, may produce profound

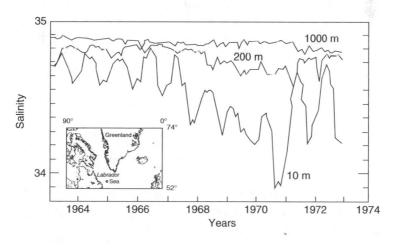

Figure 11.9   Salinity as a function of time at 10, 200 and 1000 m depth at Ocean Weather Ship Bravo in the Labrador Sea. Reproduced, with permission, from Lazier, J., 1980, Atmosphere and Ocean, 18: page 230, figure 2

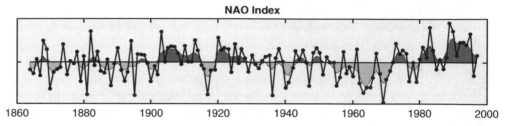

Figure 11.10 NAO is a mainly winter phenomena (Dec–Mar) and is an index of the strength of westerly winds between the Azores (40N) and Iceland (60N). The index is measured by the mean winter surface pressure difference between the Azores and Iceland. The westerly winds are always prevailing at the middle latitudes and are at their strongest each winter. NAO positive is when the winter westerlies are particularly strong, whilst NAO negative is when the westerlies are much weaker. NAO positive has storm systems developing in the Atlantic and tracking ENE towards NW Europe. We have above average rainfall in the UK especially in Northern Britain, and warmer than average winter. NAO negative the storms track northwards towards Greenland, and eastwards into the Mediterranean Sea. The UK is colder than average, with weaker westerly winds, and blocked flow in the northern N. Atlantic and Scandinavia. Reproduced from Lamont Doherty Earth Observatory, Columbia University. See plate section for a colour version of this image

changes on the deep ocean and its circulation. Computer models have shown that the addition of large amounts of fresh water from the Arctic Ocean in to the North Atlantic Ocean can significantly reduce the strength of the overturning circulation, which in turn can cause a major cooling of the North Atlantic region.

The GSA may be related to the North Atlantic Oscillation (NAO) shown in Figure 11.10. The difference in sea surface pressure between the Azores high and the Iceland low pressure gives a measure of the strength of the eastward flow over the North Atlantic. In a low index situation the eastward flow is weaker than normal, whilst in the high index state the flow is stronger than normal. In the 1960s the NAO was in a low index state, the eastward flow was weaker and the flow became 'blocked' i.e. the high pressure area moved more frequently into the region between 40° and 60°N and blocked the normal storm track in the North Atlantic. This low index situation can allow more sea ice to form in the Atlantic sector of the Arctic basin. The NAO index steadily increased from the 1970s up to 1990s. This increase in the NAO may have allowed more exchange of water between the Arctic and the North Atlantic Oceans and therefore more sea ice was brought into the North Atlantic, causing a freshening of the surface layers.

In the Pacific Ocean a Decadal Oscillation (PDO) in the winter sea surface temperature north of 20° N has been related to the surface wind circulation observed over the 20th century (see Figure 11.11). From 1900 to 1950 it was in a positive index or warm phase, with above average sea surface temperature over the Eastern Pacific Ocean and below average temperature in the

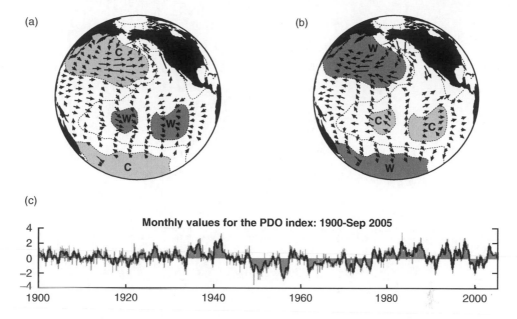

Figure 11.11 (a) Positive Pacific Decadal Oscillation is warm phase, (b) negative is cold phase. Colours are winter SST, Sea level pressure are contours, arrows are surface wind and (c) time series of the PDO. Reproduced from JIASO, University of Washington. See plate section for a colour version of this image

north-western Pacific Ocean. This pattern entered a negative phase between 1950 and 1980, when the eastern Pacific became cooler, and the north-western Pacific became warmer. This pattern reverted back to the positive phase from 1980 to the present. This oscillation is irregular and the precise reason for it is not clear. The surface wind circulation during the positive index or warm phase is associated with stronger eastward winds in the middle latitudes and a stronger Aleutian Low. In the negative index or cold phase, the surface eastward winds are weaker than normal, and the Aleutian low is weaker than normal.

The final example of decadal variability is over the Southern Ocean. In the period between 1979 and the present, an atmospheric flow pattern in the lower troposphere, known as the Southern Annual Mode (SAM), has been observed. The subtropical high pressure centres have moved further south over the whole hemisphere and the meridional pressure gradient has increased over this period. The eastward flow has subsequently increased and moved poleward over the last 30 years.

In summary, our knowledge of decadal variability in the atmosphere and the upper ocean has increased remarkably in recent years but the detailed mechanisms responsible for these changes are still not well understood.

12

Climate Change

12.1 Past climate observations

In investigating longer-term variability of the climate system, in particular the ocean-atmosphere components and their interaction, we have a problem with the available observations. The systematic collection of observations of the sea surface temperature and currents only go back 150 years, whilst deep ocean measurements of temperature and salinity go back only 100 years.

The quality of the surface observations from merchant ships was often compromised by the method of sampling but, with careful scientific analysis, corrections have been possible. Higher-quality observations of the ocean temperature and salinity across the major ocean basins were taken during the International Geophysical Year (IGY) in 1957. Some of these sections have been repeated, but only at infrequent intervals of 10 or 15 years. One of the longest continuous time series of temperature and salinity of the whole water column has been obtained at the ocean station close to Bermuda since the 1950s.

Even on land, surface temperature observations extend back about 300 years. The central England time series is one of the longest, extending from 1659 to the present. This series relies on both temperature measurements and weather diaries to obtain a reasonably reliable record. The temperature measurements were often of low quality due to poor exposure of the thermometers. Considerable judgement and care is therefore required to obtain a consistent time series from the data. Systematic measurements of the surface meteorology over the oceans commenced in 1853, but regular daily soundings of the troposphere did not commence until the 1940s.

Because the instrumental record is at best 300 years long, other methods are required to measure climate variations in the past. Scientists have developed the relatively new sciences of paleao-climatology and paleao-oceanography to provide this information. This approach uses a whole range of evidence from fossils; pollen; lake levels; deep-sea ocean sediments; glaciers and ice

The Atmosphere and Ocean: A Physical Introduction, Third Edition. Neil C. Wells.
© 2012 John Wiley & Sons, Ltd. Published 2012 by John Wiley & Sons, Ltd.

sheets to infer past variations in the climate. This is not the place to discuss all the methods and evidence that have been obtained from these sciences but some of the results which are relevant to understanding atmosphere-ocean interactions and their role in climate change will be referred to in this chapter.

One of the notable examples is from the ice cores drilled in the Antarctic and Greenland ice sheets. These ice cores, over 3000 m in length, contain records of past climate extending back to about 750 000 years before present (BP). This record extends through five major glaciations and these glacial cycles have a period of about 100 000 to 150 000 years between the ice ages and interglacial periods (see Figure 12.1). Small air bubbles trapped in the ice when it was first laid down as snow provide a time history of the composition of the atmosphere well before any anthropogenic influences occurred. In particular, they provide a unique trace of the natural variations in CO_2 and methane, as well as providing a set of baseline measurements before man's influence became detectable. The cores also provide other information on environmental conditions when the ice was first laid down. Oxygen isotopic composition, measured using air bubbles trapped in the ice, provides a record of ocean paleaotemperature and ocean volume; acidity levels record past volcanic events, whilst dust measurements provide an index of the storminess of the region. The length and the high temporal resolution of the records provide unique windows into past variations in climate. For example, past changes in the North Atlantic current can be inferred from the ocean paleaotemperature in the Greenland ice cores. One core has shown that the North Atlantic varied between warm and cold conditions on the very short time scales of 20–50 years. These rapid changes in the North Atlantic Ocean, in turn, would have had a profound influence on the northern European climate. One interesting aspect of these changes in climate is that they appear to be confined to the earlier interglacial period, known as the Eemian period, 150 000 years BP, whilst during the past 10 000 years (known as the Holocene period), the North Atlantic temperature has been very stable. The instability of climate in the northern North Atlantic is of considerable interest to oceanographers and climate scientists as this region is known to be a major control of the ocean's thermohaline circulation.

One of the problems with the ice-core data is that it is limited to Greenland and Antarctica, and therefore some climate effects will be local to these sites whilst others will be measures of global change. However, CO_2 and methane are fairly uniform throughout the atmosphere, and so the ice cap records provide excellent evidence of their global concentrations. On the other hand, the acidity levels in the Greenland ice sheet will be strongly affected by relatively local volcanic events in the Iceland area but probably only weakly affected by major low-latitude eruptions, such as the Pinatubo event in the Philippines in 1991.

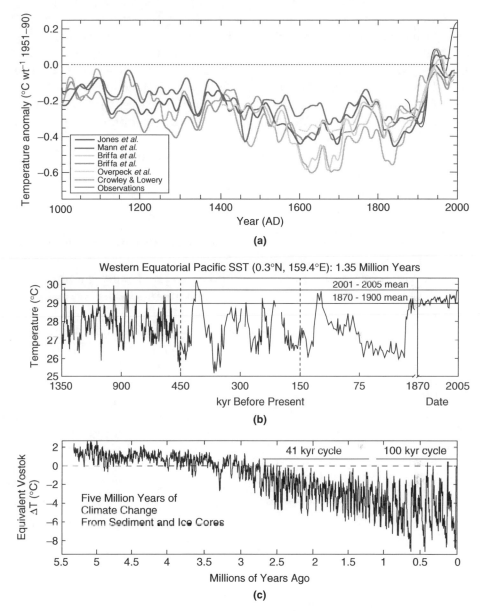

Figure 12.1   The variation of the Earth's surface temperature on three time scales: (a) estimates of global temperature for last 1000 years. Observations are based on temperature measurements and other estimates from indirect proxies. The dashed line is the global average 1951–1990. From Wikipedia Temperature record of past 1000 years. See plate section for a colour version of this image. (b) Surface temperature in the western equatorial Pacific Ocean over the last 1 000 000 years. 1870–2005 is shown for comparison. Reproduced from Hansen, J., Mki. Sato, R. Ruedy, K. Lo, D.W. Lea, and M. Medina-Elizade (2006), Global temperature change, *Proc. Natl. Acad. Sci.*, **103**, 14288–14293, doi:10.1073/pnas.0606291103, figure 5. (c) The temperature in Antarctica over 5 million years. The dashed line represents conditions at the beginning of the twentieth century. From Wikipedia Geological temperature record

The ice-core information is one source of information for the determination of past climates. Another source is the deep-ocean sediment cores, obtained from drilling programmes such as the Deep Sea Drilling Programme (DSDP) and, more recently, the Ocean Drilling Programme (ODP). Cores have been obtained in all the ocean basins (except the Arctic Ocean) and extend the record of climate change back millions of years. Because the changes in the deep ocean itself are relatively slow, it is possible to obtain a very reliable indication of climate change.

Cores from the Santa Barbara basins and the equatorial east Pacific Ocean have shown the annual deposition of diatoms and fish scales. The thickness of these fine or laminated sediments provides a record of productivity in the surface layers of the water column. This productivity, as we have mentioned in Section 10.3, is strongly influenced by El Niño; hence, the thickness of sediment or density of fish scales provides a unique record of El Niño events over the last 2000 years, a period well beyond the length of other records. These records indicate that El Niño is modulated by longer-period changes on time scales of 20–100 years. As yet, no mechanism for these variations has been established.

Finally, Figure 12.2 shows the sea level change determined by indirect palaeo-climate methods from the last glacial maximum, 18 000 years ago, to the present. The 120 m rise in sea level from 18 000 years to 8000 years ago flooded the continental shelves of the world's ocean and dramatically changed the coastline and its associated estuaries.

These examples show the diversity of information that has been gleaned from the ocean record about the past variations in climate. Figure 12.1

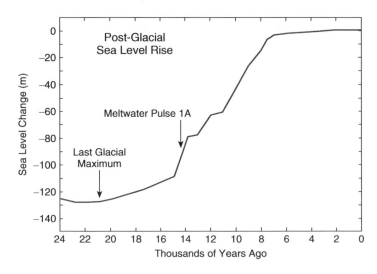

Figure 12.2 Global Sea level rise from the last glacial maximum (24,000 years ago to present). Post-glacial sea level Wikipedia

shows reconstructions of global temperature over the past five million years, inferred from palaeoclimate techniques. From this we can see that the glacial-interglacial cycles are associated with global temperature changes of about 6–8°C, whilst the Little Ice Age (1450–1850 AD) around the North Atlantic Ocean was associated with a Northern Hemisphere temperature change of less than 1°C. These long-term fluctuations in climate allow us to view the present problems of climate change in the context of natural changes in the past. The scientific understanding from the study of past variations, coupled with the intense study of the present system, should allow better predictions of climate change to be made. These predictions are based on the coupled ocean-atmosphere and climate models discussed in Chapter 10.

12.2   Mechanisms of climate change

From our cursory examination of the climate record (Figure 12.1), we see that there have been large changes in the Earth's climate both globally, associated with the waxing and waning of the ice sheets, and regionally, for instance, in the northern North Atlantic basin during the Little Ice Age. What are the mechanisms which give rise to these significant changes in the system?

First, we will consider the physical constraints on the time scales of the system, depicted in Figure 12.3. The climate system is composed of five basic components: the atmosphere; the ocean; the cryosphere, which is composed of ice sheets, mountain glaciers and pack ice; the land and its hydrology, and finally the biosphere. The biosphere includes all aspects of the living system on land, in the atmosphere and in the ocean. The latter is the most complex component and probably the least understood part of the climate system. All

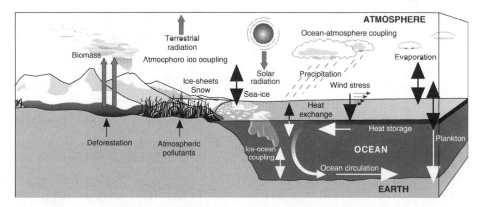

Figure 12.3   A schematic view of the Earth's climate system showing the roles of land, atmosphere, oceans, sea ice, glaciers, and ice sheets. Reproduced, with permission, from Wells, N.C., Gould, W.J. and Kemp, A.E.S., 1996, In Oceanography – An Illustrated Guide, ed. Summerhayes, C.P. and Thorpe, S.A.T. , Manson Publishing: page 42, figure 3.1. See plate section for a colour version of this image

these components interact with other components of the system to produce climate variability on a vast range of time scales.

The atmosphere provides the shortest time scale in the system. The time for a parcel of air to travel around the globe defines the global circulation time scale, which is about 10 days in the troposphere. The atmosphere transports heat, momentum and water in all its three phases, as well as radiatively active gases (such as CO_2, methane and ozone) and particulates (such as sulphates and carbon particulates) around the globe. The radiation time scale of the atmosphere (defined in Section 1.4) is a little longer, at about 30 days. The atmosphere is the most capricious part of the climate system. As we have seen, the clouds, which at any one time occupy about 50% of the surface area of the globe, are extremely important for the regulation of the Earth's surface temperature (see Chapter 2) but these clouds are quite difficult to predict, because of their complex physics discussed in Chapter 4.

By contrast, the underlying oceans have a larger range of time scales, from months to 1000's of years. The surface layers, because they are being continually driven by surface winds, solar heating and cooling by evaporation are relatively well mixed. This mixing depth is typically about 20 m in summer to over 100 m in winter. The mixed layer has a time scale of months and, as we have seen in Section 11.2, it is associated with many short-term climate interactions with the atmosphere. By contrast, the major gyre systems have time scales of years to decades. For example, the Great Salinity Anomaly in the northern Atlantic Ocean (Figure 11.9) took slightly longer than a decade to travel around the sub-polar gyre.

The thermal time scale of the gyres can be estimated from the average depth of the main thermocline. If it is assumed that the average depth is about 500 m, then this gives a time scale of about 18 years (see Section 1.4). These gyre circulations are driven by the overlying wind system and the process of Ekman pumping (Chapter 7), as well as the addition and removal of buoyancy at the sea surface. Therefore, they may hold the key to unravelling the decadal variability we observe in the climate system.

The deeper abyssal oceans provide the longest time scale of the ocean. The driving force of the abyssal oceans is focused on the high-latitude regions of the globe, wherein the abyssal water masses are formed. The circulation time scale is of the order of centuries to a millennium. For example, the North Atlantic creates about $20 \times 10^6 \, \text{m}^3 \, \text{s}^{-1}$ of North Atlantic Deep Water (NADW). If we assume that the abyssal ocean is composed entirely of NADW, and take the volume of the abyssal ocean to be about $1.0 \times 10^{18} \, \text{m}^3 \, \text{s}^{-1}$, then the time scale for the deep overturning circulation is 1600 years.

The third component is the cryosphere, wherein about 2% of the Earth's total water and 75% of the Earth's fresh water is stored. One of the shortest time scales is that of the surface pack ice which floats on the Arctic Ocean and around the Antarctic continent, and which has a thickness of a few metres.

Its growth and decay are controlled by heat fluxes and precipitation between the surface and the atmosphere, and by the ocean heat fluxes from below the ice. Because of these factors, sea ice exhibits considerable growth and decay on seasonal and interannual time scales. Furthermore, because it is driven by surface winds and ocean currents, it can transfer large quantities of fresh water from its place of formation to its place of melting, and these will need to be replaced by the ocean and atmosphere circulation. An example of this process occurs in the Antarctic where the pack ice is driven northwards from the coast by intense offshore winds, thus providing open water adjacent to the coast. This in turn allows more ice to be frozen at the coast by intense cooling of the ocean by the Antarctic winds. Pack ice is generally low in salinity due to brine removal back into the ocean during the freezing process. Hence, the ocean becomes more saline beneath the ice. The combination of high salinity and low temperatures of these waters leads to the production of the most dense water mass in the ocean: Antarctic Bottom Water.

Sea ice is also important as a regulator of ocean temperature due to its low thermal conductivity. Hence heat transfer from the ocean to the atmosphere is considerably reduced in ice-covered regions of the Arctic and Antarctic Oceans.

The mountain glaciers, ice sheets and ice shelves provide the longer time scales of the cryosphere. The growth and decay of mountain glaciers occur on time scales of decades to centuries, whilst the major ice sheets have time scales of many thousands of years. A dramatic example of the decay of glaciers is that seen in the last 150 years by most of the world's glaciers associated with the global rise in surface temperatures. In general, glaciers are in quasi-balance between the growth of ice at the head of the glacier, mainly due to snowfall accumulation over many years, and ablation processes at the lower end of the glacier. These ablation processes include surface melting and evaporation, melting at the base of the major ice sheets and calving when the ice breaks off into the sea. The growth of a glacier is therefore dependent not only on the temperature, but also on the rate of snowfall at the head. In addition, because of the vast accumulations of ice in Greenland and Antarctica (of the order of 2 km thick), there is a lowering of the Earth's lithosphere and asthenosphere by isostatic adjustment. This isostatic adjustment is seen today in Scandinavia, where the land is rising at the rate of 1 cm per year following the melting of the ice sheet over 18 000 years ago. This adjustment process brings an additional longer time scale into the dynamics of ice sheets. Figure 12.2 shows the profound changes in ocean volume associated with the melting of the ice sheets in the last 18 000 years. It is noted that global sea level rises of over 30 mm year^{-1} occurred during melt water pulse 1A over 14 000 years before present, whilst sea level rise presently is 2–3 mm year^{-1}.

The Milankovitch theory for glacial cycles implicitly relies on feedback within the climate system to amplify the relatively small changes in seasonal solar radiation incident on the atmosphere to produce large changes in the ice sheets. Milankovitch placed his emphasis on the crucial role of solar radiation in the summer-time at 60°N. When the solar radiation due to orbital effects was diminished, there would be decreased ablation of the ice in summer and therefore the ice sheet would grow slowly from one year to the next. As the area covered by the ice sheet increased, the albedo would increase, thus reducing the amount of solar radiation absorbed by the ice and further decreasing the amount of ablation. The theory does not take into account many of the physical processes which influence the growth of an ice sheet; for example, the changes in snowfall due to the increased level of the ice sheet or the influence of clouds and aerosols on the surface radiation budget of the ice sheet. Despite the theory being very simple, it has been widely accepted by the scientific community, because the analyses of deep-sea cores have shown that the main frequencies of variability are associated with these forcing frequencies of the solar radiation (see Chapter 1).

The land surface and its associated hydrology comprise the fourth component of the climate system. The air temperature above the surface is controlled by the energy budget over that surface, discussed in Chapter 5. The absorption of solar radiation will depend on the albedo of the surface; whether it is snow, ice, desert, or its type and cover of vegetation (see Table 1.3), whilst the heat loss by evaporation will depend on the availability of water stored in the vegetation or soil, in lakes and rivers, or as groundwater. The albedo is an exceptionally important variable in the climate system. The spring melt of snow cover over the continents is an example of the influence of albedo. Whilst the snow cover exists, only a relatively small proportion of the solar radiation is absorbed by the surface. At the same time, heat is lost from the surface by long-wave radiation and sublimation to the atmosphere, and therefore little or no heat is available for melting. The heat balance maintains the snow-covered ground. Now, if the snow is removed from a soil surface, the albedo will decrease from about 80% to 20% and, as a consequence, the absorbed solar radiation will increase by a factor of 4 and the surface temperature will rise. To balance the absorbed solar heat, the heat loss from the surface would increase by a similar amount, and thus the air above the surface would also be warmed rapidly. Over the continents in the spring-time, the increasing solar radiation provides the stimulus for the rapid thaw of the snow cover, which may take place over a period as short as a few days.

Albedo changes will also occur where there are major changes in land use, for instance, associated with major deforestation or with the withering of vegetation associated with drought. There is some evidence that the prolonged droughts in the Sahel region in the 1970s and 1980s involved albedo feedback. In this case the drought led to an increase in surface albedo, which in turn decreased the propensity for rainfall in this region.

The final component of the climate system is the biosphere. This component is fundamentally important, as it has been part of the process in developing the present-day composition of the atmosphere. The GAIA hypothesis takes this a stage further, by suggesting that the biosphere has regulated its own climate from early in Earth's history by changing the composition of the atmosphere.

We have already seen how land use and its associated vegetation provide an important example of albedo feedback. A second example of this albedo feedback has been hypothesised in the ocean, in connection with coccoliths. These phytoplankton, because of their carbonate shells, produce a chalky appearance to the water when they bloom in northern North Atlantic waters. The albedo of sea water is thus increased by the appearance of the bloom, which results in a reduction of the absorbed solar radiation and a cooling of the surface waters.

Vegetation is not only important in albedo feedback. It also has the capacity, with the underlying soil and hydrology, to control the energy budget of the surface and of the air above it. The tropical rainforest is an example of an ecosystem where the water and nutrients are recycled very efficiently. When the forest is removed, the water is evaporated very rapidly and, because of the poor soils, the area may revert quickly to low-quality scrubland.

In this section we have seen that, generally, there are five components which interact to produce our climate. Sometimes the interaction may involve only two major components, for example, the ENSO, which is an interaction between the tropical ocean and atmosphere (see Section 11.3). It may involve three or four components e.g. the land and its surface hydrology; the biosphere, the ocean surface and the atmosphere in phenomena such as the Sahel decadal droughts.

The mathematical climate model (see Section 10.6) provides a most powerful tool for the study of climate change. The basis of the model is a set of equations which describe the processes occurring in each of the components, together with equations which describe the transfer of quantities between each component of the system; for example, the transfer of momentum or the exchange of carbon dioxide between the atmosphere and ocean. In Chapter 10 we showed some examples of ocean, atmosphere and land models that are being used to study aspects of the climate change problem. In addition, ice-sheet models have been developed and are being actively validated from the known record. Though it would be foolish to suggest that these models are perfect, they do have sufficient authenticity to be used for prediction of the climate system. At present, perhaps the least understood component is the biosphere, particularly in the ocean. The complexity of the marine biology component, and the spatial and temporal sampling problem associated with ships and satellites, make it difficult to measure the rapid changes associated with plankton blooms over the globe. This means that, at present, a complete climate system model which has all five components interacting with each other is some way off being realised. For the present, we may have to use more limited climate models with perhaps three components of the system to understand climate change and variability.

In summary, the inherent complexity of the climate system indicates that systematic study of climate variability and change requires new methods and sustained observations to enable us to make this subject into a reliable quantitative science from which increasingly more accurate predictions can be made.

12.3   Current climate change

Temperature observations from around the Earth have been combined to calculate a global mean surface temperature. For recent history, these observations are mainly temperature recordings made from thermometers, but in order to gain an insight into historical global temperatures other sources are used such as weather diaries and ice-core data. When these global mean surface temperatures are plotted as a function of time, it becomes apparent that the surface of the Earth has been warming gradually since around 1900 (see Figure 12.4). The global temperature shows a rise of about 0.8°C from 1910 to 2005 (source: IPCC 4th assessment). This rise in temperature, however, is far from steady over the period, with the largest increase between 1910 and 1940, followed by a small cooling for 40 years, before resuming a rising trend. Recent years have been particularly warm, with 11 out of 12 years between 1995 and 2006 ranking in the warmest 12 years on record.

Although observations indicate that warming is widespread over the globe, it is interesting to note that different regions are warming at different rates, as shown in Figure 12.5. The strongest warming has been observed in the Arctic, which has been warming at almost twice the global average rate over the last 100 years. This is largely because the snow and ice, which reflects sunlight and keeps the surface cool, has decreased markedly over both land and sea. For example, summer Arctic sea ice has decreased in extent by 5% per decade since 1978 and in the last 10 years by 10%. The snow cover in the Northern Hemisphere has also declined in the last 50 years (see Figure 12.4). However, due to the thermal capacity of the ocean, it is predominantly the decrease in land-ice and snow cover which has driven the rise in temperatures over recent years. It is also worth noting that the warming signal is stronger on land than in the oceans, which indicates a lag effect in the oceans.

Consistent with the increase in global mean surface temperature, an increase in global mean sea level has been observed, with the mean sea level rising by around 2mm year^{-1} between 1961 and 2003 (see Figure 12.4). About 40% of this increase in sea level is estimated to be due to the thermal expansion of

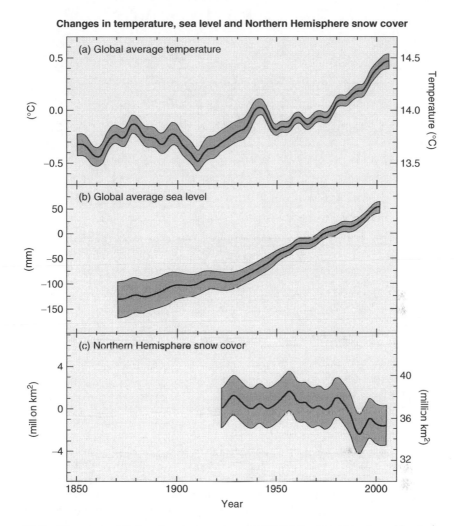

Figure 12.4 Changes in global surface temperature, global sea level and Northern Hemisphere area of snow cover (March–April), with respect to the period 1961–1990. Reproduced, with permission, from Intergovernmental Panel on Climate Change (IPCC) 2007 WG1-AR4

the oceans associated with warming, with a similar fraction associated with melting mountain glaciers. Less than 20% is associated with melting of the Greenland and Antarctica ice sheets.

Climate change over at least the past 100 years has been associated with the addition of greenhouse gases by humankind, originating from the industrial and agricultural revolutions in the late 18th and 19th centuries. The most important greenhouse gas is carbon dioxide, which has reached a level of 390 ppm (2010) from a pre-industrial value of 270 ppm. From the ice cores we know that during the Holocene carbon dioxide levels have been remarkably

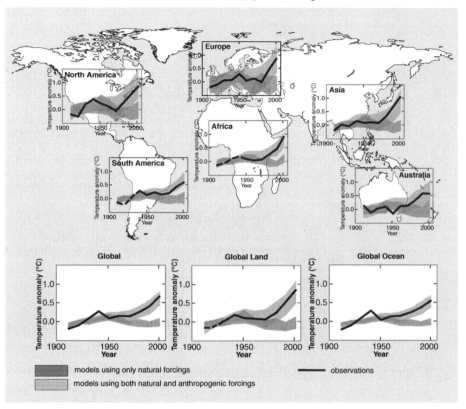

Figure 12.5   Global, ocean and continental temperature change in 20th Century from observations and climate models. Lower curves indicates models which include only natural radiative forcing due to solar and volcanic activity, whilst upper curves indicates models which include both anthropogenic (e.g. greenhouse gases) forcing and natural forcing. Reproduced, with permission, from Intergovernmental Panel on Climate Change (IPCC) 2007 WG1-AR4. See plate section for a colour version of this image

constant at the pre-industrial levels, whilst over the last 750 000 years they have varied between 180 ppm during the ice ages and 290 ppm during the inter-glacial periods.

The present climate is not in an equilibrium state, and therefore continued increases in CO_2 are likely to cause a major change in the Earth's climate. Figure 12.6 shows the radiative forcing associated with greenhouse gases, ozone and aerosols. Though it can be seen there are many factors which may change climate, greenhouse gas contribution is the most significant of all these factors. The extra radiative forcing on the Earth's surface of $1.6\,\mathrm{W\,m^{-2}}$ (2005) is small compared with the solar radiation flux at the top of the atmosphere. However, this extra flux will continue to rise over time unless the anthropogenic greenhouse gas concentrations are stabilised.

Radiative Forcing Components

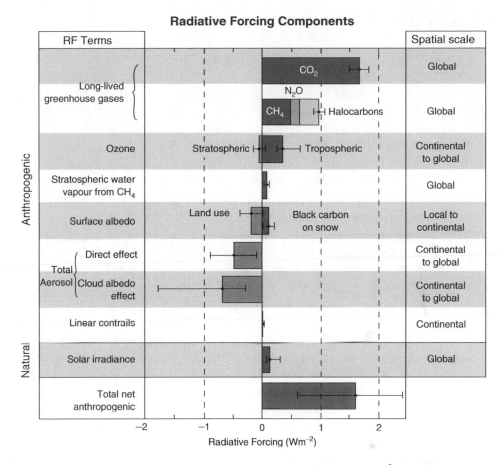

Figure 12.6 Anthropogenic and anomalous Solar Radiative Forcing ($W\,m^{-2}$) of climate system in 2005. Volcanic aerosols are not included. The net radiative forcing in 2005 is $1.6\,W\,m^{-2}$. This can be compared with a long term mean Solar Radiative forcing of $342\,W\,m^{-2}$. Reproduced, with permission, from Intergovernmental Panel on Climate Change (IPCC) 2007 WG1-AR4. See plate section for a colour version of this image

The study of the anthropogenic rise in greenhouse gases, through the IPCC process (see Section 12.4), has improved our understanding of the climate system. In particular, it has shown that the climate system is sensitive to small changes in the radiative forcing. Therefore it is likely that small changes in the output from the Sun may have a large influence on the Earth's climate and for example, the 11 year sunspot cycle is associated with a variation in solar output of $0.3\,W\,m^{-2}$. Furthermore, it is certainly known that volcanic aerosols have a major influence by reducing the surface irradiance at the Earth's surface for a few years. Changes in atmospheric composition, aerosols, and solar radiation are all important influences on the Earth's climate.

12.4   Understanding recent climate change

The mathematical climate model provides a powerful tool for the study of climate change. The model however must be able to reproduce the observed climate, before it is used for prediction (see Figure 10.4). Climate models are validated by their ability to simulate historical temperature and sea-level trends. Figure 12.5 shows a comparison of observed and simulated continental and global temperature trends for the 20th century. The lower curves indicates simulated temperatures using models which do not include anthropogenic changes in concentrations of radiatively active gases such as carbon dioxide. These models are able to simulate temperature trends from around 1900 to 1950, when changes in global mean temperatures were dominated by volcanic activity and changes in solar activity. However, these models are unable to simulate the observed increase in global mean temperatures since around 1950. In order to simulate these temperature changes it is necessary to include anthropogenic changes in radiatively active gases (upper curves). It is only the models which include anthropogenic forcings that have been able to simulate the observed geographical distribution of the warming; for example, greater warming over land than over the ocean, and the observed continental scale temperature changes. As a result of evidence such as this, most scientists are in agreement that the systematic rise in the atmospheric radiatively active gases (e.g. carbon dioxide and methane) from human activity over this period has caused the observed changes in temperature in the 20th century.

Despite a broad consensus among scientists on the causes of global warming there are many aspects of the climate system that are not well understood which lead to uncertainties in any climate predictions. Currently perhaps the least understood component is the biosphere, particularly in the ocean. The complexity of the marine biology component, and the spatial and temporal sampling problem associated with ships and satellites, makes it difficult to measure the rapid changes associated with plankton blooms over the globe.

Other uncertainties in the climate system models include the challenge of representing physical processes that occur on small scales, such as cloud formation, in mathematical models with coarse gird-spacing (see Chapter 10).

Climate scientists can estimate the uncertainties in present climate models by running many different mathematical models. For example, to understand the influence of different emissions of greenhouse gases in the future one would run the same climate model several times, changing the emissions scenario each time. If the emissions scenarios cover the range of likely possible emissions, then the difference between the results from these models gives an indication of the range of temperatures that might be expected in the future. The effect of uncertainties in the representation of physical processes in a model can be examined using a similar technique. For example, in a climate model with a coarse grid, clouds are formed when the grid-box mean relative humidity exceeds a threshold value. This threshold relative humidity is set to a value somewhat less than 100%, say 90%, to try and represent the idea that the relative humidity varies over the grid-box so that if the mean relative humidity exceeds 90% it is likely that somewhere in the grid-box the air will be saturated and clouds will form. However this threshold is not a precise value, so suppose a scientist wished to investigate the sensitivity to changing this value within a realistic range (say 85–95%). The scientist could then run several climate models, each with a different threshold. If a model does not reproduce the observed temperature trends since 1850 then one can discard that simulation. The remaining simulations will then give an indication of the sensitivity of the future climate to changing the threshold relative humidity.

12.5   Predicting future climate

Climate change predictions which are provided to decision makers and planners are made by running hundreds of different climate simulations, with different mathematical models and different choices for less defined key parameters. Once any models which do not produce the observed past climate (1860 to present) are discarded, the remaining models are combined to give probability maps of likely changes in climate under different emissions scenarios.

These models have demonstrated that global warming of the Earth's surface in the range of 2–4°C is very likely by 2100. This predicted rise in global temperature is larger than any temperature shown in Figure 12.1 over last 6 million years.

In view of this evidence for the effects of unrestrained increases in greenhouse gases in the next 100 years, the Intergovernmental Panel on Climate Change (IPCC) has taken the pragmatic approach of publishing new evidence every five years. This allows the international community of politicians, decision makers and the public to be appraised of the best evidence and advice on global warming from the scientists.

A final word on the anthropogenic addition of carbon dioxide to the atmosphere and its transfer into the ocean concerns the steady decline in the pH of the ocean that marks ocean acidification. This is raising concerns about the influence of increased levels of carbon dioxide on the biosphere, from the dissolution of coral reefs to other impacts on marine animals, including crustacea.

Advances in the all aspects of the Earth sciences are needed to guide not only this generation, but future generations, towards the maintenance of a habitable planet Earth.

Problems

Chapter 1

Problem 1

If the global albedo of the Earth increased from 0.3 to 0.31 estimate the change in radiative equilibrium temperature of the Earth.

From equation 1.4 $\frac{Q}{4}(1-\alpha) = \sigma T_e^4$

Substitute into above $Q = 1360\,\text{W m}^{-2}$; $\sigma = 5.67 \times 10^{-8}\,\text{W m}^{-2}\,\text{K}^{-4}$; $\alpha = 0.31$;

$$T_e^4 = \frac{Q(1-\alpha)}{4\sigma} = 4137566138$$

$$T_e = 253.62\,\text{K}$$

Substitute into above $Q = 1360\,\text{W m}^{-2}$; $\sigma = 5.67 \times 10^{-8}\,\text{W m}^{-2}\,\text{K}^{-4}$; $\alpha = 0.30$;

$$T_e^4 = \frac{Q(1-\alpha)}{4\sigma} = 4197530864$$

$$T_e = 254.53\,\text{K}$$

Cooling of 0.915 K with increase of albedo of 0.01 or 3.3%.

Another method is to differentiate α with respect to T_e using equation 1.3

$$\frac{d\alpha}{dT_e} = -\frac{4}{T_e}(1-\alpha) = 0.011\,\text{K}^{-1}$$

Substitute into above $T_e = 254.53\,\text{K}$ and $\alpha = 0.30$;

For a change of 0.915 K the change in α is 0.01. The change in albedo for 1 K is therefore $0.01/0.915 = 0.011$. This agrees with the earlier calculation.

The change in albedo may be caused by a change in cloud cloud. Assuming cloud covers 50% by area of the globe, then an increase in cloud cover by 6.6% would be needed to cause this change in albedo.

The planetary albedo may also be changed by the introduction of sulphate aerosols into the upper atmosphere by large volcanic eruptions.

The Atmosphere and Ocean: A Physical Introduction, Third Edition. Neil C. Wells.
© 2012 John Wiley & Sons, Ltd. Published 2012 by John Wiley & Sons, Ltd.

Problem 2

Estimate the surface temperature of the Earth's moon on the sunlit side. The moon does not have an atmosphere and therefore the surface temperature will be close to the radiative equilibrium temperature, provided the heat conduction through the surface is neglected. The sunlit area of the moon is the area of hemisphere facing the sun.

The surface area of the hemisphere is $2\pi r^2$ where r is radius of the moon. The intercepted area of the sun's radiation is πr^2.

$$\frac{Q(1-\alpha)}{2} = \sigma T_e^4$$

If $\alpha = 0.12$; $Q = 1360\,\mathrm{W\,m^{-2}}$; $\sigma = 5.67 \times 10^{-8}\,\mathrm{W\,m^{-2}\,K^{-4}}$;

$$T_e = 320.51\,\mathrm{K}$$

This calculation has assumed an average value for the solar radiation over the sunlit side of the moon.

Problem 3

Calculate the thermal capacity of an ocean (i) 2.6 m (ii) 100 m and (iii) 1000 m (iv) 3670 m deep and compare it to the thermal capacity of the atmosphere.

Thermal capacity of atmosphere = Mass/unit area × specific heat of air at constant pressure = $10,328\,\mathrm{kg/m^2} \times 1004\,\mathrm{J/kg/K} = 10,357,264\,\mathrm{J/m^2/K}$.

Mass/unit area = msl pressure/g = $1013.25 \times 100/9.81 = 10,328\,\mathrm{kg/m^2}$.

The thermal capacity of the ocean = mass/unit area × specific heat of sea water

(i)   Mass/unit area = density of sea water × depth = $1020 \times 2.6 = 2652\,\mathrm{kg/m^2}$
       Thermal capacity = $2652 \times 4000\,\mathrm{kg/m^2} \times \mathrm{J/kg/K} = 10,608,000\,\mathrm{J/m^2/K}$
       It can be seen that 2.6 m sea water has a similar heat capacity to that of the whole atmosphere. Ratio of ocean /atmosphere heat capacity = 1.024.

(ii) Heat capacity of 100 m = $(100/2.6) \times 10608000 = 408000000\,\mathrm{J/m^2/K}$
       Ratio of ocean/atmosphere = 38.5.

(iii) Heat capacity of 1000 m = 4,080,000,000
       Ratio of ocean/atmosphere = 385.

(iv) Heat capacity of 3670 m = 1.49736×10^{10}
       Ratio of ocean/atmosphere = 1445.7.

Problem 4

The sun in early Earth's history may have had a power output of 75% of the present sun. Estimate the change in the Earth's radiative equilibrium temperature, assuming the planetary albedo is the same as the present day albedo.

$$T_e^4 = \frac{0.75\,Q(1-\alpha)}{4\sigma} = 3148148148$$

$$T_e = 236.87\,\text{K}$$

$$T_e(\text{today}) = 254.53\,\text{K}$$

$$\Delta T = -17.66\,\text{K}$$

Chapter 2

Problem 1

Show that 1 decibar is approximately equal to 1 m of sea water.

The hydrostatic pressure (p) at depth h is given by: $p = \int\limits_{-h}^{0} \rho g\,dz$.

In oceanography the pressure at sea level is assumed to be zero.
If we take a typical ocean density is $1020\,\text{kg m}^{-3}$, $g = 9.81\,\text{ms}^{-2}$, and $h = 1\,\text{m}$ then $p = 1020 \times 9.81 \times 1 = 10006\,\text{Pa}$.
1 decibar $= 1/10$ bar $= 10000\,\text{Pa}$.

Problem 2

Estimate the height of the following pressure surfaces in the atmosphere.

(i) 500 hPa (ii) 200 hPa (iii) 50 hPa (iv) 10 hPa
We will use the hydrostatic equation (equation 2.5).

$$p' = p_0 \exp\left(-\frac{z'}{H}\right)$$

where p_0 is the surface pressure (1013.25 hPa) and $H = R_m \overline{T}/g$
$R_m = 287\,\text{Joules kg}^{-1}\text{K}^{-1}$ and $g = 9.81\,\text{ms}^{-2}$
Assume the average temperature $\overline{T}$ is 250 K.

$$H = 287 \times 250/9.81 = 7314\,\text{m}$$

$$p' = p_0 \exp\left(-\frac{z'}{H}\right)$$

$$\ln(p_0/p') = z'/H$$

$$z' = H \ln(p_0/p')$$

$$p' = 500 \, \text{hPa}$$

$$p_0 = 1013.25 \, \text{hPa}$$

$$z' = 7314 \times \ln(1013.25/500) = 7314 \times 0.7063 = 5,166 \, \text{m}$$

$$p' = 200 \, \text{hPa}$$

$$z' = 7314 \times \ln(1013.25/200) = 7314 \times 1.622 = 11,868 \, \text{m}$$

$$p' = 50 \, \text{hPa}$$

$$z' = 7314 \times \ln(1013.25/50) = 7314 \times 3.0089 = 22,007 \, \text{m}$$

$$p' = 10 \, \text{hPa}$$

$$z' = 7314 \times \ln(1013.25/10) = 7314 \times 4.6183 = 33,778 \, \text{m}$$

Problem 3

If the whole atmosphere warmed by 3 K, what would be the change in scale height?

$$H = \frac{R_m \overline{T}}{g} = 287 \times 250/9.81 = 7313 \, \text{m}$$

$$H = \frac{R_m \overline{T}}{g} = 287 \times 253/9.81 = 7402 \, \text{m}$$

$$\Delta H = 89 \, \text{m}$$

Chapter 3

Problem 1

The atmosphere loses heat to space by long wave radiation. Assuming a radiative equilibrium temperature of 255 K, obtain the rate of cooling for the atmosphere. The cooling is given by RHS of equation 1.3

$$\sigma T_e^4 = 5.67 \times 10^{-8} \times (255)^4 = 239 \, \text{W m}^{-2}$$

$$M \times C_p \times \frac{dT}{dt} = -239 \, \text{W m}^{-2}$$

$$M = \text{Mass/unit area} \times \text{specific heat of air at constant pressure}$$

$$= 10,328 \, \text{kg m}^{-2} \times 1004 \, \text{J/kg/K} = 10,357,264 \, \text{J/m}^2/\text{K}$$

$$M = 10,328 \, \text{kg/m}^2$$

$$C_p = 1004 \, \text{Jkg}^{-1} \text{K}^{-1}$$

$$M \times C_p = 1.0369 \times 10^7$$

$$\frac{dT}{dt} = -239/(1.0369 \times 10^7) = -2.31 \times 10^{-5} \, \text{Ks}^{-1}$$

$$\frac{dT}{dt} = -1.99 \, \text{Kday}^{-1}$$

This is an oversestimate as it ignores any direct heating of the atmosphere by solar radiation. However, it does indicate that the atmosphere tends to cool at this rate if not supplied by solar radiation.

Problem 2

Estimate the heating of the atmospheric boundary layer, extending to 1000 m above a dry insulated surface when the solar radiation incident on the surface is 800 W m^{-2} and the surface albedo is 0.30.

Assume there is no heat conducted into the ground therefore all the absorbed solar radiation is available to heat the boundary layer. The cooling of the boundary layer by radiation to space is ignored.

The absorbed radiation is $(1-0.3) \times 800 = 560 \, \text{W m}^{-2}$

$$M = \text{mass/unitarea} = \rho_a \times h = 1.2 \times 1000 \, \text{kgm}^{-2}$$

$$C_p = 1004 \, \text{Jkg}^{-1} \text{K}^{-1}$$

$$M \times C_p \times \frac{dT}{dt} = 560 \, \text{W m}^{-2}$$

$$\frac{dT}{dt} - \frac{560}{(1200 \times 1004)} = 4.64 \times 10^{-4} \, \text{Ks}^{-1} = 1.67 \, \text{K/hour}$$

Over 6 hours say from 9 am to 3 pm, the layer would increase its temperature by 10 K.

Problem 3

Estimate the heating of the upper 1 m of a lake, if 50% of the incident solar flux is absorbed in this layer. Assume incident solar radiation of 800 W m^{-2}

$$M = \text{mass/unitarea} = \rho_w \times h = 1000 \times 1 \, \text{kg m}^{-2}$$

$$C_p = 4000 \, \text{J kg}^{-1} \text{K}^{-1}$$

$$M \times C_p \times \frac{dT}{dt} = 400$$

$$\frac{dT}{dt} = \frac{400}{(1000 \times 4000)} = \times 10^{-4}\,Ks^{-1} = 0.36\,K\,hour^{-1}$$

The estimate has assumed no heat loss from the surface layer to the atmosphere, and therefore is an overestimate.

If this heating rate was maintained for 6 hours a temperature change 2.16 K would occur.

Diurnal variations of temperature in the ocean are usually less than 1 K.

Chapter 4

Problem 1

Calculate the mixing ratio, for an air parcel which has a saturated vapour pressure at 10°C of 12.27 hPa, and a total pressure of 1000 hPa.

$$q = \left(\frac{e_s}{p - e_s}\right) \times \left(\frac{m_v}{m_d}\right) = \left(\frac{12.27}{1000 - 12.27}\right) \times 0.622 = 7.729 \times 10^{-3}$$

$$q = 7.729\,gm/kg$$

Problem 2

If this water vapour condensed estimate the temperature change of the air parcel.

The latent heat of condensation is assumed to be 2500 KJoules/kg.

The heat released $= q \times L = 7.729 \times 10^{-3} \times 2500\,KJoules = 19322\,Joules$ into 1 kg of dry air.

Heat capacity of 1 kg of dry air $= C_p = 1004\,JK^{-1}$

Temperature change $= 19322/1004 = 19.24\,K$

This assumes there is no mixing of the air parcel with its surroundings.

Problem 3

If the water vapour decreased linearly from the surface at 1000 hPa to zero at 500 hPa, estimate the surface precipitation if it all condensed.

Mass of dry air $= (1000-500) \times 100/g = 50000/9.81 = 5090\,kgm^{-2}$

Mass of water vapour in column $= 0.5 \times 7.729 \times 10^{-3} \times 5090 = 19.696\,kg\,m^{-2}$

The depth of condensed water $= 19.696/\rho_w = 19.696/1000 = 1.9696 \times 10^{-2}\,\text{m} = 19.696\,\text{mm}$.

Chapter 5

Problem 1

Estimate the ocean heat divergence in Peta Watts from the equatorial region 10°S to 10°N shown in Table 5.1.

The net annual surface heat flux into the ocean is $20\,\text{W\,m}^{-2}$. The net surface heat flux into the ocean has to be transferred from the equatorial region to higher latitudes by ocean processes such as ocean currents.

To obtain the flux divergence the area of the equatorial ocean is calculated.

First the area of the total area of the equatorial band around the globe is obtained.

$$\text{Area} = 2\pi a^2 \int_{\varphi_1}^{\varphi_2} \cos\varphi\, d\varphi = 2\pi a^2 [\sin\varphi_2 - \sin\varphi_1]$$

where a is the radius of the Earth at equator $= 6.378 \times 10^6\,\text{m}$.

$\text{Area} = 2 \times 3.142 \times 6.378 \times 10^6 \times 6.378 \times 10^6 \times [0.1736 + 0.1736]$

$\text{Area} = 8.876 \times 10^{13}\,\text{m}^2$

From Figure 1.9 the ocean occupies 80% of the area.

$\text{Heat divergence} = 20 \times 0.8 \times 8.876 \times 10^{13}$

$$= 1.420 \times 10^{15}\,\text{W} = 1.420\,\text{Petawatts}.$$

This is smaller than the ocean heat divergence shown in Figure 5.5c. This indicates the difficulty of determining horizontal heat transports from observations of surface fluxes over ocean basins.

Problem 2

The freshwater transport from the South Pacific $0.02 \times 10^9\,\text{kg/s}$ whilst the export to Arctic is $0.77 \times 10^9\,\text{kg/s}$. Calculate the average freshwater input in m/year into the N. Pacific Ocean.

The net fresh water flux into the N. Pacific Ocean $= (0.77 - 0.02) \times 10^9 = 0.75 \times 10^9\,\text{kg/s}$.

Assuming density of fresh water $= 1000\,\text{kg/m}^3$ the net volume added $= 0.75 \times 10^6\,\text{m}^3/\text{s}$.

Assume the area of the North Pacific Ocean is 82.5×10^{12} m^2, hence the freshwater added $= 0.75 \times 10^6 / 82.5 \times 10^{12} = 9.09 \times 10^{-9}$ m/s.

1 year $= 365.25 \times 8.64 \times 10^4$ seconds $= 3.156 \times 10^7$ seconds.

Hence average freshwater added to N Pacific Ocean by the atmosphere (precipitation −evaporation) and runoff is 0.29 m/year.

Chapter 6

Problem 1

Estimate the volume transport of the Gulf Stream at 30°N, if the northward flow decreases from a peak of $1\,\mathrm{ms}^{-1}$ at the surface to zero at a depth of 1000 m. It is also zero at the edge of the Gulf Stream. Assume the Gulf Stream is 100 km wide.

Volume transport $=$ mean flow $\times$ cross-sectional area

Cross-sectional area $= 1000$ m $\times$ 100,000 m $= 10^8$ m^2

The mean flow over the width $= (1.0 + 0.0)/2 = 0.5\,\mathrm{ms}^{-1}$ and over the depth $= (1.0 + 0.0)/2 = 0.5\,\mathrm{ms}^{-1}$

The average cross-sectional flow $= 0.25\,\mathrm{ms}^{-1}$

Volume transport $= 0.25 \times 10^8 = 25 \times 10^{+6}$ m^{3}s^{-1} $= 25$ Sverdrups

Problem 2

Estimate the volume transport of the Hadley Cell, if the trade winds have a meridional speed of $2\,\mathrm{ms}^{-1}$ and occupy the lowest 3000 m of the atmosphere.

Assume the cell occupies the whole circumference (L) of the globe at a latitude of 20°.

$$\text{Southward volume transport} = v \times H \times L$$

$$L = 2\pi a \cos\varphi = 2.0 \times 3.142 \times 6.378 \times 10^6 \times \cos(20)$$

$$= 40074155.0 \times 0.93969 = 3.766 \times 10^7 \text{ m}$$

$$H = 3000\,\mathrm{m}$$

$$V = 2\,\mathrm{ms}^{-1}$$

$$\text{Southward volume transport} = 2 \times 3000 \times 3.766 \times 10^7$$

$$= 2.26 \times 10^{11}\,\mathrm{m}^3\,\mathrm{s}^{-1}$$

Chapter 7

Problem 1

Estimate the inertial period at $10°$, $40°$ and $80°$ latitude.

$$T = \frac{2\pi}{f} \quad where\, f = 2\Omega \sin \varphi$$

$$\varphi = 10° \, f = 2.0 \times 7.272 \times 10^{-5} \times \sin(10)$$

$$= 2.525 \times 10^{-5} T = 248800 \, \text{sec} = 69.1 \, \text{hours}$$

$$\varphi = 40° \, f = 2.0 \times 7.272 \times 10^{-5} \times \sin(40)$$

$$= 9.348 \times 10^{-5} T = 67209 \, \text{sec} \, s = 18.7 \, \text{hours}$$

$$\varphi = 80° \, f = 2.0 \times 7.272 \times 10^{-5} \times \sin(80)$$

$$= 1.432 \times 10^{-4} T = 43868 \, \text{sec} \, s = 12.185 \, \text{hours}$$

The radius of inertial circle $R = V/f$. Estimate the radius of an inertial circle for a horizontal velocity of $V = 1 \, \text{m/s}$ at the above latitudes.

$$R = \frac{V}{f} \quad where\, f = 2\Omega \sin \varphi$$

$$\varphi = 10° \, f = 2.0 \times 7.272 \times 10^{-5} \times \sin(10)$$

$$= 2.525 \times 10^{-5} \, R = 482,592 \, \text{m} = 482.6 \, \text{km}$$

$$\varphi = 40° \, f = 2.0 \times 7.272 \times 10^{-5} \times \sin(40)$$

$$= 9.348 \times 10^{-5} \, R = 10697 \, \text{m} = 10.7 \, \text{km}$$

$$\varphi = 80° \, f = 2.0 \times 7.272 \times 10^{-5} \times \sin(80)$$

$$= 1.432 \times 10^{-4} R = 6983 \, \text{m} = 6.9 \, \text{km}$$

In the lower atmosphere a typical velocity is $10 \, \text{ms}^{-1}$ and therefore R will be 10 times the above values.

Problem 2

The height of the sea surface across the Gulf Stream at 30 N increases by 1 m in an eastward direction, over a distance of 100 km. Estimate the geostrophic current and its direction.

Compared with a geopotential surface the sea surface slopes upward towards the east, and therefore the pressure on a geopotential surface is higher to the east and the west.

Hence the horizontal pressure gradient acceleration is acting towards the west and the coriolis acceleration must act towards east. The coriolis acceleration is 90 degrees to the right of the geostrophic flow and therefore the flow is to the north.

$$-f v_g = -\frac{1}{\rho}\frac{\partial p}{\partial x}$$

Hydrostatic pressure (p) on a geopotential surface is, $p = g\rho z$

where z = height above mean sea level

Substitution for p into equation 7.16

$$-f v_g = -g\frac{\partial z}{\partial x}$$

$f(30°\mathrm{N}) = 2.0 \times 7.272 \times 10^{-5} \times \sin(30) = 7.272 \times 10^{-5}\,\mathrm{s}^{-1}$

$g = 9.81\,\mathrm{ms}^{-2}$

Slope of sea surface $= \dfrac{\partial z}{\partial x} = \dfrac{1}{10^5} = 10^{-5}$

$$v_g = \frac{9.81 \times 10^{-5}}{7.272 \times 10^{-5}} = 1.349\,\mathrm{m/s}$$

Problem 3

Calculate the direction and speed of the geostrophic wind between the Azores high pressure 1030 hPa at mean sea level and 40° N and the Iceland low pressure 980 hPa at mean sea level and 60° N. Assume they are both at the same longitude and 1° latitude = 110 km. The air density = $1.2\,\mathrm{kgm}^{-3}$.

The horizontal pressure gradient acceleration acts northwards, and the coriolis acceleration acts southwards. Hence an eastward geostrophic flow is necessary to produce a southward coriolis acceleration.

$$+f u_g = -\frac{1}{\rho}\frac{\partial p}{\partial y}$$

$$-\frac{1}{\rho}\frac{\partial p}{\partial y} \simeq -\frac{1}{\rho}\frac{\Delta p}{\Delta y}$$

where Δp is pressure difference between Iceland and Azores, and Δy is the distance.

$\Delta p = 980 - 1030 = -50\,\text{hPa} = -5000\,\text{pa}; \; \Delta y = 20 \times 110 \times 1000 = 2.2 \times 10^6\,\text{m}$

$-\dfrac{1}{\rho}\dfrac{\Delta p}{\Delta y} = -\dfrac{1}{1.2}\dfrac{(-5000)}{2.2 \times 10^6} = +1.89 \times 10^{-3}\,\text{ms}^{-2}$

The mid-point value of f is the value at 50 N.

$f = 2.0 \times 7.272 \times 10^{-5} \times \sin(50) = 1.114 \times 10^{-4}\,\text{s}^{-1}$

$u_g = \dfrac{1.89 \times 10^{-3}}{1.114 \times 10^{-4}} = 16.9\,\text{ms}^{-1}$

Chapter 8

Problem 1

An observer at time $T1$ on an island notices swell waves from a storm with a 20 second period.

At later time ($T2$) the observer notices swell of 15 second period

If $T2 - T1 = 2$ hours, how far away is the storm.

L = distance of observer from storm centre.

T = time for waves to travel L

The swell waves will travel at the group velocity and therefore from Table 8.1 a 20 second wave has $C_g = 15.6\,\text{m/s}$, and a 15 second wave has $C_g = 11.7\,\text{m/s}$

$$T2 - T1 = \frac{L}{C_{g2}} - \frac{L}{C_{g1}} = L(C_{g1} - C_{g2})/C_{g1}C_{g2}$$

$$T2 - T1 = L \times 0.0214$$

$$L = (T2 - T1) \times 46.8\,\text{m}$$

$$T2 - T1 = 2\,\text{hours} = 7200\,\text{s}$$

$$L = 336930\,\text{m} = 336.9\,\text{km}$$

Problem 2

An M_2 tidal wave travels around a semi-enclosed sea of 100 m depth, with maximum elevation at the coast. Assume $f = 10^{-4}\,\text{s}^{-1}$.

Estimate the speed, the wavelength, and its width, assuming the tidal wave behaves as a Kelvin wave.

The Kelvin wave travels at $c = \sqrt{gH}$. $H = 100\,m$ and $g = 9.81\,ms^{-2}$ and hence $c = 31.3\,m/s$.

Wavelength of the wave $(L) = C \times T$ where $T =$ tidal period for the M_2 constituent is 12.4 hours.

$L = 31.3 \times 12.4 \times 3600 = 1397232\,m = 1397\,km$.

Equation 8.11 shows the elevation decays from the coast on a scale $L = c/f$.

$L = 31.3/10^{-4} = 313{,}000\,m = 313\,km$.

The elevation will reach e^{-1} or 0.36 of its maximum elevation, at 313 km from the coast.

Chapter 9

Problem 1

Estimate the kinetic energy in the following phenomena:

(i)   Gulf Stream
(ii) Antarctic Circumpolar Current
(iii) Extra-cyclone
(iv) Tropical cyclone

$$\text{K.E.} = \frac{1}{2}mv^2 \quad \text{where m(mass)} = \rho(\text{density}) \times V(\text{volume})$$

	L km	W Km	H km	Volume m³	Mass kg	V m/s	V²	K.E. (Joules)
Gulf Stream	1000	100	1	10^{14}	10^{17}	1	1	5×10^{16}
ACC	20,000	500	4	4.0×10^{16}	4.0×10^{19}	0.2	0.04	8.0×10^{17}
Extra-tropical cyclone	1000	1000	10	10^{16}	1.7×10^{16}	20	400	3.4×10^{18}
Tropical cyclone	150	150	10	2.2×10^{14}	1.6×10^{14}	50	2500	2.1×10^{17}

(i)   These are estimates of the size of the Gulf Stream at Cape Hatteras, and the length between Florida and Cape Hatteras.
(ii) The length of the Antarctic Circumpolar Current is based on a circle at latitude at 60°S.
(iii) It has been assumed that the Extra-Tropical Cyclones and Tropical Cyclones are retangular prisms. The average density of troposphere = $0.75\,kgm^{-3}$.

Problem 2

During the strong El Niño event of 1982–83 the day length increased by 1 millisecond.

Describe how the rotation of the Earth is related to day length.

Discuss the implications for the angular momentum of the atmosphere, and describe how the atmospheric winds and ocean currents may have changed to cause this change the day length.

The day length (T) is inversely proportional to the Earth's rotation angular (Ω) given by $T = \frac{2\pi}{\Omega}$. Hence a increase in day length is associated with a decrease in the rotation speed of the Earth.

The sum of the angular momentum of the solid Earth, ocean and atmosphere is a constant. A decrease in the rotation of the speed of the solid Earth, must be associated with an increase in the angular momentum of the atmosphere or ocean. If it assumed that all the angular momentum change is in the atmosphere, then it could be associated with an increase in eastward winds or a decrease in westward winds. Analyses have shown that variations in the winds of both troposphere and stratosphere during the El Niño contributed to the total change in atmospheric momentum.

However, the ocean may have made an additional contribution. For example, the eastward Antarctic Circumpolar Current may have increased in strength.

Chapter 10

Problem 1

The thermohaline circulation (THC) is important for setting the vertical density stratification of the world ocean. This can be demonstrated by a model.

The vertical stratification is determined by two processes: (i) vertical mixing and (ii) vertical advection.

To simplify we ignore the influence of salinity and consider only the effect of temperature on the density stratification. The global averaged ocean temperature is given by T.

$$w\frac{dT}{dz} = k_v \frac{d^2T}{dz^2}$$

The LHS is the vertical advection of temperature where w is the vertical velocity and the RHS is the vertical mixing of temperature where k_v is a vertical mixing coefficient.

The equation may be solved subject to the boundary conditions $T = Ts$ at $z = 0$ and $T = Ta$ in the abyssal ocean.

To further simplify the problem it is assumed that the ocean is infinitely deep.

The solution is:

$$T = (Ts - Ta)\exp\left(\frac{z}{D}\right) + Ta$$

where $D = k_v/w$

The quantity D has units of depth and when $z = -D$ the temperature difference Ts−Ta is 0.36. D therefore is a measure of the reduction of the temperature with depth.

Typical values are $Ts = 15°C$ and $Ta = 3°$ C.

By fitting this equation to the observed temperature profile D is estimated to be 500 m.

At a depth of 1500 m $z = -3D$ T is very close to Ta.

The upwelling velocity can be estimated from the transport of the THC which is 20 Sverdrups or 20×10^6 m^{3}s^{-1}.

The horizontal area of the ocean at 31×10^{12} m^{-2} and therefore w = 20 × $10^6/31 \times 10^{12} = 6.45 \times 10^{-7}$ m/s or 20 m/year.

$D = k_v/w$ and therefore $k_v = w \times D = 6.45 \times 10^{-7} \times 500 = 3.25 \times 10^{-4}$ m^{2}s^{-1}

This ocean model predicts a vertical mixing coefficient three orders higher than the molecular diffusivity. The mixing is associated with mesoscale eddies, internal waves and ocean tides.

Chapter 11

Problem 1

Discuss why it is easier to predict the climate change associated with the increase in greenhouse gases this century than predict a season or decade ahead.

The prediction of greenhouse gases a century ahead depends on the emissions by the human race this century. This in turn depends on economic growth which in the 19th and 20th Century was dependent on the use of fossil fuels. There is no strong evidence that the use of these fuels has declined and there are probably sufficient reserves to continue this growth in the first half of the century.

Based on this it is possible to make a reasonable set of predictions on the increase in greenhouse gases this century.

With our present climate models we can make predictions of the temperature change and precipitation change on a global scale with very high confidence and on a regional scale with good confidence.

Seasonal and decadal prediction using modelling methods is only in its infancy.

The ENSO prediction is one which is showing skill for a few months ahead, using climate type models. (Figure 10.10). This is a major interannual signal which strongly influences the tropics, and to some extent, the higher latitudes.

The Indian and SE Asian Monsoon however is still very difficult to predict one month ahead.

In the extra-tropics there is more difficulty because of the instability of the westerly jet stream. For example the development and breakdown of an individual 'blocking pattern' can only be predicted at best 10 days ahead.

In seasonal prediction there is a statistical relation between May sea surface temperature and the following winter atmospheric circulation in the North Atlantic which shows evidence of useful predictive skill.

Chapter 12

Problem 1

Estimate the change in sea level associated with the melting of all the ice caps and glaciers and state your assumptions.

Assume volume of ice caps and glaciers $= 28.4 \times 10^6$ km^3

If it is assumed the ocean area (A) remains the same then the sea level rise $= 28.4 \times 10^6$ km^3/A $= 28.4 \times 10^6 \times 10^9$ m^3/358×10^{12} m$^2 = 79.3$ m

Greenland Ice Cap $= 2.6 \times 10^6$ km^3 Sea level rise $= 2.6 \times 10^6 \times 10^9$/$358 \times 10^{12} = 7.26$ m

Antarctic Ice Cap $= 25 \times 10^6$ km^3 Sea level rise $= 25 \times 10^6 \times 10^9$/$358 \times 10^{12} = 69.83$ m

Glaciers $= 28.4 - 2.6 - 25.0 = 0.8 \times 10^6$ km^3 Sea level rise $= 0.8 \times 10^6 \times 10^9$/$358 \times 10^{12} = 2.23$ m

To provide a more accurate estimate one would calculate the horizontal land area from mean sea level to 79.3 m above mean sea level, and add this to the previous ocean area (A) and recalculate the sea level rise.

This process could be repeated to obtain a more accurate estimate, than the previous value.

Problem 2

The warming due to greenhouse gases is calculated at present to be 1.6 W m^{-2}

If this heat is absorbed in the ocean estimate the temperature change over one century.

Chapter 10.2 gives a straightforward method for calculating this change. First one needs to decide to what depth this heat is absorbed in the ocean.

The seasonal variations are mainly confined to upper 100 m, but with long time climate change we may expect the heat to be mixed at least to the depths of the main thermocline.

Let us choose $h = 500$ m. Then substituting for $Q = 1.6$ W m^{-2} into equation 10.1

$$C\frac{dT}{dt} = Q_T$$

where $C = \rho C_p h$

Q_T = net surface heat flux

dT/dt = temperature change with time

h = depth of surface layer

ρ = water density

C_p = specific heat at constant pressure

$C = $ density $\times\ Cp\ \times\ h = 1020 \times 4000 \times 500 = 2.04 \times 10^9 \mathrm{Jm}^{-2}$

$$\frac{dT}{dt} = \frac{1.6}{2.04 \times 10^9} = 7.84 \times 10^{-10}\ \mathrm{Ks}^{-1}$$

1 year $= 365.25 \times 8.64 \times 10^4$ s

$$\frac{dT}{dt} = 0.0247\ \mathrm{K/year}$$

Over a century the change would be 2.47 K over the global ocean.

If it was mixed to 1000 m the change would be 1.23 K and if it was only mixed to 250 m the change would 4.95 K.

This calculation assumes that there is no extra heat exported from the land, for example, by rivers or ice calving.

Secondly that the heating from the greenhouse gases remains at the current level, which is very unlikely. These estimates are therefore lower bounds on the global ocean temperature rise.

Glossary

adiabatic processes

An adiabatic process is where a fluid (gas or liquid) changes temperature, as a result of either a change in pressure or volume of the fluid, subject to no heat being removed or added to the fluid from its surroundings.

The dry adiabatic lapse rate is the change in temperature with height of a dry air parcel caused solely by the change in atmospheric pressure. There is no heat added to or removed from the air parcel.

The saturated adiabatic lapse rate is the change in temperature with height of a saturated air parcel caused solely by the change in atmospheric pressure. The latent heat of condensation provides an additional heat source and therefore it is not strictly an adiabatic process. It is often referred to as a pseudo-adiabatic process.

age of a water mass

The age of a water mass is the time elapsed since the water mass was formed by contact with the atmosphere. This can be determined by *in situ* measurements of dissolved gases, such as the family of freons whose concentrations in the atmosphere are well determined over the last 70 years.

albedo

The planetary albedo is the ratio of the incident solar radiation above the atmosphere to the reflected radiation from the Earth's atmosphere and surface. The surface albedo is the ratio of the incident solar radiation, both diffuse and direct, to the reflected radiation from the surface. It is measured across a horizontal plane above the surface.

atmospheric aerosol

A general term which refers to all particulates and droplets suspended in the atmosphere.

The Atmosphere and Ocean: A Physical Introduction, Third Edition. Neil C. Wells.
© 2012 John Wiley & Sons, Ltd. Published 2012 by John Wiley & Sons, Ltd.

atmospheric boundary layer

The region of the lower atmosphere which is influenced directly by the Earth's surface. The winds in the boundary layer are affected by the frictional forces acting across the surface, whilst the thermal and moisture properties of the air will be influenced by heat and evaporative fluxes from or across the surface.

available potential energy (APE)

This is the potential energy of the atmosphere and ocean which can be converted into the kinetic energy of the circulation. The available potential energy is dependent on horizontal variations of density (ocean) and potential temperature (atmosphere). The available potential energy is considerably smaller than the potential energy.

black body

A perfect emitter and absorber of radiation at all wavelengths is known as a black body.

climate system

The system contains the atmosphere, ocean, cryosphere, terrestial hydrology, and biosphere and their interactions.

circulation time scale

This is the time for a fluid parcel to circulate around an ocean basin. For example, a parcel of water will take about five to ten years to travel around a sub-tropical gyre in the North Atlantic Ocean.

correlation coefficient

A statistical measure of the correspondence between two variables.

cryosphere

A generic term for the frozen part of the Earth's system in all its forms on or at the Earth's surface. This includes mountain glaciers; ice sheets; sea ice and lying snow.

diabatic process

This refers to a process which transfers heat to a fluid parcel from its surroundings. For instance, radiation is a diabatic process because it can warm or cool a fluid parcel by the absorption or emission of radiation.

Earth's orbit around the Sun

The eccentricity is a measure of the ellipticity of the Earth's orbit around the Sun.

The obliquity is the tilt of the Earth with respect to the vertical through the orbital plane of the Earth around the Sun.

The longitude of the perihelion is the time of the year when the Earth is closest to the Sun. It is defined with respect to the Northern Hemisphere spring or vernal equinox. For example, 0° is the vernal equinox; 90° is the summer solstice; 180° is the autumn equinox and 270° is the winter solstice.

Earth system

Contains the climate system and its interaction with the underlying geology in particular the lithosphere and mantle.

emissivity

This is a measure of the efficiency of a body for emitting radiation. A black body is a perfect emitter of radiation and therefore has an emissivity of one.

entrainment

This is a process whereby a turbulent layer of fluid grows at the expense of a quiescent layer of fluid. For example, turbulence at the ocean surface, generated by wind and waves, will mix with quiescent fluid below and this will increase the depth of the surface turbulent layer.

equipotential surface

This is an imaginary surface around the Earth which has a constant gravitational potential energy. The acceleration due to the gravity of the Earth is always normal to the equipotential surface.

frequency spectrum or power spectrum

This is a decomposition of the energy (or variance) of a fluctuating quantity into either its frequency (*or period*) or its wavenumber (*or wavelength*).

geoid

The geoid is an equipotential surface which coincides with mean sea level over a stationary ocean. Ocean currents and tides cause deviations of sea level from the geoid.

geostationary

An imaginary point above the Earth's surface which rotates at the same angular velocity as the solid Earth. A geostationary satellite is one which is above the equator and rotates with same velocity as the Earth's surface below the satellite. The satellite will observe the same disc of the Earth at all times.

halocline

A maximum in the vertical gradient of salinity. It occurs where fresh water, either from rivers or meltwater from ice, overlies saline ocean water. The fresh water is less dense than the ocean water and therefore forms a vertical stable layer.

Indonesian maritime continent

This is a term used by meteorologists to describe the Indonesian region composed of islands and seas. It acts as a huge source of heat and water vapour and powers the atmospheric circulation in the Indian-Pacific region.

inverse analysis

This is a mathematical procedure used for calculating the ocean circulation from measured properties in the ocean. These properties may include both the normal hydrographic measurements of temperature and salinity as well as measurements of conservative chemical and radioactive tracers.

isentropic

These are surfaces in the atmosphere and ocean having a constant potential temperature.

isopycnal

These are surfaces in the ocean having a constant density.

isostatic equilibrium

A theoretical balance of all large portions of the Earth's crust floating on a denser underlying layer at a depth of about 110 km.

ITCZ

The Inter-Tropical Convergence Zone (ITCZ) is a region close to the equator, where moist trade-wind systems from both hemispheres converge. The resulting uplift causes very heavy rainfall along the convergence zone. The zone tends to move across the equator with the Sun.

lithosphere

The rigid outer layer of the Earth which comprises the Earth's crust and the solid upper part of the mantle.

Mathematical model

A set of equations which describe physical, chemical and biological processes in a system.

MODE

The acronym given to the Mid-Ocean Dynamics Experiment (MODE), an experiment to measure the properties of meso-scale ocean eddies in a 200 km × 200 km square in the North Atlantic Ocean near Bermuda during the 1970's. It was largest physical oceanography experiment of its time and was supported by the U.S.A.

ocean mixed layer

This is the term used to describe the surface layer of the ocean, which is continuously mixed by turbulent eddies, generated by wind, waves and surface cooling. It is characterised by vertically uniform properties of temperature and salinity, and may extend to depths in excess of 100 m in winter. It is also called the upper ocean boundary layer.

potential temperature

A conservative property of an incompressible adiabatic fluid. In the atmosphere this property is useful as a conservative tracer of motion in the troposphere and stratosphere on a time scale of a few days. It can also be used as a tracer of ocean circulation.

pycnocline

A maximum in the vertical gradient of density which may be caused by either temperature or salinity gradients, or both factors. The main pycnocline in the ocean closely follows the main thermocline.

Rossby waves

These are quasi-horizontal planetary waves, whose behaviour is determined by the northward gradient of the Coriolis parameter. They are named after the Swedish Meteorologist Carl Gustav Rossby.

SOFAR channel

The SOFAR (SOund Fixing And Ranging) channel is a waveguide for sound propagation in the ocean with its quasi-horizontal axis corresponding to the minimum in sound velocity. Low-frequency sound can travel many thousands of kilometres in the SOFAR channel.

supersaturation

If the water vapour pressure is e, the supersaturation (%) with respect to liquid water is $(e/e_s - 1)100$, where e_s is the saturation pressure over a plane surface of pure liquid water. The saturation pressure over saline water is slightly different from that of pure water.

A supersaturation with respect to ice is defined in a similar way.

synoptic

In meteorology this refers to a characteristic horizontal scale of motion associated with extra-tropical cyclones and anticyclones. It is typically approximately 1000 km to 5000 km in scale.

thermal equator

This is the zone of maximum sea surface temperature, which generally moves with the Sun across the equator. The ITCZ is often located on the thermal equator.

thermocline

A maximum in the vertical gradient of ocean temperature. The best example is the main thermocline which is the boundary between the warm thermocline waters and the cold abyssal waters. The warm layers are less dense than the deeper and colder layer, and hence a vertical stable layer is formed.

upwelling

Upwelling occurs where cooler thermocline waters are brought to the surface to replace the warmer surface waters. It is often observed in the trade wind regions, where winds drive the ocean surface layer away from the coast or from the equator. Tropical cyclones also cause upwelling. The upwelled waters are usually regions of high biological productivity.

vertical stability

When lighter fluid overlies heavier fluid, the vertical stability of the system is hydrostatically stable. In contrast, when heavier fluid overlies lighter fluid, the system is unstable and there is rapid vertical mixing.

General Reading

Deacon, M.B., 1971, *Scientists and the Sea – a Study of Marine Science, 1650–1900* (London: Academic Press), 445 pp.

Deacon, M.B. (ed.), 1978, *Oceanography: Concepts and History, Benchmark Papers in Geology Vol. 35* (London: Dowden, Hutchinson & Ross Inc.), 394 pp.

Defant, A., 1961, *Physical Oceanography*, Vols 1 (729 pp.) and 2 (598 pp.) (Oxford: Pergamon).

Dietrich, G., Kalle, K., Krauss, W. and Siedler, G., 1980, *General Oceanography: An Introduction*, 2nd edition (New York John Wiley & Sons, Inc.), 626 pp.

Gill, A.E., 1982, *Atmosphere-Ocean Dynamics* (London: Academic Press), 662 pp.

Grant Gross, M. and Grant Gross E., 1996, *Oceanography – A View of the Earth*, (Prentice-Hall), 236 pp.

Hess, S.L., 1959, *Introduction to Theoretical Meteorology* (New York: Holt Rinehart & Winston), 362 pp.

Hill, M.N., 1963, *The Sea: Ideas and Observations, Vol. 1. Physical Oceanography* (New York: Wiley-Interscience), 864 pp.

Holton, J.R., 2004, *An Introduction to Dynamical Meteorology*, 4th edition (Elsevier Academic Press), 535 pp.

Knauss, J.A., 1996, *Introduction to Physical Oceanography*, 2nd edition, (Prentice-Hall), 309 pp.

Maury, M.F., 1853, *The Physical Geography of the Sea* (London: Sampson Low & Son), 274 pp.

Monin, A.S., 1972, *Weather Forecasting as a Problem in Physics* (Cambridge, MA: MIT Press), 199 pp.

Monin, A.S., 1986, *Introduction to the Theory of Climate* (Dordrecht: Reidel), 261 pp.

Neuman, G. and Pierson, W., 1966, *Principles of Physical Oceanography* (Englewood Cliffs, NJ: Prentice-Hall), 545 pp.

Pond, S. and Pickard, G.L., 1983, *An Introduction to Dynamical Oceanography*, 2nd edition (Oxford: Pergamon), 329 pp.

Pugh, D.T., 1987, *Tides, Surges and Mean Sea Level*, (John Wiley & Sons, Chichester), 472 pp.

Robinson, I.S., 1995, *Satellite Oceanography* (Wiley-Praxis, Chichester), 454 pp.

Summerhayes, C.P. and Thorpe, S.A.T. (Editors), 1996, *Oceanography An Illustrated Guide* (Manson Publishing): 352 pp.

Sverdrup, H.U., 1945, *Oceanography for Meteorologists* (London: George Allen & Unwin), 246 pp.

Sverdrup, H.U., Johnson, M.W. and Fleming, R.H., 1946, *The Oceans: Their Physics, Chemistry and General Biology* (Englewood Cliffs, NJ: Prentice-Hall), 1087 pp.

Talley, L.D., Pickard, G.L., Emery, W.J. and Swift, J.H., 2011, *Descriptive Physical Oceanography*, Sixth Edition (Academic Press), 560 pp.

Tchernia, p., 1980, *Descriptive Regional Oceanography*, Pergamon Marine Series, Vol. 3 (Oxford: Pergamon), 253 pp.

Tomczak, M. and Godfrey, J.S.1994, *Regional Oceanography: an introduction* (Oxford: Pergamon), 422 pp.

Wallace, J.M. and Hobbs, P.V., 2006, *Atmospheric Science* (Elsevier Academic Press), 483 pp.

Warren, B.A. and Wunsch, C., 1981, *Evolution of Physical Oceanography* (Cambridge, MA: MIT Press), 623 pp.

Further Reading and References

Chapter 1

Goody, R.M. and Walker, J.C.G., 1972, *Atmospheres* (Englewood Cliffs, NJ: Prentice-Hall).

Henderson-Sellers, A. and Wilson, M.F., 1983, Albedo observations in climatic research, *Philosophical Transactions of the Royal Society of London*, **A309**, 285–294.

Pierrehumbert, R.T., 2011, Infra-red radiation and planetary temperature, *Physics Today*, **64**, 33–38.

Ramanathan, V., Barkstrom, B.R and Harrison, E.F., 1989, Climate and the Earth's radiation budget, *Physics Today*, **42**, 22–32.

Chapter 2

Allegre, C.J. and Schneider, S.H., 1994, Evolution of the Earth, *Scientific American*, **271**(4), 44–51.

Brancazio, P.J. and Cameron, A.G.W., 1964, *The Origin and Evolution of Atmospheres and Ocean* (New York: John Wiley & Sons, Inc.), 314 pp.

Ditmar, W., 1884, Report on research into the composition of ocean water collected by H.M.S. Challenger during the years 1873–76, in Report on the Scientific Results of the Voyage of H.M.S. Challenger. Physics and Chemistry (London: HMSO), 1, pp. 199–209, 227–235.

McDougall T.J. *et al.*, 2009, *The International Thermodynamic Equation Of Seawater 2010* (TEOS-10).

Millero, F.J. *et al.*, 2008, The composition of Standard Sea water and the definition of the Reference-Composition Salinity. Deep Sea Research, Part 1, *Oceanographic Research Papers*, **55**, 50–77 pp.

Tolstoy, I. and Clay, C.S., 1966, *Ocean Acoustics: Theory and Experiment in Underwater Sound* (New York: McGraw-Hill), 293 pp.

UNESCO, Tenth Report of the Joint Panel on Oceanographic Tables and Standards, 1981, *UNESCO Technical Papers on Marine Science*, No. 36, 24 pp.

The Atmosphere and Ocean: A Physical Introduction, Third Edition. Neil C. Wells.
© 2012 John Wiley & Sons, Ltd. Published 2012 by John Wiley & Sons, Ltd.

Chapter 3

Dopplick, T.G., 1972, Radiative heating of the global atmosphere, *Journal of Atmospheric Science*, **29**, 1278–1294.

Ivanoff, A., 1977, Oceanic absorption of solar energy, in *Modelling and Prediction of the Upper Layers of the Ocean*, ed. Kraus, E.B. (Oxford: Pergamon), Chapter 3.

Jerlov, N.G., 1976, *Marine Optics* (Amsterdam: Elsevier), 251 pp.

Kirk, J.T.O. 1994, Light and Photosynthesis in Aquatic Systems (Cambridge University Press), 509 pp.

Kondratyev, K.Ya., 1969, *Radiation in the Atmosphere* (London: Academic Press), 912 pp.

Liou, K.N., 1980, *An Introduction to Atmospheric Radiation* (London: Academic Press).

Manabe, S. and Strickler, R.F., 1964, Thermal equilibrium of the atmosphere with a convective adjustment, *Journal of Atmospheric Science*, **21**, 361–385.

Paltridge, G.W. and Platt, C.M.R., 1976, *Radiative Processes in Meteorology and Climatology* (Amsterdam: Elsevier), 318 pp.

Chapter 4

Ludlam, F.H., 1980, *Clouds and Storms* (University Park, PA: Pennsylvania State University Press), 405 pp.

Mason, B.J., 1971, *The Physics of Clouds* (Oxford: Clarendon Press), 671 pp.

Mason, B.J., 1975, *Clouds, Rain and Rainmaking* (Cambridge University Press), 189 pp.

Scorer, R.S., 1972, *Clouds of the World* (Newton Abbot, UK: David Charles), 176 pp.

Chapter 5

Broecker, W.S., 1981, Geochemical tracers and ocean circulation, in *Evolution of Physical Oceanography*, eds. Warren, B.A. and Wunsch, C. (Boston, MA: MIT Press), Chapter 15.

Brunt, D., 1932, Notes on radiation in the atmosphere, *Quarterly Journal of the Royal Meteorological Society*, **58**, 389–420.

Bryden, H. L. 2007. Heat Transport and Climate, *Encyclopedia of Ocean Sciences*, 13 pp.

Bunker, A.F., 1976, Computations of surface energy flux and annual air - sea interaction cycles in the North Atlantic, *Monthly Weather Review*, **104**, 1122–1140.

Clark, N.E., Eber, L., Laurs, R.M. , Renner, J.A. and Saur, J.F.T., 1974, Heat exchange between ocean and atmosphere in the eastern North Pacific for 1961-71. *NOAA Tech. Rep. NMFS SSRF-682*, U.S. Dept. of Commer., Washington, D.C., 108.

Lumb, F.E., 1964, The influence of clouds on hourly amounts of total solar radiation, *Quarterly Journal of the Royal Meteorological Society*, **90**, 43–56

Newell, R.E., Kidson, J.W., Vincent, D.G. and Boar, G.J., 1972, *The General Circulation of the Tropical Atmosphere, Vol. 1* (Boston, MA: MIT Press), 258 pp.

Oort, A.H. and von der Haar, T.H., 1976, On the observed annual cycle in the ocean - atmosphere heat balance over the northern hemisphere, *Journal of Physical Oceanography*, **6**, 781–800.

Priestley, C.H.B., 1959, *Turbulent Transfer in the Lower Atmosphere* (University of Chicago Press), 130 pp.

Reed, R., 1977, On estimating insolation over the ocean, *J. Phys. Oceanogr.*, **7**, 482–485.

Sellers, W.D., 1965, *Physical Climatology* (University of Chicago Press), 272 pp.

Stommel, H., 1980, Asymmetry of interoceanic fresh-water and heat fluxes, *Proceedings of the National Academy of Science, USA*, **77**, 2377–2381.

Trenberth, Kevin E., and Julie M. Caron, 2001: Estimates of meridional atmosphere and ocean heat transports. *J. Climate*, **14**, 3433–3443.

Worthington, L.V., 1981, Water masses of the world ocean - a fine scale census, in *Evolution of Physical Oceanography*, ed. Warren, B.A. and Wunsch, C. (Boston, MA: MIT Press), Chapter 2.

Chapter 6

Ganachaud, A. and C. Wunsch, 2000 Improved estimates of global ocean circulation, heat transport and mixing from hydrographic data. *Nature*, **408**, 453–456.

Lorenz, E.N., 1969, *The Global Circulation of the Atmosphere* (London: Royal Meteorological Society).

Maury, M.F., 1853, *The Physical Geography of the Sea* (London Sampson Low & Son), 274 pp.

Munk, W. and Wunsch, C., 1983a, Observing the ocean in the 1990s, *Philosophical Transactions of the Royal Society of London*, **A307**, 438–463.

Munk, W. and Wunsch, C., 1983b, The study of the ocean and the land surface from satellites, *Philosophical Transactions of the Royal Society of London*, **A309**, 243–461.

Newell, R.E. *et al.*, 1972, *The General Circulation of the Tropical Atmosphere* (Boston, MA: MIT Press).

Reid, J.L., 1981, On the mid-depth circulation of the world ocean, in *Evolution of Physical Oceanography*, ed. Warren, B.A. and Wunsch, C. (Boston, MA: MIT Press), Chapter 3.

Rossby, H.T., Voorhis, A.D. and Webb, D., 1975, A quasi-Lagrangian study of mid-ocean variability using long-range SOFAR floats, *Journal of Marine Research*, **33**, 355–382.

Rossby, H.T., Riser, S.C. and Mariano, A.J., 1983, The western North Atlantic - a Lagrangian viewpoint, in *Eddies in Marine Science*, ed. Robinson, A.R. (New York: Springer-Verlag), 66–88.

Siedler, G., Church J., and Gould, J. (Ed). 2001, *Ocean Circulation and Climate*, Academic Press, 712 pp.

Swallow, J.C., 1955, A neutral buoyancy float for measuring deep currents, *Deep Sea Research*, **3**, 74–81.

Swallow, J.C., 1971, The Aries current measurements in the western North Atlantic, *Philosophical Transactions of the Royal Society of London*, **A2 70**, 451–460.

The TWERLE Team, 1977, The TWERL Experiment, *Bulletin of the American Meteorological Society*, **58**, 936–948.

Warren, B.A., 1981, Deep circulation in the world ocean, in *Evolution of Physical Oceanography*, eds. Warren, B.A. and Wunsch, C. (Boston, MA: MIT Press), Chapter 1.

Wunsch, C., 1981, Low frequency variability of the sea, in *Evolution of Physical Oceanography*, eds. Warren, B.A. and Wunsch, C. (Boston, MA: MIT Press), Chapter 11.

Chapter 7

Cheyney, R.E. and Marsh, J.G., 1981, Seasat altimeter observations of dynamic topography in the Gulf Stream region, *Journal of Geophysical Research*, **86**, 473–483.

Ekman, V.W., 1905, On the influence of the Earth's rotation on ocean currents, *Arch. Maths. Astron. Physics.* **2**, 1–52.

Findlater, J., 1971, *Geophysical Memoirs*, No. 115 (London: HMSO), 53 pp.

McWilliams, J.C., 1976, Maps from the mid-ocean dynamics experiment. Part 1. Geostrophic streamfunction, *Journal of Physical Oceanography*, **6**, 810–827.

Stommel, H., 1958, The abyssal circulation, *Deep Sea Research*, **5**, 80–82.

Stommel, H., 1966, *The Gulf Stream* (Cambridge University Press/University of California Press), 248 pp.

Stommel, H. and Arons, A.B., 1960, On the abyssal circulation of the world's ocean. 1. Stationary planetary flow patterns on a sphere, *Deep Sea Research*, **6**, 140–154.

Swallow, J.C. and Bruce, J.G., 1966, Current measurements off the Somali coast during the SW monsoon of 1964, *Deep Sea Research*, **13**, 861–888.

Chapter 8

Cartwright, D.E., Driver, J.S. and Tranter, J.E., 1977, Swell waves at Saint Helena related to distant storms, *Quarterly Journal of the Royal Meteorological Society*, **103**(438), 655–684.

Charnock H., 1981, Air-sea interaction, in *Evolution of Physical Oceanography*, eds. Warren, B.A. and Wunsch, C. (Boston, MA: MIT Press), Chapter 17.

Flather, R.A., 1984, A numerical model investigation of the storm surge of 31 January and 1 February 1953 in the North Sea, *Quarterly Journal of the Royal Meteorological Society*, **110**(465), 591–612.

Kinsman, B., 1984, Wind Waves: *Their Generation and Propagation on the Ocean Surface* (New York: Dover Publications), 676 pp.

Pekeris, C.L. and Accad, Y., 1969, Solution of Laplace's equations for the M_2 tide in the world's ocean, *Philosophical Transactions of the Royal Society of London*, **A265**, 413–436.

Pugh, D.T., 1987, *Tides, Surges and Mean Sea Level: A Handbook for Engineers and Scientists* (Chichester: John Wiley & Sons, Ltd.), 472 pp.

Thomson, W. (Lord Kelvin), 1879, On gravitational oscillations of rotating water, *Proceedings of the Royal Society, Edinburgh,* **10**, 92 –100.

Chapter 9

Huang, R.X., 1998, Mixing and Available potential energy in a Boussinesq Ocean, *Journal of Physical Oceanography,* **28**, 669–678.

Lorenz, E.N., 1967, *The Nature and Theory of the General Circulation of the Atmosphere* (World Meteorological Organization), 161 pp.

Newell, R.E., Vincent, D.G., Dopplick, T.G., Ferruzza, D. and Kidson, J.W., 1969, The energy balance of the global atmosphere, in *The Global Circulation of the Atmosphere,* ed. Corby, G.A. (London: Royal Meteorological Society), 42–90.

Newton, C.W., 1969, The role of extratropical disturbances in the global atmosphere, in *The Global Circulation of the Atmosphere,* ed. Corby, G.A. (London: Royal Meteorological Society), 137–158.

Palmen, E. and Newton, C.W., 1969, *Atmospheric Circulation Systems* (London: Academic Press), 603 pp.

Parker, C.E., 1971, Gulf Stream rings in the Sargasso Sea, *Deep Sea Research,* **18**(10), 981–993.

Chapter 10

Dyke, P.P., 1996, *Modelling Marine Processes* (Prentice-Hall), 152 pp.

McGuffie, K. and Henderson-Sellers, A., 2005, *A Climate Modelling Primer,* 3rd edition (John Wiley & Sons), 280 pp.

Trenberth, K.E., 1992, *Climate System Modelling* (Cambridge University Press), 788 pp.

Chapter 11

Cane, M.A., 1983, Oceanographic events during El Nino, *Science,* **222**, 1189–1195.

Ghil, M., 1980, Internal climatic mechanisms participating in glaciation cycles, in *Climatic Variations and Variability: Facts and Theories,* ed. Berger, A., NATO Advanced Study Series (Dordrecht: Reidel), 539–558.

JASIN, 1983, Results of the Royal Society Joint Air-Sea Interaction Project (JASIN), *Philosophical Transactions of the Royal Society of London,* **A308**, 221–449.

Lovelock, J., 1988, *The Ages of Gaia: A Biography of our Living Earth* (Oxford: Oxford University Press), 252 pp.

Mitchell, J.M., 1975, A note on solar variability and volcanic variability, in *Physical Basis of Climate and Climate Modelling* (World Meteorological Organization), Garp 16.

Munk, W., 1989, Global ocean warming: detection by long-path acoustic travel times, *Oceanography*, **2**(2), 40–41.

Munk, W. and Wunsch, C., 1983a, Observing the ocean in the 1990s, *Philosophical Transactions of the Royal Society of London*, **A307**, 438–463.

Munk, W. and Wunsch, C. 1983b, The study of the ocean and the land surface from satellites, *Philosophical Transactions of the Royal Society of London*, **A309**, 243–464.

Namias, J., 1971, The 1968–69 winter as an outgrowth of sea and air coupling during antecedent seasons, *Journal of Physical Oceanography*, **1**, 65–81.

Namias, J., 1973, Hurricane Agnes - an event shaped by large-scale air-sea systems generated during antecedent months, *Quarterly Journal of the Royal Meteorological Society*, **99**(421), 506–519.

Philander, S.G.H., 1990, *El Niño, La Niña and the Southern Oscillation* (New York: Academic Press), 293 pp.

WMO, Garp Publication Series 16, 1975, *The Physical Basis of Climate and Climate Modelling*.

WMO, Garp Publication Series 24, 1981, *Scientific Results of the Air Mass Transformation Experiment*, 236 pp.

Chapter 12

Berger, A., 1988. Milankovitch Theory and Climate. *Review of Geophysics*, **26**(4), 624–657.

Climate Chang 2007, the *Fourth Assessment Report (AR4) of the United Nations Intergovernmental Panel on Climate Change* (IPCC).

Houghton, J.T., Jenkins, G.J. and Ephraums, J.J. (eds.), 1990, *Climate Change: The IPCC Scientific Assessment* (Cambridge University Press), 365 pp.

Lovelock, J., 1988, *The Ages of Gaia: A Biography of our Living Earth* (Oxford: Oxford University Press), 252 pp.

Trenberth, K.E., 1992, *Climate System Modelling* (Cambridge University Press), 788 pp.

Figure Sources

Figure 1.1   Allen, C.W., 1958, *Quarterly Journal of the Royal Meteorological Society*, **84**: page 311, figure 3.

Figure 1.2   Pittock, A.B. *et al.* eds, 1978, *Climate Change and Variability: A Southern Ocean Perspective*, Cambridge University Press: page 10, figure 2.1.1.

Figure 1.3   Hess, S.L., 1959, *Introduction to Theoretical Meteorology*, Holt, Rinehart & Winston: page 132, figure 9.1.

Figure 1.5   Jacobowitz, 1-I. *et al.*, 1979, *Journal of Atmospheric Science*, **36**: page 506, figure 6.

Figure 1.7   Pittock, A.B. *et al.* eds, 1978, *Climate Change and Variability: A Southern Ocean Perspective*, Cambridge University Press: page 12, figure 2.1.2.

Figure 1.8   Jacobowitz, H. *et al.*, 1979, *Journal of Atmospheric Science*, **36**: page 506, figure 7.

Figure 1.9   McLellan, H.J., 1965, *Elements of Physical Oceanography*, Pergamon Press: page 5, figure 1.2.

Figure 1.10   McLellan, H.J., 1965, *Elements of Physical Oceanography*, Pergamon Press: page 4, figure 1.1.

Figure 2.1   Friedrich, H. and Levitus, S., 1972, *Journal of Physical Oceanography*, **2**: page 516, figure 4.

Figure 2.2   Tchernia, P., 1980, *Descriptive Regional Oceanography*, Pergamon Press: page 162, figure 5.40.

Figure 2.3   Tolstoy, I. and Clay, C.S., 1966, *Ocean Acoustics: Theory and Experiment in Underwater Sound*, McGraw-Hill: page 7, figure 1.2.

Figure 2.4   Tchernia, P., 1980, *Descriptive Regional Oceanography*, Pergamon Press: page 35, figure 3.4.

Figure 2.6   Tchernia, P., 1980, *Descriptive Regional Oceanography*, Pergamon Press: page 30, figure 3.2.

Figure 2.9   Stull, R.B., 1973, *Journal of Atmospheric Science*, **30**: page 1097, figure 5.

Figure 3.1   Goody, R.M. and Walker, J.C.G, 1972, *Atmospheres*, Prentice Hall: page 25, figure 2.2.

Figure 3.2   Goody, R.M., 1964, *Atmospheric Radiation*, Oxford University Press: page 4, figure 1.1.

Figure 3.3   Ivanoff, A., 1977, Ocean absorption of solar energy in *Modelling and Prediction of the Upper Layers of the Ocean*, ed. Kraus E.B., Oxford, Pergamon: page 49, figure 5.3.

The Atmosphere and Ocean: A Physical Introduction, Third Edition. Neil C. Wells.
© 2012 John Wiley & Sons, Ltd. Published 2012 by John Wiley & Sons, Ltd.

Figure 3.5   Adetunji, J. *et al.*, 1979, *Weather*, **30**: pages 434 and 435, figures 4 and 5.

Figure 3.6   Schmetz *et al.*, 1983, *Results of the Royal Society Joint Air-Sea Interaction Project (JASIN)*, The Royal Society: page 380, figure 2.

Figure 3.7   Wallace, J.M. and Hobbs, P.V., 1977, *Atmospheric Science*, Academic Press: page 23, figure 1.8.

Figure 3.9   Manabe, S. and Strickler, R.F., 1964, *Journal of Atmospheric Science*, **21**: page 371, figure 6a.

Figure 3.10   Trenberth, K.E., 1997, *Bulletin of the American Meteorological Society*, **78**: page 206, figure 7.

Figure 3.11   Dopplick, T.G., 1972, *Journal of Atmospheric Science*, **29**: pages 1291 and 1292, figures 21 and 23.

Figure 3.12(b)   Jerlov, N.G., 1976, *Marine Optics*, Elsevier: page 140, figure 7.4.

Figure 3.13   Robinson, I.S. 1995, *Satellite Oceanography*, Wiley-Praxis Series in Remote Sensing, page 210, figure 7.5.

Figure 3.14   Pickard G.L. and Emery, W.J., *Descriptive Physical Oceanography* (5th Edition), Pergamon Press: page 45, figure 4.6, reproduced by permission of Butterworth Press.

Figure 4.3   Lorenz, E.N., 1967, *The Nature and Theory of the General Circulation of the Atmosphere*, World Meteorological Organization: 218 TP 115, pages 42 and 43, figures 14 and 16.

Figure 4.6   Mason, B.J., 1975, *Clouds, Rain and Rainmaking*, Cambridge University Press: page 21, figure 4.

Figure 4.7   Slinn, N. and George, W., 1975, *Atmospheric Environment*, **9**: page 763, figure 1.

Figure 4.8   Wallace, J.M. and Hobbs, P.V., 1977, *Atmospheric Science*, Academic Press: page 171, figure 4.16.

Figure 4.9   Wallace, J.M. and Hobbs, P.V., 1977, *Atmospheric Science*, Academic Press: page 73, figure 2.7.

Figure 4.10   Wallace, J.M. and Hobbs, P.V., 1977, *Atmospheric Science*, Academic Press: page 255, figure 5.32.

Figure 4.11   Mason, B.J., 1975, *Clouds, Rain and Rainmaking*, Cambridge University Press: page 94, figure 24.

Figure 5.1   Priestley, C.H.B., 1959, *Turbulent Transfer in the Lower Atmosphere*, University of Chicago Press: page 67, figure 18.

Figure 5.3   Ocean Heat Transport by H.L. Bryden and S. Imawaki Plate 6.1.4 in *Ocean circulation and climate*, Seidler, G., Church, J. and Gould, J., Academic Press: 715pp.

Figure 5.5   Oort, A.H. and von der Haar, T.H., 1976, *Journal of Physical Oceanography*, **6**: pages 789–795, figures 5, 8, 9 and 13.

Figure 5.7   Oort, A.H. and von der Haar, T.H., 1976, *Journal of Physical Oceanography*, **6**: page 783, figure 1.

Figure 5.8   Trenberth, Kevin E., Julie M. Caron, 2001, Estimates of Meridional Atmosphere and Ocean Heat Transports. *J. Climate*, **14**: page 3433–3443, figure 7.

Figure 5.9   Wijffels, SE., 2001, In *Ocean Circulation and Climate*, Ed: Seidler, G., Church J., and Gould J., Academic Press, International Geophysics Series, Volume 77: page 478, figure 6.2.1(a).
Figure 5.10   Charnock, H. 1996, In *Oceanography – An Illustrated Guide*, ed. Summerhayes, C.P. and Thorpe, SA.T., S.A.T. Manson Publishing: page 28, figure 2.2.
Figure 5.11   Wijffels, SE., Schmitt, R., Bryden, H. and Stigebrandt, A., 1992, In *Journal of Physical Oceanography*, **22**: page 77, figure 2.
Figure 5.12   Charnock, H. 1996, In *Oceanography – An Illustrated Guide*, ed. Summerhayes, C.P. and Thorpe, S.A.T., Manson Publishing: page 30, figure 2.5.
Figure 5.13   Tchemia, P., 1980, *Descriptive Regional Oceanography*, Pergamon Marine Series 3: page 38, figure 3.5.
Figure 5.14   Dietrich, G. *et al.*, 1980, *General Oceanography: An Introduction*, Wiley Interscience: page 197, figure 78.
Figure 5.15   Grant-Cross, M., **1987**, *Oceanography: A View of the Earth*, Prentice-Hall: page 449, figure 7.15.
Figure 5.16   Tchemia, P., 1980, *Descriptive Regional Oceanography*, Pergamon Marine Series 3: page 157, figure 5.34.
Figure 5.17   Worthington, L.V., 1981, In *Evolution of Physical Oceanography*, ed. Warren, B.A. and Wunsch, C., MIT Press: chapter 2, page 47, figure 2.2.
Figure 5.19   Broeker, W.S. and Peng, T.H., 1982, *Tracers in the Sea*, Eldigo Press: page 409, figure 8.15.

Figure 6.1   Wells, N.C., 2001, 'Ocean Circulation: General Processes' for *Encyclopaedia of Ocean Sciences*, Academic Press: page 1529, figure 1.
Figure 6.3   Rossby, H.T. *et al.*, 1983, In *Eddies in Marine Science*, ed. Robinson, A.R., Springer-Verlag: pages 68 and 70, figures 1 and 3.
Figure 6.6   van der Hoven, J., 1957, *Journal of Meteorology*, **14**: page 161, figure 1.
Figure 6.7   Rhines, P.B., 1971, *Deep Sea Research*, **18**: pages 21–26, figure 1.
Figure 6.8   Lorenz, E.N., 1969, In *The Global Circulation of the Atmosphere*, ed. Corby, GA., Royal Meteorological Society: page 11, figure 4.
Figure 6.9   The TWERLE Team, 1977, *Bulletin of the American Meteorological Society*, **58**: page 942, figure 6a.
Figure 6.10   Rossby, H.T. *et al.*, 1975, *Journal of Marine Research*, **33**: pages 366 and 369, figures 6 and 10.
Figure 6.11   Newell, R.E. *et al.*, 1969, In *The Global Circulation of the Atmosphere*, ed. Corby, G.A., Royal Meteorological Society: page 69, figures 14a and b.
Figure 6.12   Newell, R.E. *et al.*, 1969, In *The Global Circulation of the Atmosphere*, ed. Corby, GA., Royal Meteorological Society: page 65, figures 9a and b.
Figure 6.13   Charnock, H., 1995, In: *Oceanography*, Edited by S.A.Thorpe and C. Summerhayes, Manson Publishers: page 34, figure 2.9.
Figure 6.14   American Meteorological Society, 2005.
Figure 6.15(a) and (b)   Grant-Gross, M., 1987, *Oceanography: A View of the Earth*, Prentice-Hall: page 180, figure 7.5.
Figure 6.16   Wells, N.C., 2001, 'Ocean Circulation: General Processes' for *Encyclopaedia of Ocean Sciences*, Academic Press: page 1536, figure 9.

Figure 7.3   Neumann, G. and Pierson, W., 1966, *Principles of Physical Oceanography*, Prentice-Hall: page 159, figure 7.19.

Figure 7.6   Nicholls, S. *et al.*, 1983, in *Results of the Royal Society Joint Air-Sea Interaction Project (JASIP)*, The Royal Society: page 299, figure 3.

Figure 7.7   Cheney, R.E. and Marsh, J.G., 1981, *Journal of Geophysical Research*, **Cl**, 86: page 480, figure 9.

Figure 7.8   McWilliams, J.C., 1976, *Journal of Physical Oceanography*, **6**: page 813, figure 1.

Figure 7.11   Stommel, H., 1966, *The Gulf Stream*, Cambridge University Press/ University of California Press: page 92, figure 58.

Figure 7.13   Stommel, H., 1958, *Deep Sea Research*, **5**: page 81, figure 1.

Figure 7.16   Rowe, MA., 1985, University of Southampton: PhD thesis.

Figure 7.19(a)   Findlater, J., 1971, *Geophysical Memoirs*, **115**: page 2, figure 1.

Figure 7.19(b)   Swallow, J.C. and Bruce, J.G., 1966, *Deep Sea Research*, **13**: page 866, figure 2.

Figure 8.1   Kinsman, B., 1984, *Wind Waves: Their Generation and Propagation on the Ocean Surface*, Prentice-Hall, page 23, figure 1.2.1.

Figure 8.2   Grant-Gross, M., 1987, *Oceanography: A View of the Earth*, Prentice-Hall: page 208, figure 8.4.

Figure 8.3   Charnock, H., 1981, In *Evolution of Physical Oceanography*, ed. Warren, BA. and Wunsch, C., MIT Press: chapter 17, page 493, figure 17.6.

Figure 8.4   Revault-d'Allones, M. and Caulliez, G., 1980, In *Marine Turbulence*, ed. Nihoul, J.C.J., Elsevier: page 110, figure 5.

Figure 8.5   Charnock, H., 1981, In *Evolution of Physical Oceanography*, ed. Warren, BA. and Wunsch, C., MIT Press: chapter 17, page 493, figure 17.8.

Figure 8.6   Cartwright, D.E. *et al.*, 1977, *Quarterly Journal of the Royal Meteorological Society*, **103**(438): page 664, figures 4a and b.

Figure 8.7   Gill, A.E., 1982, *Atmosphere-Ocean Dynamics*, Academic Press: page 380, figure 10.3.

Figure 8.9   Dietrich, G. *et al.*, 1980, *General Oceanography*, Wiley Interscience: page 421, figure 180.

Figure 8.10   Pugh, D.T., 1987, *Tides, Surges and Mean Sea Level*, John Wiley & Sons: page 158, figure 5.7.

Figure 8.11   Flather, RA., 1984, *Quarterly Journal of the Royal Meteorological Society*, **110**(465): page 594, figure 1.

Figure 8.12   Flather, R.A., 1984, *Quarterly Journal of the Royal Meteorological Society*, **110**(465): pages 605–608, figure 5.

Figure 8.13   Dai A. and Wang J., 1999, *Journal of Atmospheric Science*, **56**: pages 3887–3888, figures 15 and 16.

Figure 9.4   Barnier, B, Medec, G. et.al. *Ocean Science*, **56**: page 9, figure 5b.

Figure 9.8   Palmen, E. and Newton, C.W., 1969, *Atmospheric Circulation Systems*, Academic Press: page 176, figure 7.4.

Figure 9.10   Palmen, F. and Newton, C.W., 1969, *Atmospheric Circulation Systems*, Academic Press: page 310, figure 10.20.

Figure 9.11(a) and (b) Richardson, P.L., 1983, In *Eddies in Marine Science*, ed. Robinson, A.R., Springer-Verlag: pages 20 and 30, figures 1 and 6.
Figure 9.12 Newell, R.E. *et al.*, 1969, *The Global Circulation of the Atmosphere*, ed. Corby, GA., Royal Meteorological Society: page 63, figure 7.
Figure 9.13 Newton, C.W., 1969, In *The Global Circulation of the Atmosphere*, ed. Corby, G.A., Royal Meteorological Society: page 138, figure 1.
Figures 9.14 Lorenz, E.N., 1967, *The Nature and Theory of the General Circulation of the Atmosphere*, World Meteorological Organization: page 82, figure 40.
Figure 9.15 Siedler, G., 2001, figure 1.2.7 as taken from Schmitz, 1996.
Figure 9.16 Bean, M. S., 1997, University of Southampton: Ph.D thesis.

Figure 10.8 Supplied by Met Office, UK.
Figure 10.9 Glantz, M.H. and Krenz, J.H., 1992, Human components of the system, In *Climate System Modelling*. Ed: Trenberth K.E. Cambridge University Press: Plate 2.
Figure 10.10 Supplied by ECMWF, Reading, UK.

Figure 11.2 WMO, 1981, Garp Publication Series, 24: page 167, figure 6.1.
Figure 11.3 Guymer, T.H. et al., 1983, *Philosophical Transactions of the Royal Society of London*, **A308**: page 265, figures 10 and 11.
Figure 11.4 Smagorinsky, J., 1953, *Quarterly Journal of the Royal Meteorological Society*, **79**(341): page 362, figure 7.
Figure 11.5 Namias, J., 1973, *Quarterly Journal of the Royal Meteorological Society*, **99**(421): page 509 and 517, figures 2 and 10.
Figure 11.7 Philander, G., 1983, *Nature*, **302**: page 297, figure 3.
Figure 11.8 Wells, N.C., Gould, W.J. and Kemp, A.E.S., 1996, In *Oceanography – An Illustrated Guide*, ed. Summerhayes, C.P. and Thorpe, S.A.T., Manson Publishing: page 54, figure 3.22.
Figure 11.9 Lazier, J., 1980, *Atmosphere and Ocean*, **18**: page 230, figure 2.
Figure 11.10 Lamont-Doherty Earth Observatory, Columbia University.
Figure 11.11 JISAO, University of Washington.

Figure 12.1(a) Wikipedia Temperature record of past 1000 years.
Figure 12.1(b) Hansen, J., Sato, Mki., Ruedy, R., Lo, K., Lea, D.W. and Medina-Elizade M., (2006), Global temperature change, *Proc. Natl. Acad. Sci.*, **103**: pages 14288–14293, doi:10.1073/pnas.0606291103, figure 5.
Figure 12.1(c) Wikipedia Geological temperature record.
Figure 12.2 Post-glacial sea level Wikipedia.
Figure 12.3 Wells, N.C., Gould, W.J. and Kemp, A.E.S., 1996, In *Oceanography – An Illustrated Guide*, ed. Summerhayes, C.P. and Thorpe, S.A.T., Manson Publishing: page 42, figure 3.1.
Figures 12.4–12.6 *Intergovernmental Panel on Climate Change (IPCC) 2007 WG1-AR4.*

Appendices

A Standard International (SI) Units

Quantity	Symbol	Name	Definition
Force	N	Newton	$kg\,ms^{-2}$
Pressure	Pa	Pascal	$kg\,m^{-1}s^{-2}$
Energy/Work	J	Joule	$kg\,m^2s^{-2}$
Power	W	Watt	$kg\,m^2s^{-3}$
Temperature	K	Kelvin	
Celsius Temperature	°C	degree Celsius	K-273.15

B SI Unit Prefixes

Name	Symbol	Multiplying factor 10^N, where N is given below
peta	P	15
tera	T	12
giga	G	9
mega	M	6
kilo	k	3
hecta	h	2
deca	da	1
deci	d	−1
centi	c	−2
milli	m	−3
micro	μ	−6
nano	n	−9
pico	p	−12
femto	f	−15
atto	a	−18

The Atmosphere and Ocean: A Physical Introduction, Third Edition. Neil C. Wells.
© 2012 John Wiley & Sons, Ltd. Published 2012 by John Wiley & Sons, Ltd.

Index

The Atmosphere and Ocean: A Physical Introduction, Third Edition. Neil C. Wells.
© 2012 John Wiley & Sons, Ltd. Published 2012 by John Wiley & Sons, Ltd.